Innovations and New Processes in the Olive Oil Industry

Innovations and New Processes in the Olive Oil Industry

Sebastián Sánchez Villasclaras
Juan Francisco García Martín

MDPI

Basel • Beijing • Wuhan • Barcelona • Belgrade • Novi Sad • Cluj • Manchester

Sebastián Sánchez Villasclaras
Chemical, Environmental and Materials Engineering
University of Jaen
Jaen
Spain

Juan Francisco García Martín
Chemical Engineering
University of Seville
Seville
Spain

Editorial Office
MDPI AG
Grosspeteranlage 5
4052 Basel, Switzerland

This is a reprint of articles from the Special Issue published online in the open access journal *Processes* (ISSN 2227-9717) (available at: www.mdpi.com/journal/processes/special_issues/olive_oil_processes).

For citation purposes, cite each article independently as indicated on the article page online and using the guide below:

Lastname, A.A.; Lastname, B.B. Article Title. *Journal Name* **Year**, *Volume Number*, Page Range.

ISBN 978-3-7258-1972-0 (Hbk)
ISBN 978-3-7258-1971-3 (PDF)
https://doi.org/10.3390/books978-3-7258-1971-3

Cover image courtesy of Sebastián Sánchez Villasclaras and Juan Francisco García Martín

Contents

About the Editors

Sebastián Sánchez Villasclaras

Sebastián Sánchez was a professor of Chemical Engineering at the University of Granada (Spain). Since 1993, he has been a full professor of the Department of Chemical, Environmental and Materials Engineering of University of Jaen (Spain). From September 2017 to June 2021, he was the director of the 'Center for Advanced Studies of Olive Groves and Olive Oil'. His research interests are in the areas of 'Olive oil technology', 'Use of by-products and residues from olive oil industry', 'Gas absorption in system with chemical reaction', and 'Biofuels production' at the University Institute of Research in Olive Grove and Olive Oils. He has extensive experience in the production of virgin olive oils and management of by-products. From 1993 to the present, Professor Sánchez has been the head of the Research Group 'Bioprocesses' (TEP-138, Andalusia Government, Spain).

Juan Francisco García Martín

Professor Juan Francisco García Martín's thesis (University of Jaén) was awarded two Spanish Research Environment Prizes. Afterwards, Professor Juan Francisco García Martín took postdoctoral positions at the University of Granada, University College Dublin, and Spanish National Research Council. Additionally, he completed research stays at the Universities Paul Sabatier Toulouse III and Claude Bernard Lyon I. He was also a lecturer at the Universities of Jaén, Granada, and Málaga. Currently, his research interests focus on the use of byproducts of the food industry and food quality. Since 2016, he has been an associate professor at the University of Seville. In addition, Professor García Martín is head of the Research Group 'Obtaining Biofuels' (AGR-155, Andalusia Government, Spain) and he is in the top 2% of the most cited scientists in the high scientific impact list published annually by Stanford University.

processes

Editorial

Innovations and New Processes in the Olive Oil Industry

Sebastián Sánchez Villasclaras [1,2,*] and Juan Francisco García Martín [2,3,*]

1 Department of Chemical, Environmental and Materials Engineering, University of Jaén, 23071 Jaen, Spain
2 University Institute of Research on Olive Groves and Olive Oils, GEOLIT Science and Technology Park, University of Jaén, 23620 Mengibar, Spain
3 Departamento de Ingeniería Química, Facultad de Química, Universidad de Sevilla, 41012 Seville, Spain
* Correspondence: ssanchez@ujaen.es (S.S.V.); jfgarmar@us.es (J.F.G.M.)

The olive oil industry, one of the largest industries in the Mediterranean basin and in other countries around the world, is fundamentally composed of olive groves, olive oil mills, pomace oil extraction plants, and oil refineries. Additionally, there are other types of closely related industries, such as those producing table olives and margarine and the industries based on physicochemical transformation to obtain high value-added products for the pharmaceutical, cosmetic, nutritional, and dietary sectors [1].

Innovative processes and new technologies have been implemented in olive oil mills, oil refineries, and transformation industries in the last decade to improve the olive oil extraction and its quality, as well as to minimise and valorise the generated wastes. In this sense, biofuels (bioethanol, biodiesel, biogas, synthetic gas . . .) and high value-added products (tyrosol, hydroxytyrosol, squalene, oleuropein, tocopherols and others) can be obtained from byproducts and residues from the olive oil and pomace oil extraction processes.

This Special Issue "Innovations and New Processes in the Olive Oil Industry" brings together high-quality research studies addressing challenges faced in the olive oil industry and related industries, along with comprehensive overviews and in-depth technical research papers addressing recent progress in olive oils and table olives production and the management of the different olive oil industry wastes.

With regard to the management of olive oil mill wastes, Kurtz et al. studied the benefits and drawbacks of applying olive mill wastewater (vegetation water, tissues of olive fruits, and process water used in different stages of the olive oil extraction) in irrigated olive orchards [2]. Used in semi-arid areas with sandy loam soils, olive mill wastewater brings some benefits to soils (accumulation of K+, Ca^{2+}, PO_4^{3-}, organic carbon and increased biological activity and diversity). Notwithstanding, it has also negative effects (accumulation of phenolic compounds, increased soil hydrophobicity, and salinity). Most of these effects depend on the dose of wastewater applied, mainly in the topsoil.

In relation to olive oil quality, the control of temperature and time during olive oil production is of major importance since olive oil quality is strongly dependent on it [3]. In this sense, Plasquy et al. discuss the advantages of the cold storage of olive fruit before the olive oil extraction process [4]. Cold storage is not just a method to prolong the quality of the harvested olives, but it is also a temperature management procedure. These authors highlight the most important factors during cold storage and their influence on both fruit and oil produced. The same authors also studied the adjustment of the temperature of the intact fruit of three different olive varieties ('Arbequina', 'Cobrançosa' and 'Gordal') before grinding at both the laboratory level and a pilot plant [5]. In their work, Plasquy et al. demonstrated the feasibility of bringing the fruit to the ideal temperature for malaxing without applying excessive heat and within a period according to the characteristics of an industrial washing equipment.

During the malaxation stage, certain olive varieties, such as 'Hojiblanca' and 'Picual', give rise to pastes from which it is difficult to separate the oil, leading to low extraction yields. To improve oil extraction in these pastes, one alternative is the addition of natural

Citation: Sánchez Villasclaras, S.; García Martín, J.F. Innovations and New Processes in the Olive Oil Industry. *Processes* **2024**, *12*, 1570. https://doi.org/10.3390/pr12081570

Received: 19 July 2024
Accepted: 23 July 2024
Published: 26 July 2024

microtalcs. In their work, Sánchez et al. added a natural microtalc of great purity ($CaCO_3$ concentration less than 6 wt.%) and small average particle size ($\phi \leq 2.1$ μm) in the malaxation stage during the production of virgin olive oils from the two aforementioned olive varieties on an industrial scale at two olive mills [6]. The high-purity natural microtalc not only increased the extraction yields but also improved the virgin olive oil quality, especially in relation to antioxidant compounds. To be specific, Sánchez et al. found increases of 10.4% in phenolic compounds and of 21.5% in the tocopherols when adding the natural microtalc into the malaxer, which enhance the oxidative stability of the virgin olive oils. In relation to the increases in industrial yield in the extraction process, these results agree with those obtained by Caponio et al. [7,8].

Another alternative to improve the malaxation stage is the use of a sono-heat-exchanger. Clodoveo et al. presents it as a suitable solution to respond to the needs mapped in the community of millers because the sono-heat-exchanger is a continuous system and a potential substitute for the current malaxers [9]. It is based on ultrasonic waves, which induce cavitation (the phenomenon responsible for the mechanical action of ultrasound) and penetrate the entire thickness of the olive paste, all without representing a decrease in or elimination of the minor components in the olive oils. This ultrasonic device aims at favouring the extraction yield and increasing virgin olive oil quality by enhancing its content of phenolic compounds without causing undesired sensorial defects, thus effectively eliminating the bottleneck represented by the traditional malaxer. This conclusion was also obtained and supported by other research groups [10,11].

With regard to the olive oil itself, there is a current demand for innovative and specialised products of high nutritional quality and with various sensory characteristics, such as flavoured oils (which correct technical name is food seasoning). Following this, Chéu-Guedes et al. flavoured extra virgin olive oils of the 'Madural' variety from Trasos Montes region of Portugal with different aromatics herbs and condiments (flower of salt and bay leaf, garlic, rosemary, and dehydrated lemon peel), with the aim of studying the influence of these aromatic herbs and condiments on the physicochemical parameters of the oils [12]. The authors concluded that the flavourings did not affect the fatty acid profile of the extra virgin olive oils, while the impact on quality, purity, oxidative stability and microbiological spoilage was complex. In this work, it is worth highlighting the excellent study carried out on the composition profile of the fractions of phenolic compounds, tocopherols and sterols in flavoured olive oils. Concerning total phenolic compounds, an increase can be observed with respect to the monovarietal oil, perhaps due to the contribution of the phenolic compounds present in the composition of bay leaves, rosemary and garlic. These results agree with what different research groups indicated for garlic and rosemary [13,14].

Studies of advanced techniques and methods for high value-added product and bioenergy production within the olive oil industry are also included in this Special Issue. With regard to the former, Lozano et al. studied the use of the pomace obtained with two-outlet decanters within a biorefinery scheme to produce value-added products along with solids that can be used for energy or adsorbent production [15]. The aim was to develop a novel, efficient and environmentally friendly extraction process based on a combination of hydrothermal treatments with liquid/liquid extractions. Under mild extraction conditions (30 °C and 30 min), significant amounts of bioactive polyols (1126 mg/L inositol), saccharides (15,960 mg/L D-glucose, 385 mg/L D-xylose, 5550 mg/L D-fructose, 165 mg/L lactose, and 248 mg/L sucrose) and phenolic compounds (4792 mg/L) were extracted. Furthermore, Fainassi et al. studied the surface-active and emulsifying properties of various aqueous ethanolic extracts from the pomace obtained in the process with a two-outlet decanter [16]. The objective was to produce stable oil-in-water emulsions with smaller droplet sizes and for an extended period of storage using these extracts as sole emulsifiers.

Concerning bioenergy production, Cubero-Cardoso et al. [17] and Fernández-Rodríguez et al. [18] studied the production of methane from olive mill byproducts by anaerobic digestion. Cubero Cardoso et al. broaden the knowledge behind the processes of anaerobic degradation of olive oil solid waste under mesophyll conditions operating in a semi-continuous regime. These authors assess the instability of the degradation of this complex biomass and determine the relationship between the decrease in methane production and the microbial population present in the waste [17]. On the other hand, Fernández-Rodríguez et al. study the anaerobic digestion bioprocess of the waste generated in the extraction process for cold-pressed olive (a mixture of wastewater and solid waste). In order to assess the potential influence of the ripening level of the olives on the performance of the anaerobic digestion of the produced waste, the authors collected olives of the 'Picual' variety at two stages; i.e., green olives and olives in veraison. The highest methane yield and energy output result were detected in the waste from cold-pressed green olives [18]. On the other hand, this work concludes that the use of microwaves as pretreatment improved the solubilisation of organic compounds and the hydrolysis of proteins to NH^{4+}-N, and also increased the methane yield and the maximum methane production rate during the co-digestion process, a result consistent with what can be found in the available literature for other wastes [19].

Finally, in relation to closely related industries, Demir et al. assess the effect of growing regions and processing methods on the phenolic compound profile in table olives of the 'Gemlik' variety (very extended in Türkiye), concluding that both regions and processing methods have statistical significant effects on the phenolic composition of table olives [20], which is in agreement with the results of Pistarino et al., who also considered the olive-growing region as a key factor [21]. In this study, six regions of Türkiye were considered. The main phenolic compounds found were 3-hydroxytyrosol (from 4.58 to 168.21 mg/kg), 4-hydroxyphenyl (from 0.76 to 97.58 mg/kg), luteolin 7-glucoside (from 0.32 to 58.64 mg/kg), tyrosol (from 1.57 to 47.24 mg/kg) and luteolin (from 0.17 to 53.56 mg/kg). The results in hydroxytyrosol and tyrosol are quite similar to those obtained by Malheiro et al. [22].

Funding: This research received no external funding.

Acknowledgments: The Guest Editors are grateful to all the authors that have contributed to this Special Issue, all the reviewers for their excellent work in evaluating the submitted manuscripts and the editorial staff of *Processes*.

Conflicts of Interest: The authors declare no conflicts of interest.

References

1. *IOC/T.15/NC No 3/Rev.17/2021*; Trade Standard Applying to Olive Oils and Olive Pomace Oils. International Olive Council: Madrid, Spain, 2021; pp. 1–17.
2. Kurtz, M.P.; Dag, A.; Zipori, I.; Laor, Y.; Buchmann, C.; Saadi, I.; Medina, S.; Raviv, M.; Zchori-Fein, E.; Schaumann, G.E.; et al. Toward balancing the pros and cons of spreading olive mill wastewater in irrigated olive orchards. *Processes* **2021**, *9*, 780. [CrossRef]
3. García, J.M.; Streif, J. The effect of controlled atmosphere storage on fruit and oil quality of 'Gordal' olives. *Gartenbauwissenschaft* **1991**, *56*, 233–238.
4. Plasquy, E.; García Martos, J.M.; Florido, M.d.C.; Sola-Guirado, R.R.; García Martín, J.F. Cold storage and temperature management of olive fruit: The impact on fruit physiology and olive oil quality—A review. *Processes* **2021**, *9*, 1543. [CrossRef]
5. Plasquy, E.; García Martos, J.M.; Florido Fernández, M.D.C.; Sola-Guirado, R.R.; García Martín, J.F. Adjustment of olive fruit temperature before grinding for olive oil extraction. Experimental study and pilot plant trials. *Processes* **2021**, *9*, 586. [CrossRef]
6. Sánchez, S.; Olivares, I.; Puentes, J.G.; Órpez, R.; La Rubia, M.D.; Pacheco, R.; García Martín, J.F. Use of natural microtalcs during the virgin olive oil production process to increase its content in antioxidant compounds. *Processes* **2022**, *10*, 950. [CrossRef]
7. Caponio, F.; Squeo, G.; Difonzo, G.; Pasqualone, A.; Summo, C.; Paradiso, V.M. Has the use of talc an effect on yield and extra virgin olive oil quality? *J. Sci. Food Agric.* **2016**, *96*, 3292–3299. [CrossRef] [PubMed]
8. Caponio, F.; Monteleone, J.I.; Martellini, G.; Summo, C.; Paradiso, V.M.; Pasqualone, A. Effect of talc addition on the extraction yield and quality of extra virgin olive oils from 'Coratina' cultivar after production and during storage. *J. Oleo Sci.* **2014**, *63*, 1125–1132. [CrossRef] [PubMed]
9. Clodoveo, M.L.; Crupi, P.; Corbo, F. Olive sound: A sustainable radical innovation. *Processes* **2021**, *9*, 1579. [CrossRef]
10. Jiménez, A.; Beltrán, G.; Uceda, M. High-power ultrasound in olive paste pretreatment. Effect on process yield and virgin olive oil characteristics. *Ultrason. Sonochem.* **2007**, *14*, 725–731. [CrossRef]
11. Bejaoui, M.A.; Beltran, G.; Aguilera, M.P.; Jimenez, A. Continuous conditioning of olive paste by high power ultrasounds: Response surface methodology to predict temperature and its effect on oil yield and virgin olive oil characteristics. *LWT Food Sci. Technol.* **2016**, *69*, 175–184. [CrossRef]

12. Chéu-Guedes, M.H.; La Rubia, M.D.; Sánchez, S.; Ramos, N.; Pacheco, R. Characterization of flavoured olive oils of 'Madural' variety. *Processes* **2023**, *11*, 205. [CrossRef]
13. Gambacorta, G.; Faccia, M.; Pati, S.; Lamacchia, C.; Baiano, A.; La Notte, E. Changes in the chemical and sensorial profile of extra virgin olive oils flavored with herbs and spices during storage. *J. Food Lipids* **2007**, *14*, 202–215. [CrossRef]
14. Soares, V.P.; Fagundes, M.B.; Guerra, D.R.; Leães, Y.S.V.; Speroni, C.S.; Robalo, S.S.; Emanuelli, T.; Cichoski, A.J.; Wagner, R.; Barin, J.S.; et al. Ultrasound assisted maceration for improving the aromatization of extra-virgin olive oil with rosemary and basil. *Food Res. Int.* **2020**, *135*, 109305. [CrossRef] [PubMed]
15. Lozano, E.J.; Blázquez, G.; Calero, M.; Martín-Lara, M.Á.; Pérez-Huertas, S.; Pérez, A. Optimizing the extraction process of value-added products from olive cake using neuro-fuzzy models. *Processes* **2024**, *12*, 317. [CrossRef]
16. Fainassi, F.; Taarji, N.; Benkhalti, F.; Hafidi, A.; Neves, M.A.; Isoda, H.; Nakajima, M. Emulsion formation and stabilizing properties of olive oil cake crude extracts. *Processes* **2021**, *9*, 633. [CrossRef]
17. Cubero-Cardoso, J.; Fernández-Prior, Á.; Ramiro-Garcia, J.; Trujillo-Reyes, A.; Caballero-Guerrero, B.; Rodríguez-Gutiérrez, G.; Fermoso, F.G. Microbial population dynamics during unstable operation of a semicontinuous anaerobic digester fed with a mild-treated olive mill solid waste. *Processes* **2023**, *11*, 2724. [CrossRef]
18. Fernández-Rodríguez, M.J.; Cubero-Cardoso, J.; de la Lama-Calvente, D.; Fernández-Prior, Á.; Rodríguez-Gutiérrez, G.; Borja, R. Anaerobic digestion of the residue (combination of wastewater and solid waste) from a new olive-oil manufacturing process based on an olive cold-pressing system: Kinetic approach and process performance. *Processes* **2022**, *10*, 2552. [CrossRef]
19. Liu, J.; Zhao, M.; Lv, C.; Yue, P. The effect of microwave pretreatment on anaerobic co-digestion of sludge and food waste: Performance, kinetics and energy recovery. *Environ. Res.* **2020**, *189*, 109856. [CrossRef]
20. Demir, C.; Yildiz, E.; Gurbuz, O. Profile phenolic compounds in Spanish-style and traditional brine black olives ('Gemlik' cv.) provided from different regions of Türkiye. *Processes* **2023**, *11*, 2412. [CrossRef]
21. Pistarino, E.; Aliakbarian, B.; Casazza, A.A.; Paini, M.; Cosulich, M.E.; Perego, P. Combined effect of starter culture and temperature on phenolic compounds during fermentation of 'Taggiasca' black olives. *Food Chem.* **2013**, *138*, 2043–2049. [CrossRef]
22. Malheiro, R.; Sousa, A.; Casal, S.; Bento, A.; Pereira, J.A. Cultivar effect on the phenolic composition and antioxidant potential of stoned table olives. *Food Chem. Toxicol.* **2011**, *49*, 450–457. [CrossRef] [PubMed]

processes

Article

Toward Balancing the Pros and Cons of Spreading Olive Mill Wastewater in Irrigated Olive Orchards

Markus Peter Kurtz [1], Arnon Dag [2], Isaac Zipori [2], Yael Laor [3], Christian Buchmann [1], Ibrahim Saadi [3], Shlomit Medina [3], Michael Raviv [4], Einat Zchori-Fein [5], Gabriele Ellen Schaumann [1,*] and Dörte Diehl [1]

1 iES Landau, Group of Environmental and Soil Chemistry, Institute for Environmental Sciences, University of Koblenz-Landau, Fortstraße 7, 76829 Landau, Germany; kurtz@gmx.de (M.P.K.); buchmann@uni-landau.de (C.B.); diehl@uni-landau.de (D.D.)
2 Gilat Research Center, Institute of Plant Sciences, Agricultural Research Organization—Volcani Institute, Gilat 85280, Israel; arnondag@volcani.agri.gov.il (A.D.); matabsor@volcani.agri.gov.il (I.Z.)
3 Newe Ya'ar Research Center, Institute of Soil, Water and Environmental Sciences, Agricultural Research Organization—Volcani Institute, Ramat Yishay 30095, Israel; laor@volcani.agri.gov.il (Y.L.); saadi@volcani.agri.gov.il (I.S.); shmedina@volcani.agri.gov.il (S.M.)
4 Newe Ya'ar Research Center, Institute of Plant Sciences, Agricultural Research Organization—Volcani Institute, Ramat Yishay 30095, Israel; mraviv@volcani.agri.gov.il
5 Newe Ya'ar Research Center, Institute of Plant Protection, Agricultural Research Organization—Volcani Institute, Ramat Yishay 30095, Israel; einat@volcani.agri.gov.il
* Correspondence: schaumann@uni-landau.de

Citation: Kurtz, M.P.; Dag, A.; Zipori, I.; Laor, Y.; Buchmann, C.; Saadi, I.; Medina, S.; Raviv, M.; Zchori-Fein, E.; Schaumann, G.E.; et al. Toward Balancing the Pros and Cons of Spreading Olive Mill Wastewater in Irrigated Olive Orchards. *Processes* **2021**, *9*, 780. https://doi.org/10.3390/pr9050780

Academic Editors: Sebastián Sánchez Villasclaras and Juan Francisco García Martín

Received: 31 March 2021
Accepted: 25 April 2021
Published: 28 April 2021

Abstract: The controlled application of olive mill wastewater (OMW) as a by-product of the olive oil extraction process is widespread in olive oil-producing countries. Therefore, a sustainable approach necessarily targets the positive effects of soil resilience between successive annual applications to exclude possible accumulations of negative consequences. To investigate this, we applied 50, 100, 100 with tillage and 150 m^3 OMW ha^{-1} y^{-1} for five consecutive seasons to an olive orchard in a semi-arid region and monitored various soil physicochemical and biological properties. OMW increased soil water content with concentration of total phenols, cations, and anions as well as various biological and soil organic matter indices. Soil hydrophobicity, as measured by water drop penetration time (WDPT), was found to be predominantly in the uppermost layer (0–3 and 3–10 cm). OMW positively affected soil biology, increased the activity and abundance of soil arthropods, and served as a food source for bacteria and fungi. Subsequent shallow tillage reduced the extent of OMW-induced changes and could provide a simple means of OMW dilution and effect minimization. Despite potentially higher leaching risks, an OMW dose of 50–100 m^3 ha^{-1} applied every two years followed by tillage could be a cost-effective and feasible strategy for OMW recycling.

Keywords: olive mill wastewater; phenolic compounds; bait-lamina; Collembola; biodegradation; water re-use

1. Introduction

Olive mill wastewater (OMW) is the liquid by-product obtained from three-phase (solid, aqueous, oil) olive oil extraction systems. Up to 30 million m^3 OMW is generated annually in the Mediterranean region [1]. OMW consists of vegetation water, tissues of olive fruits, and process water used in different stages of the extraction. Therefore, it has a high biological and chemical oxygen demand, high concentrations of soluble phenolic compounds (Total Phenols; TP), oil residues, as well as the residual solid content (Total Suspended Solids) and, therefore, it cannot be processed by conventional sewage treatment plants [2].

Numerous studies advised the controlled spreading of OMW on cultivated soils as a viable recycling approach. It was further suggested that OMW could be considered as a useful, low-cost soil amendment and fertilizer [2]). Specially, in regions having low soil organic matter (SOM) content and consequently low soil fertility, OMW can serve as a readily available organic matter (OM) source. Annual allowed application rates vary in different countries with, for example, 50 m^3 ha^{-1} y^{-1} for Italy and Israel [3,4] and 30 m^3 ha^{-1} y^{-1} for Catalonia [5]. These recommendations were formulated to avoid issues associated with discharging OMW into the public sewage network, from sealed pipes to the collapse of pumping stations and sewage treatment plants. To overcome such negative effects, different approaches have been considered including engineered-oriented technologies (flotation, anaerobic digestion) or recycling (co-composting the OMW with other organic solid wastes) [6]. None has been applied on a large scale, as these approaches failed to overcome the existing challenges due to high costs for infrastructure, instruments, chemicals and transportation, low efficiency, low intake capacities, or the final treated wastewater did not fulfill the respective water regulations (e.g., [2])

OMW spreading in olive orchards has the potential to be a cost-effective recycling solution, since soil plays a key role for the transport and biodegradation of pollutants by filtering, storing, and transforming OMW constituents. From an economic and logistic point of view, on-site application of OMW near mills directly after its generation from October to January would be preferred. However, an optimal dose and application practice should be based on local variables related to climatic conditions, soil properties, and crop management [7,8]. Assuming a linear relationship between the magnitude of OMW-induced changes and OMW dose would be an oversimplification and has been disproved for processes such as phenolic compound degradation [5] or degree of weed germination inhibition [9,10]. Recently, the practice of shallow tillage following OMW application was considered by Levy et al. [11] and Zipori et al. [12] with the purpose of avoiding possible negative effects on soil physical and hydraulic properties. Evidently, reservations about tillage are based on its possible drawbacks such as a disturbed faunal community, disrupted soil aggregates, and reduced nutrients and soil organic carbon [13].

Phytotoxicity induced by the accumulation of phenolic substances [9], soil hydrophobicity [14], salinization, and acidification [15] are prominent adverse effects caused by OMW application. Still, only a few studies trace such effects with a particular focus on small-scale spatial resolutions. Recent findings from irrigated olive orchards in the semi-arid region of Israel suggest that such adverse effects may be overlooked unless the soil profile is analyzed at a high resolution. For example, Peikert et al. [14] found persistent hydrophobicity, higher electrical conductivity (EC), and sorption capacity toward agrochemicals, mainly in the very topsoil layer (0–3 cm). Steinmetz et al. [16] identified hydrophobicity and adverse effects toward soil biology exclusively in the upper soil layers (0–3 cm). Moreover, OMW toxicity toward soil biota is poorly understood. In ecotoxicological assessments (OMW from an evaporation pond, [17]) and short-term field studies (150 m^3 OMW ha^{-1}, [8]), OMW was pointed out to stress soil biota. Due to the degradation of OMW-derived phenolic compounds, it is likely that adverse effects disappear with time, as shown for recurring soil microbiological activity [18] or seed germination [9,19]. Therefore, the aims of this study were (i) to characterize soil alterations at increased resolution across soil profiles by assessing OMW-induced effects on soil surface hydrophobicity and other physicochemical properties, at annual doses in a range somewhat above what is considered to be practical; and (ii) to assess residual effects of OMW application on soil biota and microbiota as sensitive indicators to soil resilience between successive annual applications. For this, we hypothesized the following:

1. A relatively low OMW dose in the range of 50 m^3 ha^{-1} y^{-1} balances the low degradation rates of OMW organic residues expected during the cold and wet winter season, and the soil can recover between consecutive winter applications;
2. Toxic effects of OMW toward soil biota will disappear between successive winter applications; and
3. Soil tillage following OMW application will enhance soil biodegradation rates of OMW constituents and therefore reduce the negative effects of OMW on soil physicochemistry and biology.

To test our hypotheses, we conducted a five-year field study at an olive orchard (cv. 'Leccino') on Loess soil in a semi-arid region and analyzed at increased resolution the spatio-temporal effects of OMW before and after successive annual winter applications. Specifically, we assessed soil surface hydrophobicity, analyzed physicochemical properties in aqueous extracts of soil samples from five depths, determined the amount and thermal stability of soil organic matter, and as biological parameters, bait lamina consumption and abundance of soil invertebrates using pitfalls were studied to detect OMW-derived toxic effects.

2. Materials and Methods

2.1. Study Area and Sampling Design

The field study was conducted between the years 2012 and 2016 at the Gilat Research Center of the Israeli Agricultural Research Organization, Northern Negev (31°20′ N, 34°40′ E). The orchard (cv. 'Leccino'; 7-year-old) is typical for intensive olive cultivation featuring a high mature tree density (450 trees ha^{-1}). The row spacing in this rectangular orchard is 7 m, while it is 3.5 m for the tree spacing in the row. No fertilizers were applied during the 5 years of the OMW application; irrigation was generally performed from March to October with a Kc of 0.55 relative to Penman ET_0, resulting in an average annual amount of 650 mm.

OMW was applied to the soil using a spreading tank at different annual doses of 0 (control), 50, 100, and 150 m^3 ha^{-1} y^{-1}. Additionally, a fifth treatment included an annual application of 100 m^3 ha^{-1} followed by a shallow tillage to a depth of 5 cm, using a hand rototiller, three weeks after OMW application after the upper layer was dried. OMW was applied shortly after the end of the olive milling season (January–February). All five treatments were designed in five replicates in a randomized block design (a total of 25 plots; more details in [11]).

Soil samples were taken at a depth of 0–10 cm twice a year, 1–2 months before OMW application (autumn) and 1–2 months after application (spring). A more detailed field survey ("detailed survey") was conducted in January 2014 (shortly before the winter application of 2014) and included soil samples from 0–3, 3–10, 10–20, 20–40, and 40–60 cm depth.

Three unified soil samples were taken from each plot at a perpendicular distance of 1.5 m from the drip irrigation line (at an area that does not receive irrigation water), which is located in the center between two trees. The detailed survey of January 2014 was conducted at a distance of 80 cm from the drip line (also at an area that does not receive irrigation water). This distance was chosen based on soil hydrophobicity.

2.2. Soil and OMW Analyses

Soil water content was determined gravimetrically at 105 °C according to ISO 11456 [18]. Loss on ignition (LOI_{550}) was determined by igniting the sample to 550 °C (>4 h). Additionally, soil samples were air-dried, gently manually ground in a mortar to destroy larger aggregates, and sieved (<2 mm). A representative sample of the control plots (no OMW) was used to measure soil texture, bulk density, and effective cation exchange capacity [19]. Soil pH and EC from all treatments were measured in a 1:5 aqueous suspension of soil and water (shaken for 24 h and filtered through 0.45 μm filter) [19]. From these extracts, TP was determined after Box (1983) using the Folin–Ciocalteu reagent (Sigma-Aldrich, Germany) (cf. [20]) and caffeic acid as a reference (gallic acid for the detailed survey using a caffeic acid–gallic acid conversion factor of 1.2). Dissolved organic carbon (DOC) was measured using a multi N/C 2100 analyzer (Analytik Jena, Germany). Ultraviolet absorbance (UVA) at 254 nm was measured and used to calculate the specific ultraviolet absorbance, SUVA (=UVA/DOC × 100; L mg^{-1} m^{-1}). Soluble ion contents, F^-, Cl^-, NO_2^-, NO_3^-, PO_4^{3-}, and SO_4^{3-} were analyzed using an ion chromatograph (881 Compact IC pro, Metrohm, Switzerland). Total cation contents (Na^+, K^+, Mg^{2+}, Ca^{2+}, Mn^{2+}, and Fe^{2+}) were analyzed by inductively coupled plasma optical emission spectroscopy (Agilent 720 ICP-OES) in microwave-assisted reverse aqua regia (HCl + $3HNO_3$) extraction at a pH < 2.

To characterize soil water repellency, water drop penetration time (WDPT) was determined directly in the field. One drop of tap water (100 µL) was placed onto the soil surface in 20 cm (4 cm for the detailed survey) distance intervals, and the time of complete penetration of each droplet was counted and classified from wettable to completely water repellent according to Bisdom et al. [21].

Thermogravimetric analysis and differential scanning calorimetry of 10–20 mg soil samples (sieved to 1 mm) were conducted for the detailed survey on a TG DTA/DSC Apparatus STA 449F3 Jupiter (NETZSCH, Germany) coupled with the mass spectrometer MS 403 Aëolos II (NETZSCH, Germany). The sample was heated from 20 to 1000 °C at a heating rate of 10 °C min^{-1}. An empty crucible was used as a reference. Smoothed raw data of the thermogram, a heat flux curve, and mass spectrometer signals (18 for H_2O and 44 for CO_2) were evaluated (more details in Peikert et al. [22] and Tamimi et al. [23]). According to Plante et al. [24], the thermogram was analyzed for two fractions: labile organic matter (LOI_{labile}, calorific value of the labile fraction, CV_{labile}), and recalcitrant organic matter ($LOI_{recalitrant}$, $CV_{recalcitrant}$). The weight loss and the calorific value in the labile and recalcitrant fractions were summed up to determine the part caused by organic matter (LOI_{TGA}, CV_{TGA}). Then, the thermal stability of a sample was defined by the relative amount of organic matter that oxidizes with respect to the total weight loss (i.e., $LOI_{recalcitrant}/(LOI_{recalcitrant} + LOI_{labile})$.

OMW analyses were conducted on filtered samples (0.45 µm). Analyses included pH, EC, TP, DOC, ion contents, as well as thermogravimetry.

2.3. Biological Activity and Invertebrates

Invertebrates were sampled using pitfall traps before (2012 and twice 2014) and after (2012 and 2014) OMW application. Within each plot, two pitfall traps were set to level with the soil surface and placed in 40 cm distance (at an area that does receive irrigation water) from the irrigation line in the middle between two trees. Pitfall traps were filled to one-third with a 1:2 propylene-glycol and tap-water solution, which was emptied weekly for two weeks. Catches were transferred to 70% ethanol and determined on family level subsequently and summed up for both traps.

Biotic degradation processes were estimated before OMW application for the detailed survey using bait-lamina sticks (terra protecta GmbH) according to Kratz [25]. The 16 apertures of these perforated PVC strips were filled with a mixture of 70% cellulose, 27% wheat bran, and 3% activated carbon. Parallel to the irrigation line, 32 sticks were placed equidistantly distributed in 80 cm distance on both sides of the plot. Sticks were placed after soil sampling, removed after 14 days, and empty apertures were counted.

2.4. Data Analysis

Statistical analysis and plotting were done using *R* Statistics [26]. Each parameter was analyzed according to Lane [27] and Hothorn et al. [28], finding a suitable generalized linear model analysis of deviance [29]. Differences between treatments were identified (significance levels $p = 0.05$, 0.01, and 0.001) applying Bonferroni-adjusted Tukey tests post-hoc. The invertebrate community structure was analyzed using nonmetric multidimensional scaling [30]. The community structure was related to treatment and environmental parameters from the first soil sampling layer as described by Oksanen et al. [31].

3. Results and Discussion

3.1. Olive Mill Wastewater (OMW) and Untreated Soil Characteristics

The OMW and soil used in this study was already characterized by Zipori et al. [11] as well as Levy et al. [3] and is typical for OMW characterized in the Mediterranean region [6,10]. OMW was acidic (pH 4.4 ± 0.3) and loaded with conductive ions and inorganic material (EC 12.1 ± 1.2 dS m^{-1}, K^+ being most abundant with 5.3 ± 0.9 g L^{-1}). Total phenols (2.7 ± 0.3 g L^{-1}) were prominent in its organic fraction (Biological oxygen demand of 35.8 ± 10.0 g L^{-1}). Additionally, organic parameters were analyzed solely in the detailed survey and can be seen in Table 1. The soil in Gilat is a sandy loam with 50% sand, 35% silt, and 15% clay, a pH of 8.2, and an effective cation exchange capacity of 33 mmolc kg^{-1}.

Table 1. Selected properties of the OMW used in 2012 and 2013 (preceded the detailed survey). Other properties of the OMW used during 2012–2016 are found in Zipori et al. [11]. LOI, loss on ignition; CV, calorific value.

Parameter		OMW	
		2012	2013
SUVA	L mg C^{-1} m^{-1}	0.11 ± 0.01	0.13 ± 0.01
LOI	%	58 ± 0.9	64 ± 0.5
LOI_{labile}	%	48.9 ± 0.6	51.9 ± 0.7
$LOI_{recalcitrant}$	%	9.1 ± 0.4	12.2 ± 0.3
CV_{LOI}	kJ g^{-1}	19.7 ± 1.6	19.0 ± 1.3
CV_{labile}	kJ g^{-1}	4.4 ± 0.2	4.1 ± 0.2
$CV_{recalcitrant}$	kJ g^{-1}	9.4 ± 0.5	9.1 ± 0.6

3.2. Soil Hydrophobicity

The first WDPT recording was done in 2013, after two successive years of winter OMW applications. During the detailed survey, the control plots exhibited a distance-dependent hydrophobic profile (Figure 1a). Between 0–48 cm and 128–200 cm distance from the irrigation line, the soil was classified as wettable to slightly water repellent according to the arbitrary classification of Bisdom et al. [21]. However, at a distance of 48–128 cm, the soil had a median WDPT of 50.5 ± 70 s and can be classified up to strongly water repellent. OMW application increased WDPT most substantially after an application of 100 m^3 OMW ha^{-1} with a median of 90 ± 111 s ($p < 0.001$). In all OMW treatments, single spots showed WDPTs above 600 s, which is considered as severely water-repellent soil.

The repellency classes (subplots in Figure 1) show that the strongest hydrophobizing effects were found on 100 m^3 OMW ha^{-1} y^{-1} treated plots, which was followed by 50 and 150 m^3 OMW ha^{-1} y^{-1} treated plots, whereas the tilled plots with an application of 100 m^3 OMW ha^{-1} were even less hydrophobized than the control plots. This hydrophobicity effect was measured during summer in the following three successive years of winter OMW applications. The hydrophobicity effect was not intensified over the years but rather diminished substantially. In 2014 (Figure 1b), water repellency was higher near the irrigation line (100 and 150 m^3 ha^{-1} y^{-1}), and also at 48 cm distance but only at the 150 m^3 ha^{-1} y^{-1}. Plots receiving an OMW application rate of 150 m^3 OMW ha^{-1} y^{-1} showed high WDPT at 48 cm with a total of 50% of all spots of WDPT ≥ 5 s. The water repellency and its spatial distribution diminished in the following year (Figure 1c). However, only OMW-treated plots, also tilled ones, showed WDPT above 5 s in the last assessment (Figure 1d).

The noticeable distance-dependent spatial distribution of water repellency is mainly due to the field microtopography. Starting at the drip line, which is arranged at the tree trunks, there is a slight downward slope in the 50 cm parallel to the drip line. The area between the tree lines is flat. In these sinks, OMW, as well as natural organic matter such as olive leaves or fruits, accumulate and cause repellency. Harman et al. [32] found that the microtopography strongly influences soil hydraulic conductivity and soil organic matter contents in semi-arid shrubland. Steinmetz et al. [15] observed a similar spatial distribution with higher WDPT in control as well as OMW-treated plots in the first 60 cm ("irrigation zone") compared to the region between 80 and 200 cm ("dry zone") in the same study region where our study was conducted but in a different field experiment. However, 18 months after the application of 140 m^3 OMW ha^{-1} y^{-1}, they found no increased water repellency compared to control plots. This suggests that the additional time of six months might have been sufficient to degrade OMW-derived organic compounds. Similar to our study, Mahmoud et al. [33] found persistent water repellency after repeated OMW applications in winter.

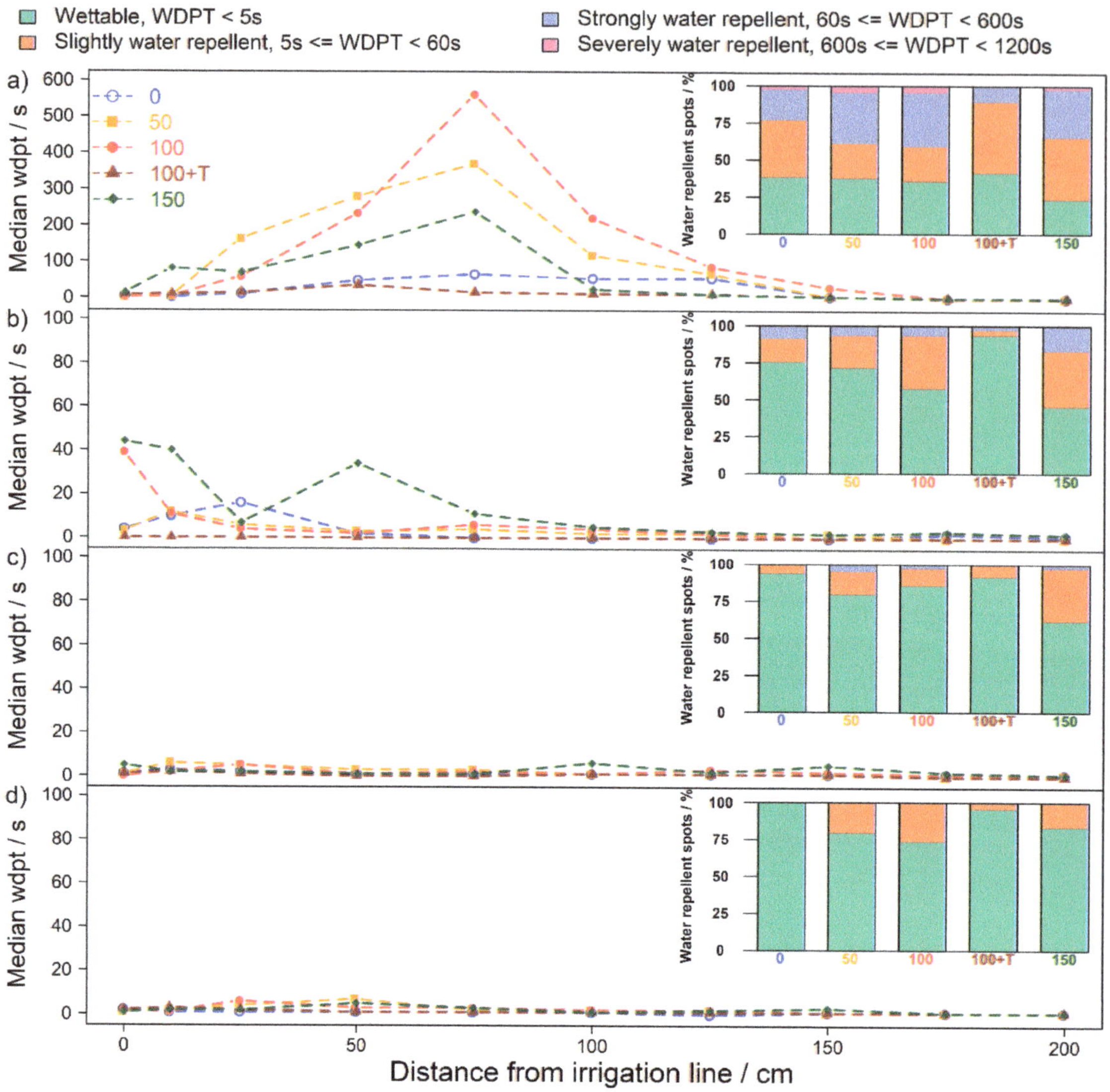

Figure 1. Median values of water drop penetration time (WDPT) measured in distance steps of 20 cm from the drip irrigation line. (**a**) 2013; (**b**) 2014; (**c**) 2015; (**d**) 2016. Subplot (stacked bars) shows the frequency distributions of four hydrophobicity categories ranging from wettable to strongly water repellent. OMW application rates are 0 (control), 50, 100, and 150 m^3 ha^{-1} y^{-1} and 100 m^3 ha^{-1} y^{-1} followed by shallow tillage (100 + T).

Tillage reduced the occurrence of water repellency and its spatial distribution as the natural and OMW-derived organic compounds were incorporated into deeper soil layers and crusts destroyed. However, the simple addition of natural and OMW-derived organic compounds does not solely lead to the expression of water repellency in soils. It can be caused by the natural organic compounds in the soil if they form a specific molecular arrangement triggered by the water content [34]. The critical water content of a soil is suggested to define a transition zone between a water repellent and a wettable state [35] and is a function of the soil organic carbon [36]. Due to their significantly lower soil organic carbon, only the control plots had critical water content below the measured water content in the upper layer, which could additionally lead to increased water repellency. Soil hydrophobicity is especially problematic in rainfed olive orchards, as it will increase the water run-off and reduce the amount of rain that will be absorbed by the soil. Based on the observations shown in Figure 1a, it was decided to focus the detailed survey at a distance of 80 cm.

3.3. Effects of OMW Application on Soil Properties

The net effect (delta) of the OMW treatments relative to the control plots is presented in Figure 2. Following OMW application, a short-term increase in soil water content was observed at the top 0–10 cm layer (Figure 2a). This increase was highest in tilled plots with a surplus of up to 3%, 2% after an application of 100 and 150 m^3 OMW ha^{-1} y^{-1} and 1% after 50 m^3 OMW ha^{-1} y^{-1}. Interestingly, higher soil water content by 1.8% was still observed at the 150 m^3 ha^{-1} y^{-1} treatment ($p < 0.05$) in the detailed survey (Figure 3a) just before the winter application of 2014. Neither irrigation water (starting in March) nor the cumulative amount of rain (226.5 mm between the previous application of 2013 and the detailed survey), which are both the same for the entire experimental plot, could be the source for differences in water contents among treatments. On the other hand, crust formation after OMW application due to fast drying could be responsible for this effect by covering the exchange surface and reducing drying kinetics [37]. Such an effect of crusts formed following OMW application has been seldom referred to in the literature, although biological soil crusts are known to positively affect the water balance in semi-arid ecosystems by modifying hydrological processes [38]. Moreover, the contribution of OMW to soil organic matter can increase the soil water-holding capacity [39]. This effect must be investigated further during spring and summer months, as any water surplus could be beneficial in this semi-arid ecosystem. Finally, although not measured in this study, improved soil aggregation, which might be a result of OMW application [40], may lead to higher water infiltration rates, which in turn increases the water content.

TP ($p < 0.001$) and DOC ($p < 0.05$) generally increased at the top 0–10 cm soil layer in OMW-treated plots (Figure 2b,c, respectively). This increase was OMW dosage-dependent in the detailed survey (Jan 2014) also in a depth of 10–20 cm (TP, $R^2 = 0.23$, $p < 0.001$, DOC, $R^2 = 0.07$, $p < 0.05$; not shown). The significant difference and dose-dependent correlation were even stronger at a depth of 0–10 cm. In the tilled plots, TP was also higher compared to the non-tilled control treatment down to a depth of 20–40 cm ($p < 0.05$). DOC as well as TP accumulated over the years, whereby this effect was strongest after an OMW application rate of 150 m^3 OMW ha^{-1} y^{-1} (1.9 mg TP L^{-1} y^{-1}); only soil samples before an OMW application were evaluated and linearly correlated to the OMW application rate. In general, TP was high in the upper soil layer (e.g., control plot in the detailed survey 28 ± 9 mg kg^{-1}) and then ranged between 14 ± 2 mg kg^{-1} and 17 ± 3 mg kg^{-1} in the beneath layers Figure 3b). The same trend was observed for dissolved organic carbon (DOC, correlation coefficient with TP was 0.74, $p < 0.001$, data not shown).

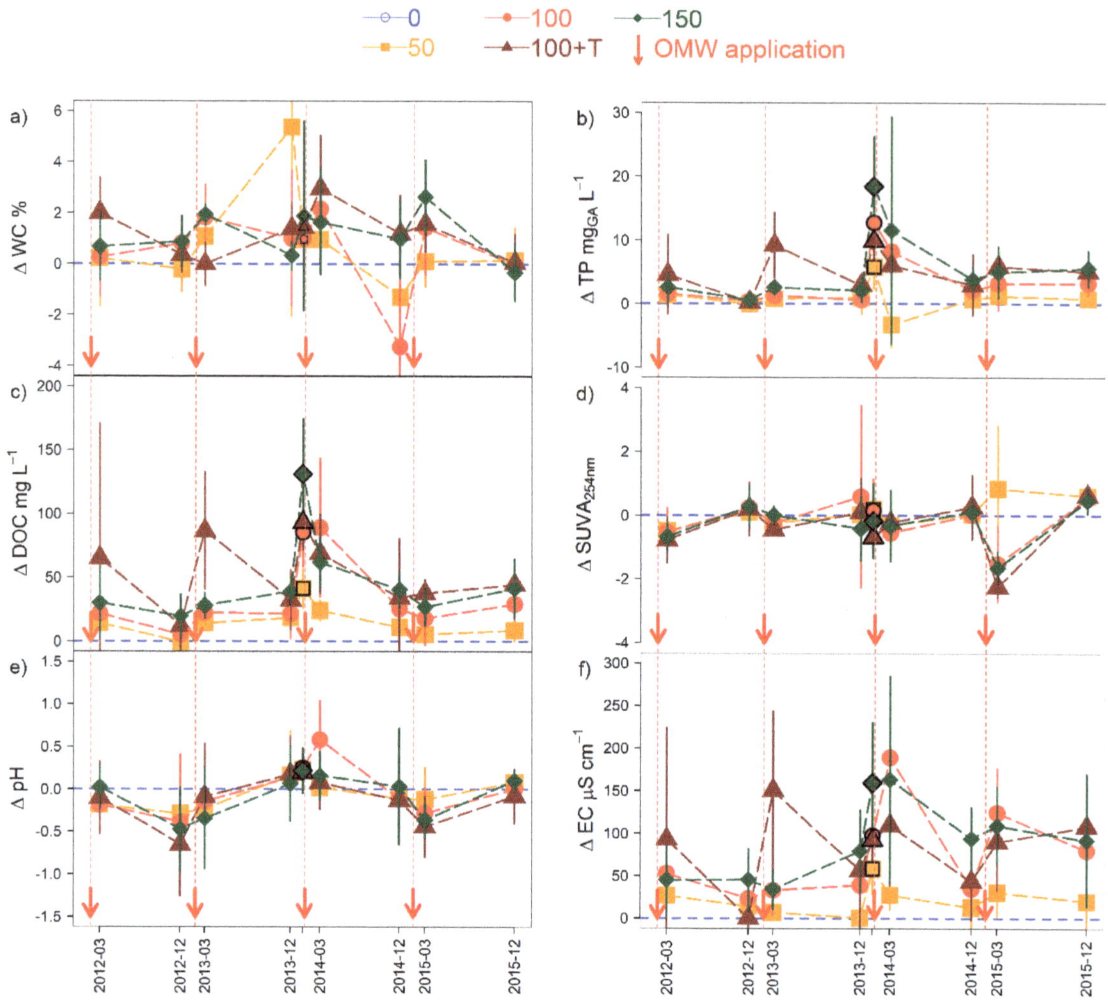

Figure 2. Evolution of physicochemical soil properties at a depth of 0–10 cm, expressed as the difference between OMW-treated and control plots. (**a**) Gravimetric water content; (**b**) total phenolic compounds (TP); (**c**) specific UV absorbance (SUVA); (**d**) pH; (**e**) electrical conductivity (EC); and (**f**) dissolved organic carbon (DOC). Bold red arrows with dashed lines mark OMW application dates. OMW application rates are 0 (control), 50, 100, and 150 m^3 ha^{-1} y^{-1} and 100 m^3 ha^{-1} y^{-1} followed by shallow tillage (100 + T).

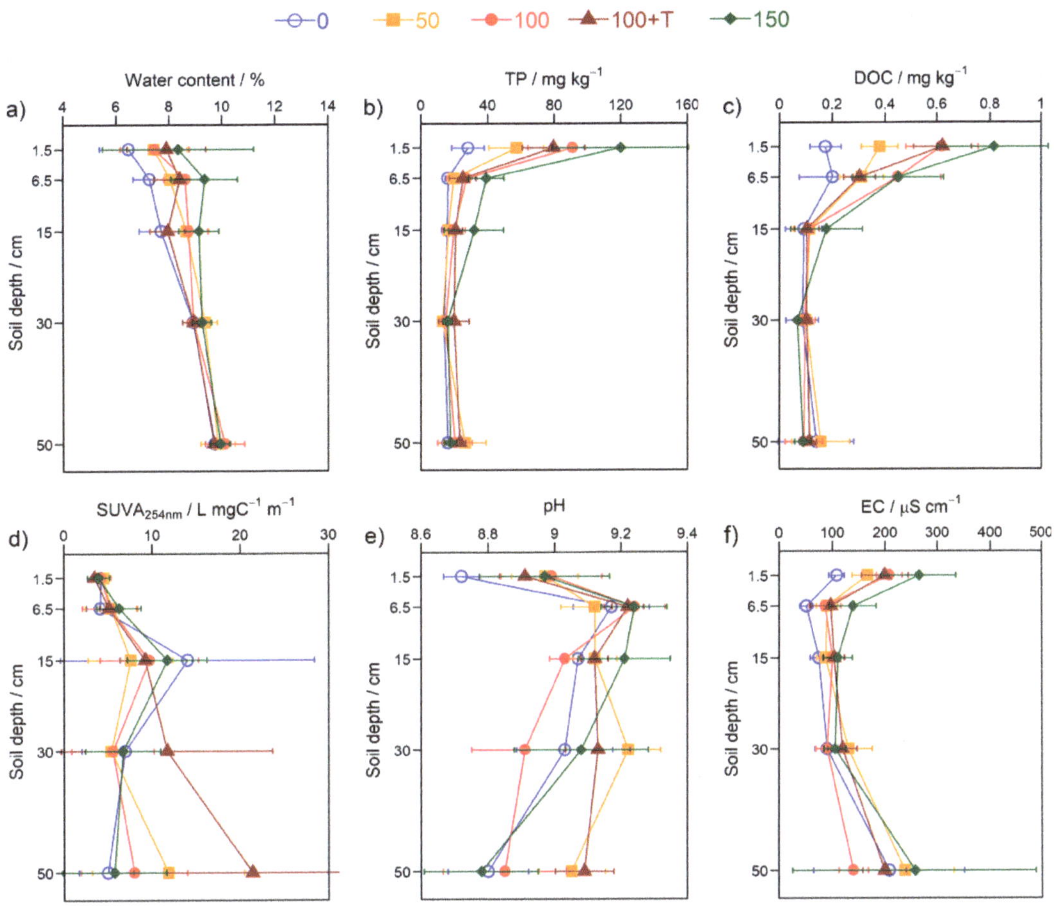

Figure 3. Depth soil profiles of (**a**) water content; (**b**) Total Phenols (TP); (**c**) dissolved organic carbon (DOC) (**d**) specific UV absorbance (SUVA); (**e**) pH; (**f**) electrical conductivity (EC) of the detailed survey before the third application of OMW. OMW application rates are 0 (control), 50, 100, and 150 m^3 ha^{-1} y^{-1} and 100 m^3 ha^{-1} y^{-1} followed by shallow tillage (100 + T).

OMW application affected the quality of soluble organic compounds expressed as SUVA254 toward higher aromaticity solely following tillage in a depth of 40–60 cm ($p < 0.001$, Figure 3c). At a depth of 15 cm, OMW application favored a lower aromaticity compared to the control. During the whole study period, no differences in the top layer could be detected. At the end of the study period, no effect on the balance of soil H^+ could be detected (Figure 3d). However, in the detailed survey, soil pH in the control soil ranged from 8.7 to 9.2 with its minimum at 0–3 cm and maxima at 3–10 cm (Figure 3e). Twelve months after the two first annual OMW applications, soil pH increased to 9.0 ± 0.2 in all treated plots solely in the upper soil layer (0–3 cm; $p < 0.01$). Previous reports show the opposite or no effect after OMW application (e.g., Lanza et al. [41]; or the review by Mekki et al. [17]). However, the mineralization of OMW-derived carbon releases free OH^- ions, which in turn could lead to a ligand exchange with ions brought by OMW. Such effects are also reported in other studies on OMW land application [42–44].

A strong effect of OMW application was found for soil EC (Figure 3f). Only the lowest application rate of 50 m^3 OMW ha^{-1} y^{-1} showed no increase compared to water-treated plots at the end of the study. In the detailed survey, the depth dependence of EC showed a characteristic c-shape in all treatments. In a depth of 0–3 (110 ± 14 µS cm^{-1}) and 40–60 cm (210 ± 100 µS cm^{-1}), we found the highest EC in the control, while at a depth of 3–10 cm (48 ± 9 µS cm^{-1}), it was lowest. OMW application did not affect this shape but increased the values especially in the upper soil layers ($p < 0.001$) and application rate dependent at 0–3 cm (adj. $R^2 = 0.65$, $p < 0.001$). This relation became weaker with increasing depth (3–10 cm, adj. $R^2 = 0.48$, $p < 0.001$; 10–20 cm, adj. $R^2 = 0.23$, $p < 0.001$). Ca^{2+} was the most abundant cation in each layer of the control plots between 130 and 180 mg kg^{-1} (Figure 4), which was followed by K^+ (43 to 130 mg kg^{-1}, Figure 4a). While PO_4^{3-} and NO_2^- were distributed equally within the soil layer (11.5 ± 1.2 mg kg^{-1} and 5.0 ± 0.3 mg kg^{-1}), NO_3^- was high in the upper soil layer (11 ± 16 mg kg^{-1}) and lower in the layers beneath (5.3 ± 0.7 mg kg^{-1}). Conversely, SO_4^{3-} increased with increasing depth. These relations changed after OMW application. The absolute and relative effect of annual application of 150 m^3 OMW ha^{-1} y^{-1} was by far highest for K^+ (590 ± 100 mg kg^{-1}). Even in a depth of 10–20 cm, plots receiving this application rate of OMW contained K^+ contents of 160 ± 110 mg kg^{-1} while control plots contained only 54 ± 21 mg kg^{-1}. Up to this depth, OMW application increased K^+ contents linearly with the applied application rate. Similar differences and application rate dependencies were found for Mn^{2+}, Na^+, Cl^-, and PO_4^{2-} (Figure 4). For sulfate, this was only found between 0 and 3 cm, while calcium, magnesium, and iron showed this relation only in depth of 3–10 cm. Linear relations were always stronger and more significant when the tillage treatment was not incorporated into the model. K^+, Mg^{2+}, Ca^{2+}, Fe^{2+}, and Mn^{2+} as well as SUVA were elevated in tilled plots at a depth of 40–60 cm compared to the control and all other OMW treatments, but these trends were not significant.

OMW spreading led to an application rate-dependent increase of cations as well as anions in the soil. Ions were contributed by OMW itself but also mobilized from the solid soil phase as shown especially for Fe^{2+} and Mn^{2+}. Aharonov-Nadborny [45], who investigated the leaching of soil cations after OMW application, suggested that cation mobilization is mainly due to the formation of cation–organic complexes with the OMW-derived dissolved organic matter and cation exchange. In loess soil, they found a high leaching potential for Mn, which can be confirmed in our study. Especially tillage increases this risk, as we found higher cation contents in the lowest investigated soil layer.

Even the lowest application rate of 50 m^3 OMW ha^{-1} lead to significant physicochemical changes 12 months after two annual OMW applications. TP, DOC, and EC as well as most anions (Cl^-, PO_4^{3-}, SO_4^{3-}) and cations (Na^+, K^+, Ca^{2+}, Mg^{2+}) accumulated linearly with the OMW application rate. This was attenuated but still significant for the application of 100 m^3 OMW ha^{-1} plus tillage. In the same study area, 18 months (6 months more compared to our study) after two annual OMW applications of 70 and 140 m^3 ha^{-1} y^{-1}, pH, EC, all cations (except K^+), C_{org}, TP, WDPT, as well as bait consumption were unaffected [15]. It is assumed that increasing temperatures and increasing soil humidity stimulate soil resilience. Moreover, OMW-derived changes disappeared already within three weeks after OMW application of 140 m^3 ha^{-1} y^{-1} in the summer in this study region [8]. Such seasonal effects were also shown for sandy clay loam treated with OMW 40 m^3 ha^{-1} y^{-1} and 80 m^3 ha^{-1} y^{-1} [11] and 150 m^3 ha^{-1} y^{-1} for clay loam [3,4]. These works highlight the unfavorable environmental conditions resulting in low biological activity and a risk for leaching when spreading OMW during winter, as done in our study.

Notably, the high resolution of soil profiles described in the present study revealed more OMW-derived effects; e.g., when analyzing the differences between treatments based on the weighted average of the three layers together (0–3, 3–10, and 10–20 cm), significant differences for all parameters became insignificant or were weaker (e.g., no difference between 0 and 50 m^3 OMW ha^{-1} y^{-1} could be detected for TP or the significant application rate effect on K^+ disappeared). Some significances already disappeared when using the weighted average of only the depths 0–3 and 3–10 cm (e.g., Mn^{2+}, Fe^{2+}, SO_4^{3-}, SUVA, thermogravimetric-derived parameter).

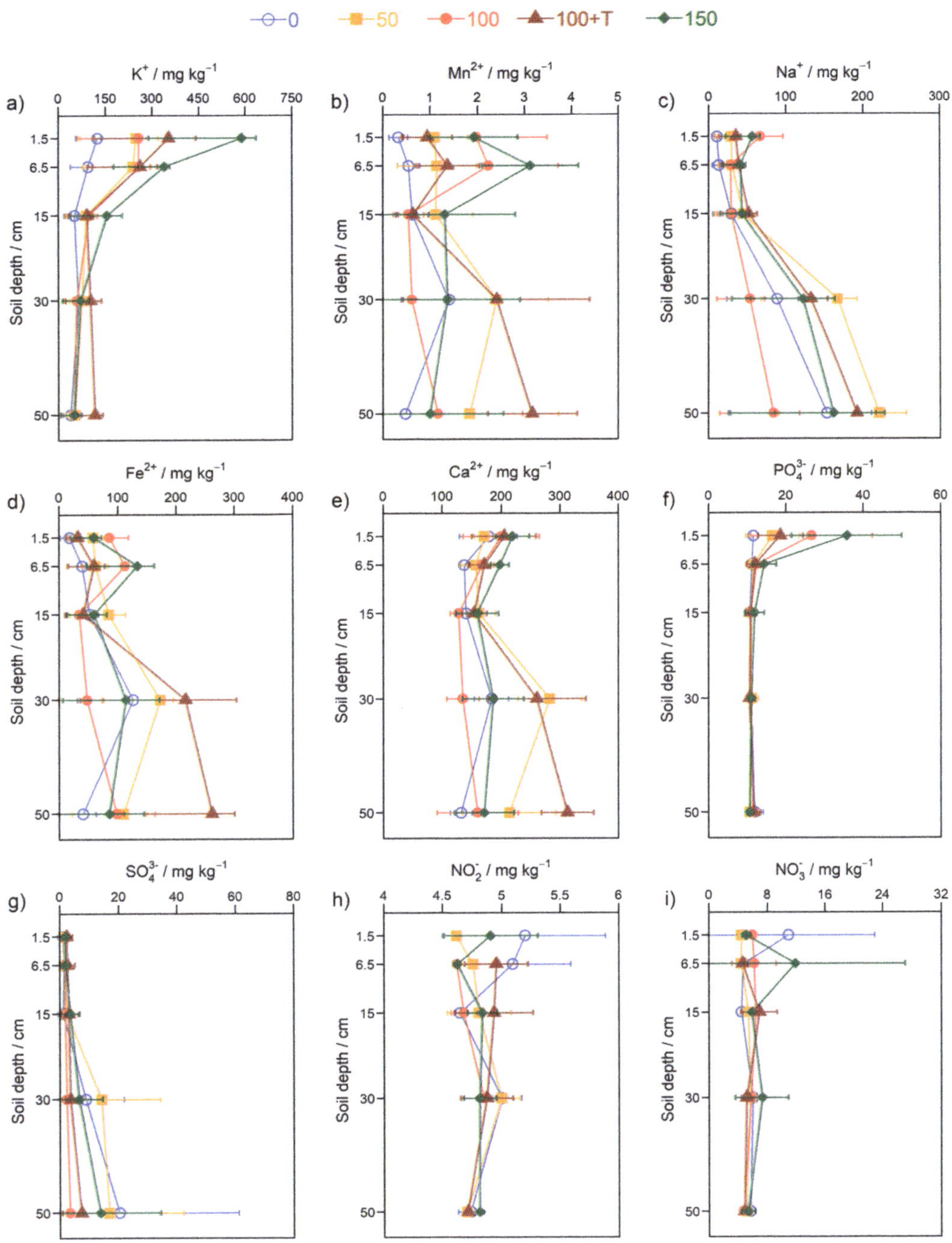

Figure 4. Depth soil profiles of K^+ (**a**), Mn^{2+} (**b**), Na^+ (**c**), Fe^{2+} (**d**), Ca^{2+} (**e**), PO_4^{3-} (**f**), SO_4^{3-} (**g**), NO_2^- (**h**), and NO_3^- (**i**) of the detailed survey before the third OMW application. OMW application rates are 0 (control), 50, 100, and 150 $m^3\ ha^{-1}\ y^{-1}$ and 100 $m^3\ ha^{-1}\ y^{-1}$ followed by shallow tillage (100 + T).

3.4. Changes in Thermal SOM Properties after OMW Application

During the thermoanalysis of soil samples from the detailed survey, temperatures of CO_2 evolution below 550 °C ranged from 286 ± 1 °C to 508 ± 6 °C. LOI_{labile} was identified by the thermogram and ranged from 286 ± 1 °C to 398 ± 3 °C (Figure 5a). $LOI_{recalcitrant}$ ranged from 398 ± 3 °C to 508 ± 6 °C (Figure 5b). Only the application of 100 m^3 OMW ha^{-1} y^{-1} with tillage significantly increased the LOI_{labile} compared to the control in both upper soil depths of 0–3 and 3–10 cm ($p < 0.01$). $LOI_{recalcitrant}$ decreased after an application of 50 m^3 OMW ha^{-1} y^{-1} in a depth of 0–3 cm and in all OMW treatments except 100 m^3 OMW ha^{-1} y^{-1} tillage compared to the control in a depth of 3–10 cm ($p < 0.001$), while this was true only for the application of 50 m^3 OMW ha^{-1} y^{-1} for the above soil layer. The application of OMW reduced the thermal stability in all OMW treatments compared with the control ($p < 0.001$, Figure 5c) in both soil layers except for the lowest application rate of 50 m^3 OMW ha^{-1} y^{-1} (3–10 cm). The calorific value (CV_{labile}, Figure 5d) of the labile fraction decreased after an application of 100 m^3 OMW ha^{-1} y^{-1} ($p < 0.01$).

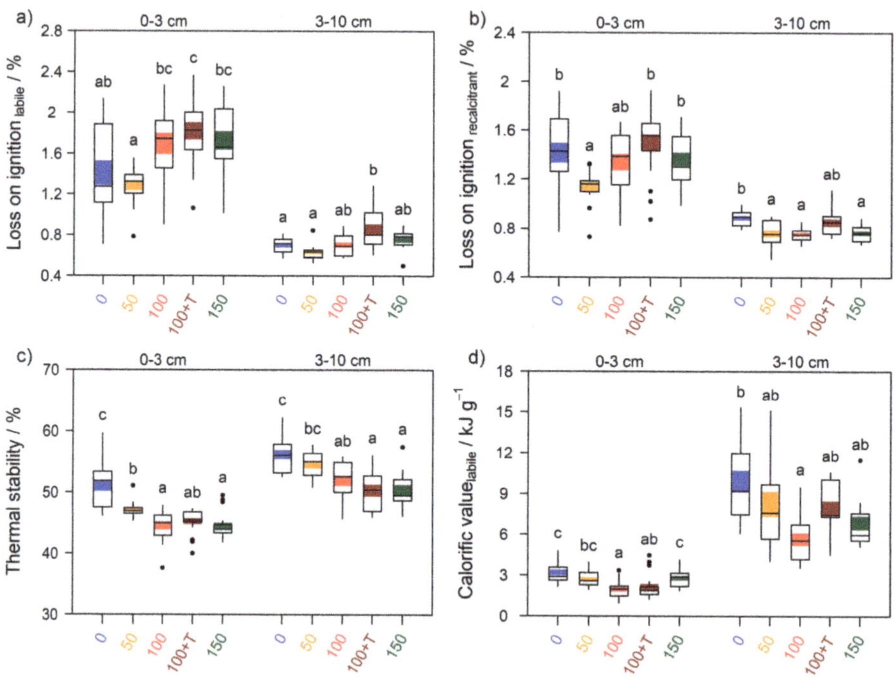

Figure 5. Thermogravimetric analysis of soil organic matter in soil depths of 0–3 and 3–10 cm: (**a**) mass loss on ignition of the labile temperature range; (**b**) mass loss on ignition of the recalcitrant temperature range; (**c**) thermal stability index; (**d**) calorific value of the labile fraction. Boxplots show the median and ranges from the 25th to the 75th percentile; the whisker length was set to 1.5 times the interquartile range. OMW application rates are 0 (control), 50, 100, and 150 m^3 ha^{-1} y^{-1} and 100 m^3 ha^{-1} y^{-1} followed by shallow tillage (100 + T). Colored overlays indicate the standard deviation of the mean. Mean values within one soil depth followed by the same letter were not statistically different ($p \leq 0.05$).

The reduced $LOI_{recalcitrant}$ could be due to a priming effect, which was also identified as responsible by Tamimi et al. [23]. By adding easily degradable exogenous substrate (OMW) to the soil as an energy source for microorganisms, co-metabolized enzymes are synthesized by K-strategists capable of degrading recalcitrant OM [46]. This and the OMW-OM could also be one source for increased LOI_{labile} in OMW-treated plots, as the degraded recalcitrant OM is converted to labile OM. Another explanation for the increased labile organic carbon in tilled treatments could be due to residual plant materials and roots brought into the soil matrix through this management practice [47].

3.5. *Effects of OMW on Soil Biological Activity and Invertebrates*

In the detailed survey conducted on January 2014 after two OMW applications and just before the third winter application, the median biological activity in terms of consumed bait material showed a distinct depth dependency (Figure 6). Moreover, OMW application positively affected soil biological activity in the upper soil layer (0–0.5 cm). Overall, a significant effect on soil biological activity could be found for OMW application rates of 100 m^3 OMW ha^{-1} y^{-1} with and without tillage.

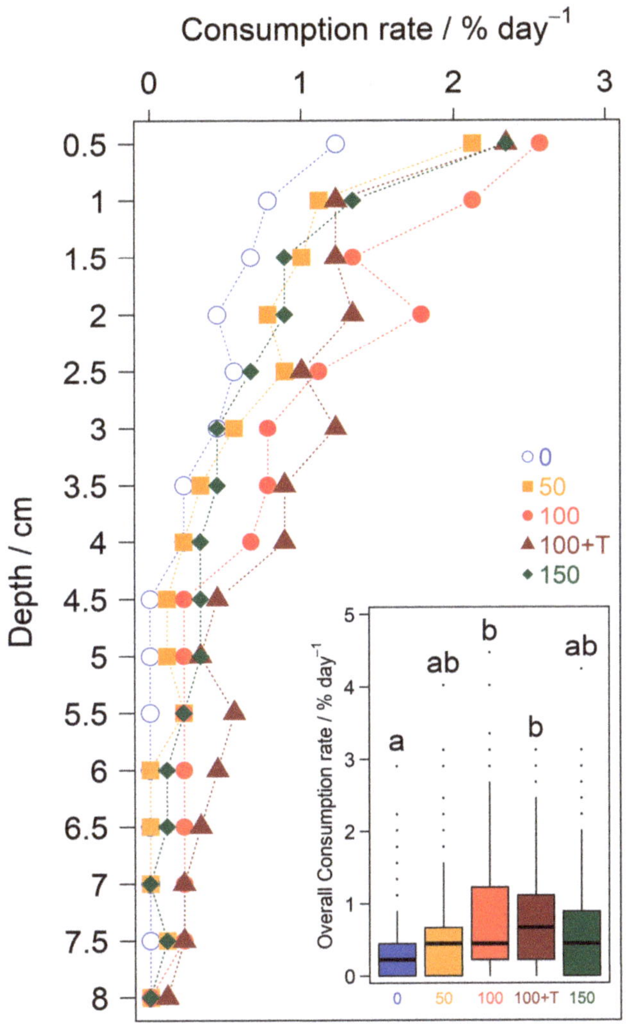

Figure 6. Soil biological activity as consumed bait material per day. The subplot shows the median and ranges from the 25th to the 75th percentile, the whisker length was set to 1.5 times the interquartile range. OMW application rates are 0 (control), 50, 100, and 150 m^3 ha^{-1} y^{-1} and 100 m^3 ha^{-1} y^{-1} followed by shallow tillage (100 + T). Dots represent outliers. Different small letters indicate significant differences between OMW applications ($p < 0.05$).

During the detailed survey, the abundance of Collembola (Figure 7) was not affected by OMW. In addition to grubs, Coleoptera, Arachnida, Dermaptera, and Diplopoda were not affected and different compared with the control. During the whole study period, no clear pattern was found.

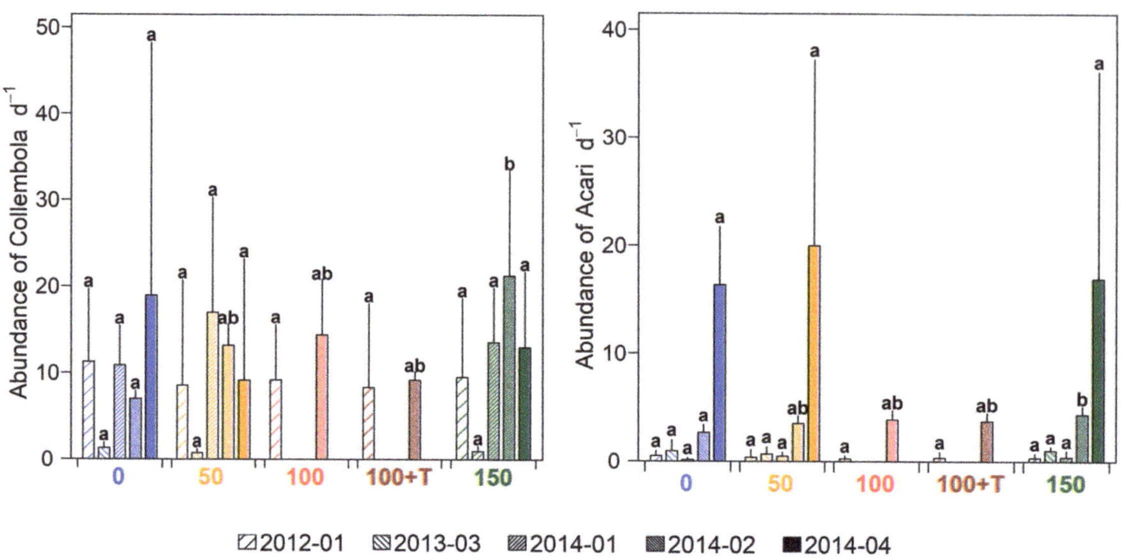

Figure 7. Abundance of Collembola and Acari per sampling day in five scoring events. OMW application rates are 0 (control), 50, 100, and 150 m^3 ha^{-1} y^{-1} and 100 m^3 ha^{-1} y^{-1} followed by shallow tillage (100 + T). Different small letters indicate significant differences within one OMW application rate ($p < 0.01$).

OMW application, independent of the application rate, created temporarily biologically favorable conditions resulting in higher biological activity and higher abundances of Collembola and Acari. Steinmetz et al. [15] did not find any effect on bait consumption after winter application but rather an adverse effect when applying OMW in summer due to high TP contents. To our best knowledge, the only field study available dealing with effects of OMW toward soil invertebrates was performed in a neighboring field, and the authors monitored the effects three weeks after the application of OMW in summer [8]. In their study, neither negative nor positive effects toward Collembola were found. The latter was surprising, as a high application rate of 150 m^3 OMW ha^{-1} y^{-1} was applied four days before the first invertebrate sampling. Even more, positive effects on Acari abundance were discovered three weeks after the OMW application. It is to be noted that the abundance was much lower in this study compared to ours. During six sampling events, Kurtz et al. [8] found 56 individuals of Collembola and 716 individuals of Acari, whereas in this study, 4598 and 1268 individuals were found for Collembola and Acari during the detailed survey, respectively. This difference is mostly attributed to the sampling season (hot and dry summer vs. wet and mild winter) and sampling method (soil extraction with Berlese–Tullgren funnels vs. pitfalls). Conditions of ideal humidity, temperature, and food supply could be the reason for such high abundances due to induced swarming [48]. This is highlighted by the increased consumption of bait material, which indicates a stronger activity of soil mesofauna [49] in all depths down to 8 cm. OMW is known to increase the number of bacteria and fungi in soil [41,50], which in turn are preferred feeding sources for Collembola [51]. Moreover, Hentati et al. [16] showed that the standard soil arthropod Folsomia candida favors some soils from OMW ponds (with an EC > 1600 µS cm^{-1}) compared to a farmland 'reference' soil without OMW treatment. These results strongly suggest that the soil fauna is insensitive or even prefers OMW-derived edaphic changes (i.e., phenols, carbon input, high salt contents, or hydrophobic films) after a time period of one year. The toxic effects of OMW-derived phenolic compounds disappeared through the degradation or polymerization of these OMW constituents. Soil fauna is known to stimulate microbial activity and modify the recalcitrance persistence of soil carbon pools [52]. Wang et al. [53] found that high numbers of Collembola transfer soil C from recalcitrant to more labile sources. Additionally to the priming effect, this can explain the decreased recalcitrant carbon in the lower soil layer as OMW application significantly increases soil biological activity.

3.6. Practical Implications

Considering national and international recommendations of 30–80 m^3 OMW ha^{-1} y^{-1}, the area available for spreading in olive orchards is sufficient to absorb the entire amount of OMW produced from olive oil production. In non-irrigated orchards, where olive production is in the range of 2–3 tons ha^{-1}, and the resulting OMW is ca. 2.5–3.5 m^3 ha^{-1} accordingly, the area available for spreading (e.g., 50% of the land) means a low dose of only 5–7 m^3 ha^{-1} y^{-1}, or 10–14 m^3 ha^{-1} every second year. Even in intensive irrigated orchards, where olive production is much higher, in the range of 10 tons ha^{-1}, and the resulting OMW is ca. 10–12 m^3 ha^{-1} accordingly, the area available for spreading (e.g., 50% of the land) means a dose of 20–22 m^3 ha^{-1} y^{-1}, or 40–44 m^3 ha^{-1} every second year. These volumes are much lower than the above-mentioned recommendations, indicating that spreading OMW in olive orchards is a sustainable practice. Potential annual contribution of OMW to soil nutritional status is significant and in line with the idea of turning waste into a resource. The application of 50 m^3 OMW ha^{-1} from our study would provide 318 kg K_2O ha^{-1} y^{-1} (annual recommendation in e.g., Israel is 300 kg K_2O ha^{-1} y^{-1}) and save around 1500 € ha^{-1} y^{-1} [11].

4. Conclusions

OMW application in semi-arid areas with sandy loam soils has positive effects (accumulation of K^+, PO_4^{3-}, organic carbon, and increased biological activity and diversity) but also negative consequences (accumulation of phenolic substances, increased soil hydrophobicity, and salinity). Most of these effects are OMW dose-dependent and most substantial in the topsoil. The positive effects underline the strategy to apply low application rates of OMW in alternating locations. Especially tillage after OMW application needs to be further investigated as we found indicators (SUVA, K^+, Mn^{2+}) for higher leaching risks after this management practice. Since tillage 'dilutes' the OMW within the soil, the displacement of OMW constituents favors biological degradation as the effect of UV degradation is negligible in deeper soil layers. Although the strongly reduced soil hydrophobicity after tillage did not affect soil biology and even improved soil physicochemical properties, further investigations on higher leaching risks after this management practice are mandatory. A simple field kit or annual sampling to determine soil EC, sum of phenolic substances, and water drop penetration time would be sufficient for a rough estimation if another OMW application could be performed.

The overall results of this study indicate that applying the recommended doses of 30–80 m^3 OMW ha^{-1} y^{-1} in olive orchards at a semi-arid region pose no clear negative effects on soil chemical and physical properties. The only measured negative effect is the temporary increase in soil surface hydrophobicity. To overcome this potential effect, shallow tillage may be applied subsequently after OMW spreading.

Author Contributions: Conceptualization: M.P.K., G.E.S., D.D. and A.D.; methodology, M.P.K., G.E.S., I.Z., Y.L., D.D., S.M., E.Z.-F. and A.D.; software, M.P.K.; validation, M.P.K., A.D. and Y.L. formal analysis, M.P.K., G.E.S., A.D. and D.D.; investigation, M.P.K., I.S., S.M., M.R., E.Z.-F., I.S.; resources, G.E.S., A.D., I.S.; data curation, M.P.K., C.B., A.D., Y.L., I.Z. and D.D.; writing—original draft preparation, M.P.K., C.B.; and D.D.; writing—review and editing, M.P.K., C.B., A.D., Y.L. and D.D.; visualization, M.P.K.; supervision, D.D. and G.E.S.; project administration, G.E.S. and A.D. funding acquisition, G.E.S. and A.D. All authors have read and agreed to the published version of the manuscript.

Funding: This study was supported by the Deutsche Forschungsgemeinschaft (TRILAT-OLIVEOIL, SCHA 849/13 and also part of the MERC foundation (Research Project M31-019).

Institutional Review Board Statement: Not applicable.

Informed Consent Statement: Not applicable.

Data Availability Statement: The data presented in this study are available on request from the corresponding author. The data are not publicly available at the time of publication due to an ongoing data transfer to the corresponding author. An openly accessible link to the raw data will be provided in a timely manner.

Acknowledgments: We want to thank the researchers in Gilat and Landau for fruitful discussions and help in the laboratory. Finally, we thank Yulia Subbotin and Silvia Zypman for help in Gilat and Avi for hospitality in (Hapoel!) Be'er Sheva.

Conflicts of Interest: The authors declare no conflict of interest.

References

1. Elkacmi, R.; Bennajah, M. Advanced oxidation technologies for the treatment and detoxification of olive mill wastewater: A general review. *J. Water Reuse Desalination* **2019**, *9*, 463–505. [CrossRef]
2. Laor, Y.; Saadi, I.; Raviv, M.; Medina, S.; Erez-Reifen, D.; Eizenbergc, H. Land spreading of olive mill wastewater in Israel: Current knowledge, practical experience, and future research needs. *Isr. J. Plant Sci.* **2011**, *59*, 39–51. [CrossRef]
3. Levy, G.J.; Dag, A.; Raviv, M.; Zipori, I.; Medina, S.; Saadi, I.; Krasnovski, A.; Eizenberg, H.; Laor, Y. Annual spreading of olive mill wastewater over consecutive years: Effects on cultivated soils' physical properties. *Land Degrad. Dev.* **2018**, *29*, 176–187. [CrossRef]
4. Aboud, T. *Online Presentation: Estimation of Environmental Pollution Potential from Olive Mill Wastewaters in Israel & Integrated Solutions for Olive Mill Wastewater with Sewage Treatment Plants and Policy;* Agro-Ecology Department, North District, Ministry of Environmental Protection: Bet Dagan, Israel, 2012.
5. Mohawesh, O.; Albalasmeh, A.; Al-Hamaiedeh, H.; Qaraleh, S.; Maaitah, O.; Bawalize, A.; Almajali, D. Controlled Land Application of Olive Mill Wastewater (OMW): Enhance Soil Indices and Barley Growth Performance in Arid Environments. *Water Air Soil Pollut.* **2020**, *231*, 1–12. [CrossRef]
6. Sierra, J.; Martí, E.; Garau, M.A.; Cruañas, R. Effects of the agronomic use of olive oil mill wastewater: Field experiment. *Sci. Total Environ.* **2007**, *378*, 90–94. [CrossRef]
7. Barbera, A.; Maucieri, C.; Cavallaro, V.; Ioppolo, A.; Spagna, G. Effects of spreading olive mill wastewater on soil properties and crops, a review. *Agric. Water Manag.* **2013**, *119*, 43–53. [CrossRef]
8. Kurtz, M.P.; Peikert, B.; Brühl, C.; Dag, A.; Zipori, I.; Shoqeir, J.H.; Schaumann, G.E. Effects of Olive Mill Wastewater on Soil Microarthropods and Soil Chemistry in Two Different Cultivation Scenarios in Israel and Palestinian Territories. *Agriculture* **2015**, *5*, 857–878. [CrossRef]
9. Buchmann, C.; Felten, A.; Peikert, B.; Muñoz, K.; Bandow, N.; Dag, A.; Schaumann, G.E. Development of phytotoxicity and composition of a soil treated with olive mill wastewater (OMW): An incubation study. *Plant Soil* **2015**, *386*, 99–112. [CrossRef]
10. Piotrowska, A.; Iamarino, G.; Rao, M.A.; Gianfreda, L. Short-term effects of olive mill waste water (OMW) on chemical and biochemical properties of a semiarid Mediterranean soil. *Soil Biol. Biochem.* **2006**, *38*, 600–610. [CrossRef]
11. Zipori, I.; Dag, A.; Laor, Y.; Levy, G.J.; Eizenberg, H.; Yermiyahu, U.; Medina, S.; Saadi, I.; Krasnovski, A.; Raviv, M. Potential nutritional value of olive-mill wastewater applied to irrigated olive (*Olea europaea* L.) orchard in a semi-arid environment over 5 years. *Sci. Hortic.* **2018**, *241*, 218–224. [CrossRef]
12. Bronick, C.; Lal, R. Soil structure and management: A review. *Geoderma* **2005**, *124*, 3–22. [CrossRef]
13. Peikert, B.; Schaumann, G.; Keren, Y.; Bukhanovsky, N.; Borisover, M.; Garfha, M.; Shoqeric, J.; Dağ, A. Characterization of topsoils subjected to poorly controlled olive oil mill wastewater pollution in West Bank and Israel. *Agric. Ecosyst. Environ.* **2015**, *199*, 176–189. [CrossRef]
14. Tamimi, N.; Diehl, D.; Njoum, M.; Marei, A.; Schaumann, G.E. Effects of olive mill wastewater disposal on soil: Interaction mechanisms during different seasons. *J. Hydrol. Hydromech.* **2016**, *64*, 176–195. [CrossRef]

15. Steinmetz, Z.; Kurtz, M.P.; Dag, A.; Zipori, I.; Schaumann, G.E. The seasonal influence of olive mill wastewater applications on an orchard soil under semi-arid conditions. *J. Plant Nutr. Soil Sci.* **2015**, *178*, 641–648. [CrossRef]
16. Hentati, O.; Oliveira, V.; Sena, C.; Bouji, M.S.M.; Wali, A.; Ksibi, M. Soil contamination with olive mill wastes negatively affects microbial communities, invertebrates and plants. *Ecotoxicology* **2016**, *25*, 1500–1513. [CrossRef]
17. Mekki, A.; Dhouib, A.; Sayadi, S. Review: Effects of olive mill wastewater application on soil properties and plants growth. *Int. J. Recycl. Org. Waste Agric.* **2013**, *2*, 1–7. [CrossRef]
18. *ISO 11465 Soil Quality: Determination of Dry Matter and Water Content on a Mass Basis: Gravimetric Method*; International Organization for Standardization: Geneva, Switzerland, 1993.
19. *ISO 11277 Soil Quality—Determination of Particle Size Distribution in Mineral Soil Material—Method by Sieving and Sedimentation*; International Organization for Standardization: Geneva, Switzerland, 2020.
20. Box, J. Investigation of the Folin-Ciocalteau phenol reagent for the determination of polyphenolic substances in natural waters. *Water Res.* **1983**, *17*, 511–525. [CrossRef]
21. Bisdom, E.; Dekker, L.; Schoute, J. Water repellency of sieve fractions from sandy soils and relationships with organic material and soil structure. *Geoderma* **1993**, *56*, 105–118. [CrossRef]
22. Peikert, B.; Schaumann, G.E.; Bibus, D.; Fischer, J.; Braun, U.; Brunkhardt, J. Effects of olive oil mill wastewater on chemical, microbiological, and physical properties of soil incubated under four different climatic conditions. *Biol. Fertil. Soils* **2016**, *53*, 89–102. [CrossRef]
23. Tamimi, N.; Schaumann, G.E.; Diehl, D. The fate of organic matter brought into soil by olive mill wastewater application at different seasons. *J. Soils Sediments* **2017**, *17*, 901–916. [CrossRef]
24. Plante, A.F.; Fernández, J.M.; Leifeld, J. Application of thermal analysis techniques in soil science. *Geoderma* **2009**, *153*, 1–10. [CrossRef]
25. Kratz, W. The bait-lamina test. *Environ. Sci. Pollut. Res.* **1998**, *5*, 94–96. [CrossRef] [PubMed]
26. R Core Team. *R: A Language and Environment for Statistical Computing*; R Foundation for Statistical Computing: Vienna, Austria, 2020.
27. Lane, P.W. Generalized linear models in soil science. *Eur. J. Soil Sci.* **2002**, *53*, 241–251. [CrossRef]
28. Hothorn, T.; Bretz, F.; Westfall, P.; Heiberger, R.M.; Schuetzenmeister, A. Multcomp: Simultaneous Inference in General Parametric Models. *R Package Version* **2014**, *50*, 1–2.
29. Nelder, J.A.; Wedderburn, R.W.M. Generalized Linear Models. *J. R. Stat. Soc. Ser. A* **1972**, *135*, 370. [CrossRef]
30. Venables, W.N.; Ripley, B.D. *Modern Applied Statistics with S*, 4th ed.; Statistics and Computing; Springer: New York, NY, USA, 2002; ISBN 978-0-387-95457-8.
31. Oksanen, J.; Blanchet, F.G.; Kindt, R.; Legendre, P.; Minchin, P.R.; O'Hara, R.B.; Simpson, G.L.; Solymos, P.; Stevens, M.H.H.; Wagner, H. *Vegan: Community Ecology Package*; International Centre for Research in Agroforestry (ICRAF): Nairobi, Kenya, 2015.
32. Harman, C.J.; Lohse, K.A.; Troch, P.A.; Sivapalan, M. Spatial patterns of vegetation, soils, and microtopography from terrestrial laser scanning on two semiarid hillslopes of contrasting lithology. *J. Geophys. Res. Biogeosci.* **2014**, *119*, 163–180. [CrossRef]
33. Mahmoud, M.; Janssen, M.; Haboub, N.; Nassour, A.; Lennartz, B. The impact of olive mill wastewater application on flow and transport properties in soils. *Soil Tillage Res.* **2010**, *107*, 36–41. [CrossRef]
34. Dekker, L.W.; Ritsema, C.J. How water moves in a water repellent sandy soil: 1. Potential and actual water repellency. *Water Resour. Res.* **1994**, *30*, 2507–2517. [CrossRef]
35. Dekker, L.W.; Doerr, S.H.; Oostindie, K.; Ziogas, A.K.; Ritsema, C.J. Water Repellency and Critical Soil Water Content in a Dune Sand. *Soil Sci. Soc. Am. J.* **2001**, *65*, 1667–1674. [CrossRef]
36. Goebel, M.-O.; Bachmann, J.; Reichstein, M.; Janssens, I.A.; Guggenberger, G. Soil water repellency and its implications for organic matter decomposition—Is there a link to extreme climatic events? *Glob. Chang. Biol.* **2011**, *17*, 2640–2656. [CrossRef]
37. Jeguirim, M.; Dutournié, P.; Zorpas, A.A.; Limousy, L. Olive Mill Wastewater: From a Pollutant to Green Fuels, Agricultural Water Source and Bio-Fertilizer—Part 1. The Drying Kinetics. *Energies* **2017**, *10*, 1423. [CrossRef]
38. Chamizo, S.; Cantón, Y.; Rodríguez-Caballero, E.; Domingo, F. Biocrusts Positively Affect the Soil Water Balance in Semiarid Ecosystems: The Role of Biocrusts in the Local Water Balance. *Ecohydrology* **2016**, *9*, 1208–1221. [CrossRef]
39. Mahmoud, M.; Janssen, M.; Peth, S.; Horn, R.; Lennartz, B. Long-term impact of irrigation with olive mill wastewater on aggregate properties in the top soil. *Soil Tillage Res.* **2012**, *124*, 24–31. [CrossRef]
40. Mao, J.; Nierop, K.G.J.; Dekker, S.C.; Dekker, L.W.; Chen, B. Understanding the mechanisms of soil water repellency from nanoscale to ecosystem scale: A review. *J. Soils Sediments* **2018**, *19*, 171–185. [CrossRef]
41. Lanza, B.; Di Serio, M.G.; Di Giovacchino, L. Long-Term Spreading of Olive Mill Wastewater on Olive Orchard: Effects on Olive Production, Oil Quality, and Soil Properties. *Commun. Soil Sci. Plant Anal.* **2017**, *48*, 2420–2433. [CrossRef]
42. Chaari, L.; Elloumi, N.; Mseddi, S.; Gargouri, K.; Ben Rouina, B.; Mechichi, T.; Kallel, M. Changes in Soil Macronutrients after a Long-Term Application of Olive Mill Wastewater. *J. Agric. Chem. Environ.* **2015**, *4*, 1–13. [CrossRef]
43. Di Bene, C.; Pellegrino, E.; Debolini, M.; Silvestri, N.; Bonari, E. Short- and long-term effects of olive mill wastewater land spreading on soil chemical and biological properties. *Soil Biol. Biochem.* **2013**, *56*, 21–30. [CrossRef]
44. Vella, F.M.; Galli, E.; Calandrelli, R.; Cautela, D.; Laratta, B. Effect of Olive Mill Wastewater Spreading on Soil Properties. *Bull. Environ. Contam. Toxicol.* **2016**, *97*, 138–144. [CrossRef] [PubMed]

45. Aharonov-Nadborny, R.; Tsechansky, L.; Raviv, M.; Graber, E. Mechanisms governing the leaching of soil metals as a result of disposal of olive mill wastewater on agricultural soils. *Sci. Total. Environ.* **2018**, *630*, 1115–1123. [CrossRef]
46. Blagodatskaya, E.; Kuzyakov, Y. Mechanisms of real and apparent priming effects and their dependence on soil microbial biomass and community structure: Critical review. *Biol. Fertil. Soils* **2008**, *45*, 115–131. [CrossRef]
47. Youjin, L.; Zifang, W.; Ming, G.; Chaofu, W. Effects of conservation tillage on organic carbon, nitrogen and enzyme activities in a hydragric anthrosol of Chongqing, China. *Energy Procedia* **2011**, *5*, 30–36. [CrossRef]
48. Baweja, K.D. Studies of the Soil Fauna, with Special Reference to the Recolonization of Sterilized Soil. *J. Anim. Ecol.* **1939**, *8*, 120. [CrossRef]
49. Helling, B.; Pfeiff, G.; Larink, O. A comparison of feeding activity of collembolan and enchytraeid in laboratory studies using the bait-lamina test. *Appl. Soil Ecol.* **1998**, *7*, 207–212. [CrossRef]
50. Gargouri, K.; Masmoudi, M.; Rhouma, A. Influence of Olive Mill Wastewater (OMW) Spread on Carbon and Nitrogen Dynamics and Biology of an Arid Sandy Soil. *Commun. Soil Sci. Plant Anal.* **2014**, *45*, 1–14. [CrossRef]
51. Hopkin, S.P. *The Biology of the Collembola (Springtails): The Most Abundant Insects in the World*; Natural Histology Museum: London, UK, 1997.
52. Fox, O.; Vetter, S.; Ekschmitt, K.; Wolters, V. Soil fauna modifies the recalcitrance-persistence relationship of soil carbon pools. *Soil Biol. Biochem.* **2006**, *38*, 1353–1363. [CrossRef]
53. Wang, M.; Zhang, W.; Xia, H.; Huang, J.; Wu, Z.; Xu, G. Effect of Collembola on mineralization of litter and soil organic matter. *Biol. Fertil. Soils* **2017**, *53*, 563–571. [CrossRef]

processes

MDPI

Review

Cold Storage and Temperature Management of Olive Fruit: The Impact on Fruit Physiology and Olive Oil Quality—A Review

Eddy Plasquy [1,*], José María García Martos [1], María C. Florido [2], Rafael Rubén Sola-Guirado [3] and Juan Francisco García Martín [4,*]

1 Department of Biochemistry and Molecular Biology of Plant Products, Instituto de la Grasa Spanish National Research Council, 41013 Seville, Spain; jmgarcia@cica.es
2 Department of Crystallography, Mineralogy and Agricultural Chemistry, Escuela Técnica Superior de Ingeniería Agronómica, Universidad de Sevilla, 41013 Seville, Spain; florido@us.es
3 Department of Mechanics, University of Cordoba, 14014 Cordoba, Spain; ir2sogur@uco.es
4 Departamento de Ingeniería Química, Universidad de Sevilla, 41012 Seville, Spain
* Correspondence: eddy.plasquy@telenet.be (E.P.); jfgarmar@us.es (J.F.G.M.)

Abstract: Cold storage of olive fruit has been the subject of study for over more than 50 years. From the 1990s on, an increasing amount of knowledge is build-up about the impact of the conservation on the physiological response of the fruit as well as on the quality of the extracted oil therefrom. This review offers a comprehensive synopsis of this research, discusses the most important influential factors and summarizes the results on the influence of the studied parameters on both the fruit and the oil. Currently, changing climatic conditions, new harvesting techniques and a more demanding consumer market are triggering the need to broaden this strict focus on conservation. A more dynamic view on the effects of temperature from the moment the fruit is harvested up to the oil extraction process, reveals the necessity to manage this crucial influential factor more diversely. An overview of how this management can take form is structured through a focus on the different phases of the postharvest processing and the widely different harvesting scales. Future prospects of research are presented based on the actual state of the art of cold storage research as well as on the necessities that come forward from a broader fruit temperature management perspective.

Keywords: olive fruit; heat transfer; physicochemical analysis; postharvest; thermal treatment; temperature management

Citation: Plasquy, E.; García Martos, J.M.; Florido, M.C.; Sola-Guirado, R.R.; García Martín, J.F. Cold Storage and Temperature Management of Olive Fruit: The Impact on Fruit Physiology and Olive Oil Quality—A Review. *Processes* **2021**, *9*, 1543. https://doi.org/10.3390/pr9091543

Academic Editor: Elwira Sieniawska

Received: 4 August 2021
Accepted: 27 August 2021
Published: 30 August 2021

1. Introduction

The International Olive Council (IOC) defines virgin olive oils as "the oils obtained from the fruit of the olive tree (*Olea europaea* L.) solely by mechanical or other physical means under conditions, particularly thermal conditions, that do not lead to alterations in the oil, and which have not undergone any treatment other than washing, decantation, centrifugation and filtration." [1]. This definition underlines the importance and the strict relation between the fruit on the one side, and the obtained oil on the other. The quality of the latter depends to a large extent on the quality of the former. At the same time, the definition specifies the need to restrict to a minimum the treatments that can be applied before and during the extraction process and explicitly points to the importance of controlling the temperature during the processing.

Virgin olive oils are differentiated into Extra virgin olive oil (EVOO), virgin olive oil (VOO), ordinary virgin olive oil (OVOO) and lamp oil (LOO) [1]. EVOO is regarded as the olive oil with the highest quality. It has a free acidity, expressed as oleic acid, of no more than 0.8 g per 100 g (0.8%) besides other characteristics that correspond to those fixed for each category in the IOC standards. In the case of VOO, a maximum level of 2.0% free fatty acids (FFA) is tolerated and 3.5% in the case of OVO. LOO's have a FFA of more than 3.3%. Moreover, an evaluation by skilled tasters is mandatory and based on the absence (extra virgin) or the weak presence (virgin) of sensory defects and presence of three positive attributes: fruitiness, bitterness and pungency.

The applied harvesting method and the environmental conditions during the pre-processing storage of olive fruit are crucial factors in maintaining the quality of the fruit. For more than 30 years, both aspects have been studied intensively using the official physicochemical parameters and sensory evaluations that determine the quality of the oils and the impact of the applied treatments. The growing economic importance of premium olive oils further triggered fine-tuned studies in which more sophisticated variables and analyses were introduced to differentiate between olive oils that comply with the official EVOO standards and those that aspire to an even higher quality.

The storage of olive fruit has been investigated using different varieties, temperatures, conservation times and controlled atmospheres. The focus on the cold storage of olive fruit was initially driven by the need of finding a solution to a logistic problem, namely the managing of an increasing amount of fruit that needed to be processed as quickly as possible. Storing the fruit under optimum conditions was presented as an alternative that allows the use of the available equipment, avoiding significant investments. Later, various studies probed the possibility of modifying the specific characteristics of the olive oil, such as the bitterness and content of polyphenols, by cooling or heating the fruit just before or after being ground. More recently, attention is brought to the possibilities to fine-tune the temperature of the paste before entering the malaxer to optimize not only the malaxing process but also to correct too high or too low fruit temperatures. The latter necessity is becoming more and more relevant as the industry copes with fruit harvested at field temperatures that are too high for ideal extraction. At the same time, high field temperatures can induce fermentation processes during transport, especially when ventilation within the container is reduced.

The published studies in which the control of the temperature during the post-harvest and pre-processing stages came to play a crucial role thus not only increased during the last decade but also brought the possibilities of thermal modification more to the fore. Shifting from cold storage for several weeks to the necessity to control and correct the fruit or paste temperature during its processing, opened various lines of research in which the possibilities to dynamically modify the fruit temperature before their processing are a central focus. Simultaneously, new technologies are needed to incorporate this new challenge in the actual olive fruit harvesting and processing.

This review aims to present a comprehensive overview of the available literature concerning cold storage of olive fruit along with an outline of the advantages that dynamic control of the olive fruit temperature can offer and the unavoidable technological challenges such a shift inevitably implies. After a historical overview of the progress in this field, the most important independent variables that interfere in the cold storage of olive fruit are discussed as well as the actual state of the art concerning the chemical and physiological responses of olive fruit and the physicochemical characteristics of the oil extracted from cold-stored olives. In a third chapter, the focus is brought towards fruit temperature management as a novel framework to establish new technological challenges and opportunities.

2. Cold Storage: An Historical Overview

The first studies concerning the conservation of olive fruit were published in the 1960s. Researchers of the University of California envisioned the processing of table olives and studied the ripening and the storage of this fruit [2–4]. Their first aim was to define the lowest permissible temperature for storage while avoiding chilling injuries. Secondly, they explored the possibilities to apply a controlled atmosphere during storage. The experiments led to the conclusion that altering the oxygen and carbon dioxide concentrations around the fruit increased their susceptibility to chilling injury [3]. From 1965 on, the storage of olive fruit for oil extraction was tackled by Italian researchers working in the Instituto di Industrie Agrarie dell'Universitá di Perugia [5,6]. Aware of the deterioration of the olive fruit during storage they focused on two causes: the fruit constituents and the contamination by lipolytic microorganisms. The volatile and phenolic compounds of olive fruit processed immediately after harvest and stored at ambient temperature for 10 days were compared [7] and pointed to the loss of these components because of hydrolytic enzymatic mechanisms. The potential solutions varied from inactivation of fruit enzymes, dehydration, the use of inert gas atmospheres or ammonia; immersion into salt and acidic solutions, or treatment with fungicides or antibiotics [5–8]. However, and quite remarkably given the earlier published work [2–4], cold storage was not mentioned as a possible treatment. Meanwhile, at the University of California, the use of Controlled Atmospheres (CA) at various temperatures was further evaluated in order to maintain the quality of the fruit to extend the post-harvest life of 'Manzanillo' olives and underlined the potential to store olive fruit at 5 °C at 2% O_2 for up to about 12 weeks [9].

It is only from the 1990s that an increasing interest in the cold storage of olive fruit for oil extraction took form, resulting in a continuous increasing in published papers on the topic from the 90s up to date (Figure 1).

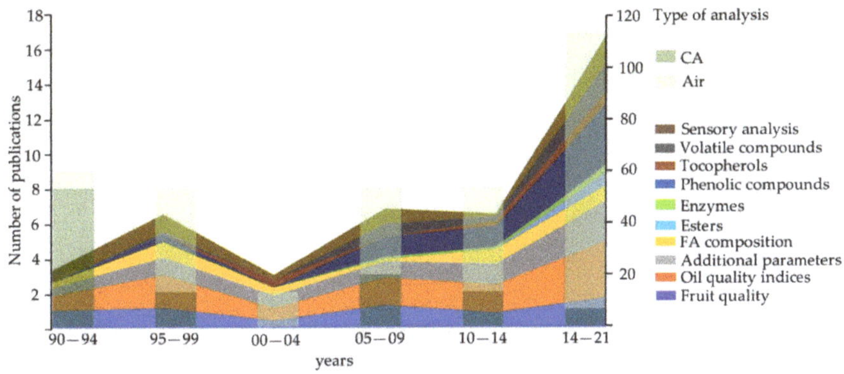

Figure 1. Number and type of works of analysis, under air and controlled atmosphere, published between 1990 and 2021.

The pivotal work of García and Streif [10], published in 1991, was followed with 8 more publications in the following 4 years, all belonging to a research group of the Instituto de la Grasa (CSIC) of Sevilla. Their focus included the effect of cold storage and CA during cold storage on the fruit as well on the extracted oil [11–14]. Fruit quality parameters and oil quality parameters (FFA, peroxides, K232 and K270, oxidative stability and sensory analysis) were measured, while more specific publications indicated shifts in polar compound concentrations [15], and changes in polyphenol content [16] when stored under different conditions. Whilst the advantages for cold storage became more and more clear at laboratory scale, attention was brought to the limitations of the used storage system especially when storage over more than 15 days in containers of 400 kg was envisioned [17,18]. In the following years, the experiments were scaled up to an industrial level and carried out on two different varieties, 'Blanqueta' and 'Villalonga' [19–22]. The results confirmed the advantages of cold storage and its potential for improving the processing. At the same time, it underlined the need for specific requirements, such as an adequate cooling installation to store the fruit that cannot be processed immediately, the capability to renew the air to avoid the accumulation of CO_2, specific material, and machinery to handle the boxes and the supply of undamaged fruit. In the knowledge that each degree higher than 5 °C implies a significant reduction of energy costs, an experiment at a pilot-scale demonstrated that bringing the temperature from 5 to 8 °C would result in significant differences in fruit ripening and olive oil quality [20].

Meanwhile, Italian researchers built on the work of [7] and specified that to optimize the quality of the oil in relation to the level of ripeness and storage time, the latter could be stored up to a maximum of 6 days and even shorter when dealing with ripe olive [23]. A local variety from northeast Italy was used to focus on the effect of hypoxic and CO_2 enriched atmospheres on olive ripening and oil quality [24]. As an alternative to cold storage, Koprivnjak et al. assessed an ancient conservation method from Croatia that consisted of keeping the olive fruit in seawater, a method that was previously mentioned by other authors [5,6], only to conclude that although the basic physicochemical quality indicators did not change significantly, the sensory characteristics were affected dramatically [25].

The pioneering work of the group from the Instituto de la Grasa inspired researchers to design storage experiments with other varieties, modifying the temperature and the atmosphere. Significant intercultivar differences were observed between 'Sevilla' and 'Mission' cultivars on the hand, and 'Manzanillo' and 'Ascolano' on the other hand [26]. While the former could be stored up to 8 weeks at 5 °C, the latter only lasted 4 weeks at the same temperature. A study on the 'Koroneiki' variety and designed a fine-tuned study in which the fruit was kept at 0, 5 and 7.5 °C while modifying the atmosphere at the same time [27]. They confirmed the earlier conclusions that cold storage at 7.5 °C resulted in low quality and could largely be attributed to the development of fungus [20]. In line with the observations of Maxie, storage at 0 °C was neither recommended because of the risk of chilling injuries [3]. In the following decades, the potential to cold storage olive fruit was evaluated and confirmed by Californian [26,28], Australian [29]; Israelian [30], Italian [31–35], Spanish [36–46], Portuguese [47], Egyptian [48], Tunisian [49–54], Turkish [55], Iranian [56,57] and Croatian [58] research centers, which used most of the time the local cultivars for their experiments.

Thus far, a total of 38 varieties has been the subject of an experiment related to the conservation and cold storage of olives (Figure 2).

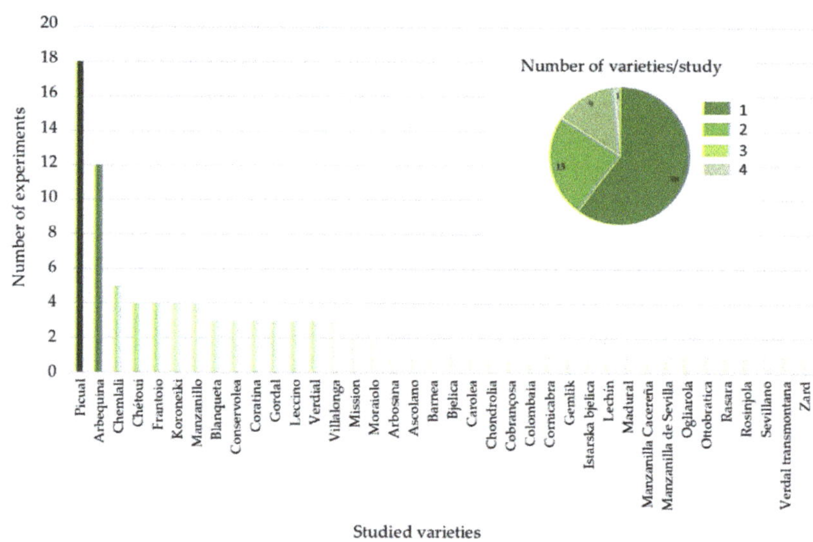

Figure 2. Varieties used in cold storage experiments and the number of varieties per study.

The 'Picual' and 'Arbequina' stand out as the most studied varieties with 18 and 12 experiments, respectively, followed by 'Chemlali' (5), 'Chétoui' (4); 'Frantoio' (4), 'Koroneiki' (4) and 'Manzanillo' (4) cultivars. More than half of the varieties appeared only in one experiment (22). 60% of the experiments were realized with 1 cultivar, and 24 and 14% with, respectively, 2 and 3 cultivars. Only one study took 4 varieties into account (Figure 2).

The more than 60 published studies since the pivotal work of García et al. [10] are not evenly distributed over time but came in three waves (Figure 1). The first one was in the 1990s, during which the prospect for cold storage was evaluated for time spans up to 90 days and where the majority of the studies assessed the possibilities of CA. A second boost of studies arose from 2005 on, when CA became less prominent, and the mean maximum storage time was reduced to 30 days. Between 2015 and 2021 the number of publications doubled, while only marginal attention was given to CA, and the storage time was further reduced. From the second wave on, the type of analysis became not only more differentiated with an increasing focus on the phenolic and volatile compounds (Figure 1) but also more focused on the harvesting method. The last is in large part due to the progressive mechanization and the introduction of overhead mechanical harvesters in high-density orchards.

The broadening and the deepening of the research were mainly performed at the laboratory level, dealing with small quantities of fruit to keep the explanatory variables under strict control. Even though a few experiments were performed at a pilot-plant level and industrial level, cold storage has never been incorporated into the daily workings of an actual mill. The absence of thermodynamic studies to calculate the energy cost and economic studies to estimate the full size of a cold storage investment at an industrial level is in this respect striking but not surprising. At the time the alternative offered by cold storage was published, new extraction lines were already on the market, allowing mills to increase their milling capacity easily.

The prospect to modify the sensory characteristics of the extracted oil through thermal treatment of the olive fruit was suggested once it became clear that the amount of polyphenols in the oil was susceptible to temperature. However, experiments at laboratory as well as pilot plant scale, revealed that controlling the conditions to attain a desired level of bitterness was extremely difficult, if not impossible, given the influence of the fruit cultivar, their ripeness and their origin. At the same time, an inadequate application provoked a denaturalization of the fruit structure and undesirable emulsions during the malaxing [38,59–63].

Recently, the advantages of cold storage for small producers were brought to the attention. Keeping harvested olives at 5 °C for a short time, offered a solution to preserve the quality of the fruit until the necessary quantity for transport to the mill was reached. Combined with a gentle harvesting method, the refrigeration method made it possible to extract high-quality oil without disproportionate investments [45,46]. Cooling the fruit at the farm for a short term opens up new possibilities to revalorize the advantages of cold storage. Meanwhile, climatological changes and earlier harvests, especially of those varieties that are planted in high-density orchards, trigger new queries in which the management of the fruit temperature is coming to the fore as a key factor. The accumulated knowledge concerning cold storage did form a crucial element to address these new challenges.

3. Explanatory Variables

3.1. Storage Temperature

In 80% of the studies, the temperature for the cold storage experiments was kept at 4–6 °C, following the recommendation given in earlier works to respect the 5 °C limit to avoid chilling injuries [3,4,9]. However, the control temperature varied significantly between 15 and 25 °C in 75% of the cases, considered as ambient temperature (Figure 3).

This wider range is partly explained because several studies took the mean maximum day and night temperatures into account or registered the temperatures over longer storages times. A minority of studies used intermediate temperatures of 8 °C [17,18,20] and 10 °C [30,48].

Most of the published studies reported that the quality of the fruit and/or extracted oil were significantly lower when compared to a lower temperature over the same storage time. Yet, it was also suggested that 'Picual' and 'Koroneiki' cultivar can be stored at 10 °C or even at room temperature for 9 days without much reduction in oil quality [30]. A few studies looked at the possibilities to keep the fruit at lower temperatures. Black 'Manzanillo' olives were stored at 0, 2.2 and 5 °C for 6 weeks but, although no chilling injuries were observed, the oil extracted after 4 weeks of storage lost its quality as EVOO [28]. Mechanically and manually harvested olives of the 'Arbequina' variety, kept at 3 °C for 21 days neither presented chilling injuries [39]. 'Manzanilla de Sevilla' and the 'Manzanilla Cacereña' varieties were stored at 2 °C for 11 days in the best conditions without chilling damage [43].

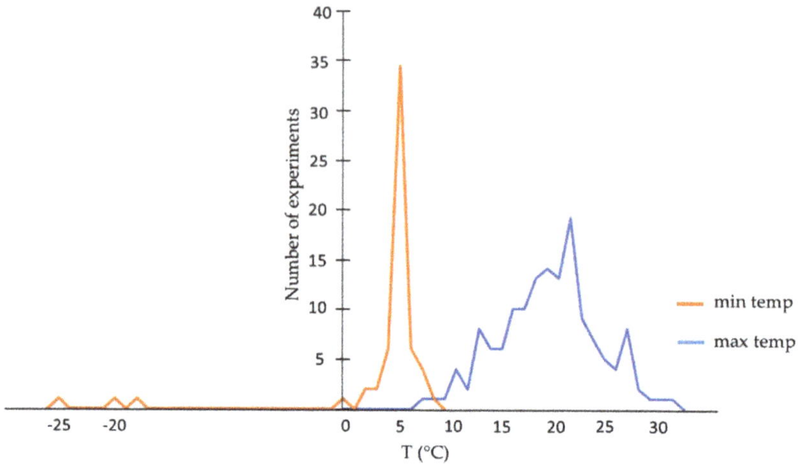

Figure 3. Temperatures use during several experiments of cold storage.

Storage at 0 °C was detrimental due to the destruction of the natural antioxidants present in the olive fruit, leading to a diminishing of the oxidation resistance [27]. Interestingly, oils obtained from 'Arbequina', 'Koroneiki' and 'Mission' cultivars that were kept at −4 °C for up to 3 weeks, maintained the standard indices used for assessing virgin olive oil quality [64]; however, the amounts of FFA in the studied samples were not indicated in that study. After 24 h at −18 °C, a negative effect was found on the oxidative stability of VOO due to olive fruit freezing [65]. It is suggested that without thawing out the fruit before malaxation, the activity of oxidative enzymes is reduced. Olive freezing significantly modified the phenolic and volatile profiles of VOO from 'Picual' and 'Arbequina' cultivars [66]. Similar results were obtained after storing olive fruit at −25 °C for 6 months [67]. However, they underlined that the extracted oils met the standard commercial parameters for EVOO's while indicating that the sensorial defects may also be related to microbial activity. It was suggested that an olive sanitization treatment could possibly overcome this problem. The 'Koroneiki' cultivar was kept for up to 60 days at −18, 5 and 20 °C, showing no negative effect of storage of olive samples at −18 °C in the extracted oil based on the levels of acidity and peroxide values [57]. After 7 days stored at 4 °C, the two Croatian cultivars, 'Istarska Bjelica' and 'Rosinjola', preserved the concentration of most phenolic compounds at levels that were comparable with fresh oil [58]. A decrease in sensory quality was observed after storage at room temperature and −20 °C, leading to the conclusion that, under production conditions, refrigeration is the most suitable option when prolonged fruit storage is taken into consideration [58].

3.2. Storage Time

The storage time is a variable of major importance when the effects of cold storage on the physiology of the fruit as well as the quality of the extracted oils therefrom are examined (Figure 4). The effect of the storage time on the quality parameters, and especially on the FFA, is profoundly studied and confirmed for a wide range of varieties.

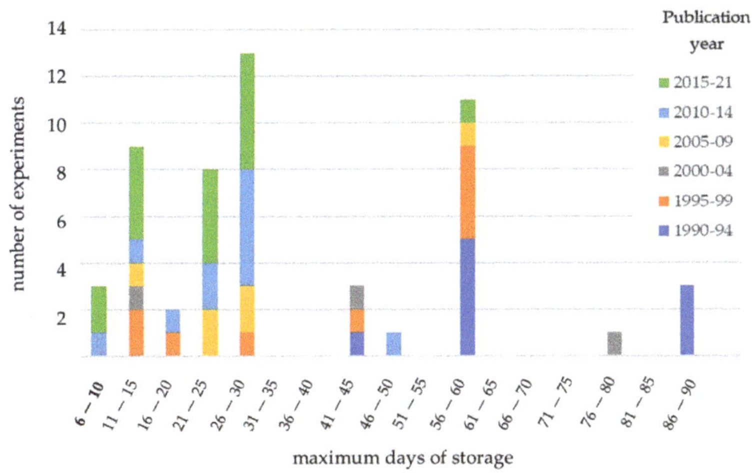

Figure 4. Maximum days of storage considered in the experiments between 1990 and 2021.

When in the 1990's the interest in cooling olive fruit took form, timespans up to 3 months were used to follow the official quality parameters of the olive oil [10,17,27,68,69]. Once it became clear that cold storage for more than one month resulted in problematic fruit quality and deficient oils, the maximum storage time was reduced and the observations were made within shorter intervals [29–31,45,46,49,51–53,70]. Monitoring the amount of polyphenolic and volatile compounds in the stored oils, triggered even more precise observations for only several weeks and between days [34,40–42,58]. More recently, the effect of short-term storage of less than one day is acquiring attention as a tool to modulate the aroma profile of the oil [71].

3.3. *Use of Controlled Atmosphere*

The prospect to control the atmosphere to enlarge the storage time while guaranteeing the quality of the fruit was extensively tested for table olives as well as for mill olives. Increasing the concentration of CO_2 and/or the decrease of O_2 allows, on the one hand, control of the development of fruit maturation and pathogenic proliferation and can delay the physiological changes due to chilling sensitive crops on the other [72]. On the other hand, these atmospheres can at the same time induce other physiological disorders and even aggravate chilling injuries for specific commodities.

The early experiments at Davis were designed to evaluate the effects of controlled atmosphere (CA) on olive fruit. Olives of the 'Manzanillo' and 'Mission' cultivars kept at 3.8 °C (39 °F) stored under low O_2 and high CO_2 concentrations were severely injured fruit due to chilling damage [2]. At the same time and based on color change, there was a clear indication of delaying ripening. 'Manzanillo' olives stored under different temperatures and levels of 5% of CO_2 caused internal injuries, while storage at 2% O_2 retarded fruit softening and color changes [9]. Two Greek varieties for table olives, stored at 5 °C, developed skin injuries after prolonged storage at high CO_2 concentrations, with or without low O_2. A significant difference was observed between the varieties, with 'Chondrolia' being highly sensitive to chilling injury and 'Conservolea' having the potential to be stored at 7.5 °C for up to 3 weeks under CA conditions [68,69].

The possibility of applying CA to optimize the cold storage of mill olives was one of the main objectives of the research group of the Instituto de la Grasa of Seville (Figure 1). Their experiments all led to the conclusions that CA induced physiological disorders that facilitated fruit decay [10–15]. Oil extracted from these fruits resulted in poor quality levels, characterized by high acidy levels and sensorial defects. The results were confirmed by other authors, reporting the best results after 30 days under air at 5 °C [27]; the lowest panel test scores were obtained for oils extracted from fruit kept during 11 days under low O_2 and high CO_2 concentrations [62]; several conventional and nonconventional analysis led to the conclusion that CA did not produce any remarkable advantage over storage at 5 °C [31,32]. In the light of this evidence, the results obtained after the storage of Turkish 'Gemlik' olives under CO_2 are quite surprising, as the treatment protected their physical properties up to 25 days, whilst the quality parameters were not affected, and no significant effect on the phenolic and fatty acid composition of the extracted oil were detected [55].

3.4. Cultivar

Even though the intervarietal comparison among different experiments is often complicated, if not impossible, due to other interfering factors such as different temperature, ripeness or storage conditions, sufficient studies do illustrate the significant effect of the genetic factor on the quality of the fruit and the oils extracted from olives that were cold-stored. 'Villalonga' and 'Blanqueta' varieties showed significant differences between de levels of FFA of both cultivars when stored for more than 20 days at 5 °C [19]. 'Manzanillo' and 'Ascolano' could be stored up at 5 °C for 2 and 4 weeks, while 'Mission' and 'Sevillano' varieties could be kept in good fruit condition up to 6–8 weeks [26]. Comparing the Portuguese cultivars 'Cobrançosa', 'Madural' and 'Verdeal Transmontana', 'Verdeal Transmontana' deteriorated the most rapidly, while 'Cobrançosa' showed the best results in oil quality after cold storage op 14 days [47]. A significant difference in the bitterness intensity was observed among the 'Manzanilla', 'Picual' and 'Verdial' varieties when stored at 5 °C for 8 weeks [38]. Differences in the levels of decay as well as FFA were reported for three Italian cultivars ('Coratina', 'Olgiarola leccese' and 'Leccino') [32], while the effect of storage temperature and time on the oil quality of mechanically harvested olives from three varieties, 'Koreneiki', 'Picual' and Barnea' were compared [30]. Significant differences between the 'Manzanilla de Sevilla' and the 'Manzanilla Cacereña' were observed in a similar study [43]. Differences on both the fruit as well as oil characteristics of the cultivars studied were confirmed between the Italian varieties 'Carolea' and 'Ottobratica' [34] and the 'Arbequina', 'Picual' and 'Verdial' cultivars [45,46].

3.5. Loading Unit

The cooling process of fresh fruits and vegetables involves a complex interaction between the thermophysical properties of the commodity (heat generation due to respiration, specific heat and thermal conductivity), the kind of packaging and palletization, flow field parameters (airflow rate and cooling temperature) and the accessibility of the cooling air to the produce [73–77]. The storage of olives implies keeping the fruit in containers that reduces the physical and biological deterioration to a minimum. The capacity of these containers, or the quantity of fruit stored per loading unit, plays a critical role in the storage efficiency of the olives as the incidence of postharvest losses increased together with the container's capacity [18,22]. After 15 days at 5 °C, the decay incidence approached 40% and at 30 days, and half of the container was rotten. The temperature in the inner zone of a container of 400 kg of 'Picual' never went below 25 °C. Olives stored in boxes with a capacity of 64 kg, attained over the same period an incidence below 20% and showed an internal temperature of 6 °C. The incidence of the ones stored in boxes of 2 and 6 kg, stayed below 10% after 45 days of storage. Olives kept in 400 kg containers gave oils with significantly higher acidity values than the ones from containers for 2, 6, 64 kg before 45 days of storage. Similarly, no appreciable differences were reported in the sensorial analysis. Based on these observations, the use of boxes with a capacity of 22 kg of fruit was recommended.

The effect of the layer thickness on the quality and the composition of the minor components was equally studied [40]. For up to 3 weeks, olive batches (ripening index of 4), were stored at 10 and 20 °C in plastic boxes (capacity of 50 kg of olives) filled in different layers, from monolayer, 10 cm, 20 cm to 60 cm thickness. The VOO quality parameters were, besides being affected by the storage temperature and time, also significantly influences by the thickness of the layer. After 5 days of storage, the fruit stored with a 60 cm thickness reached the upper limit for the EVOO (0.8%). The suggestion that olives can be stored for 15 days when placed in one layer at 10 °C, does not offer a workable solution but instead, underline one of the bottlenecks in the cold storage of olive fruit namely the crucial importance of the loading unit.

The handling of many boxes leads to large, if not unsurmountable logistic problems once dealing with several tons. A precooling treatment to the harvested fruit before dumping them in a container for conservation was suggested and investigated [77]. In a first experiment, the evolution of the internal temperature of a container of 400 kg was assessed with and without a precooling treatment at 5 °C. In a second experiment, the effects of a short cooling treatment at −18 °C to attain rapidly the desired 5 °C was tested. In the container with fruit at a field temperature of 18 °C, the mean internal temperature descended towards 10 °C during the first week, followed by a more stabilized period with temperatures between 8.8 and 9 °C. The container with precooled fruit followed an inverse behavior. Starting at 5 °C a slow rise was set in, amounting to 7.6 °C on day 14. The examined parameters of the extracted oils from both treatments showed levels that corresponded to the EVOO category, although the olives that underwent a precooling treatment in small boxes, did not experience a significant increase in the acidity level after 14 days of storage. A comparison with the published data [50] was complicated as neither the intactness nor the state of the ripening of the fruit was provided. Notwithstanding, the observed low acidity levels, suggest that additional factors must have exerted a detrimental effect on the fruit. While fruit at a field temperature of 18 °C is rather low as compared with the day temperatures at the start of the harvesting campaign when fruit temperatures above 25 °C are far from unusual.

Recently, the focus is brought to the short-term cold storage (16–18 h) in containers (perforated plastic bins), either filled with 250 kg, or half-filled with 125 kg, and perforated plastic boxes with 20 kg of fruit [77]. The surface-to-volume ratio (SVR) was 7, 10 and 16, respectively. Unfortunately, the used ambient temperature was only 13.5 ± 1 °C. Thus, the study evaluated the effect of a 5–6 °C fall in temperature. At ambient temperature, the core temperature of the boxes with an SVR of 10 and 16, attained equilibrium in 6 h, while the temperature in bins with 250 kg (SVR 7) showed an increase which reached over 3 °C after 18 h. Regarding the cooled containers, all three reached thermal equilibrium, although the cooling rate depended on the SVR. No significant differences were reported for the oil characteristics (quality parameters, phenolic compounds and VOCs). The concentration of CO_2 was significantly higher in the refrigerated storage cell, while the O_2 was significantly lower, showing that although the lower temperature reduced cellular respiration, large quantities of CO_2 accumulated in olives stored in containers (≅1%) while O_2 decreased (≅20%).

3.6. Harvesting System

In general, the published articles on the effects of the cold storage conditions on the quality of the olive fruit and the extracted oil used hand-picked olives for their experiments or gave no information on the harvesting method used. It is only with the rise of mechanization and specifically the introduction of straddle machines in super-intensive olive orchards, that the attention was brought to the degree of damage that the used harvesting method could provoke (Figure 5).

The effect of the used harvesting method on the fruit characteristics and the oil quality was brought to the fore [30]. The study reported quality parameters of the oil (FFA, peroxides, polyphenol content) after storage up to 23 days at 4, 10 °C and ambient temperature. It was concluded that, at least for 'Picual', the modernization did not reduce the olive's capacity for storage when the results of [20] were taken as a reference. The potential damage during mechanical harvesting was minimized as well for the 'Koroneiki' variety, which kept the values suitable for EVOO up to 24 days when stored at 4 °C. A comparative study to evaluate the effect of the harvesting system and cold storage on

'Arbequina' olive fruit and the chemical composition of the extracted oil, revealed that the used adapted wine grape harvester induced internal damage in the fruit that led to more rapid decay, softening and higher weight losses during storage [39,70]. Consequently, the chemical composition of the extracted oils was modified during storage, rendering inferior quality, a decline in oxidative stability and lower phenolic and volatile compounds. A similar study focused on the phenolic and volatile compounds and concluded that the cold storage of mechanically harvested olives was only effective for one of the studied cultivars, 'Manzanilla Cacereña' [43]. The damage caused by the grape straddle harvester provoked in the 'Manzanilla de Sevilla' cultivar a quick increase in decay and consequently in the acidity levels of the extracted oils, up to the level that the EVOO category was lost. In general, the levels of the phenolic and volatile compounds in the oils obtained from mechanically harvested olives were significantly lower than manually harvested. Few in-depth studies have been performed on the effect of trunk shakers and hand-held machines. One study [35] used five different harvesting systems: (1) gentle manual harvesting; (2) manual harvesting using hand-held combs to detach the olives and nets placed under the tree; (3) using two hand-held pneumatic combs with telescopic handles to detach the olives, while collecting them on nets on the ground; (4) mechanical harvesting with a trunk shaker and a reversed umbrella to collect the olives; (5) a straddle machine. The results left no doubt that gentle hand-harvesting caused the least damage, whereas the trunk shaker and the straddle machine the greatest. Stored at ambient temperature (18 ± 2 °C), the observed fruit damage explained most of the worsening of oil quality across the studied harvesting systems and storage duration.

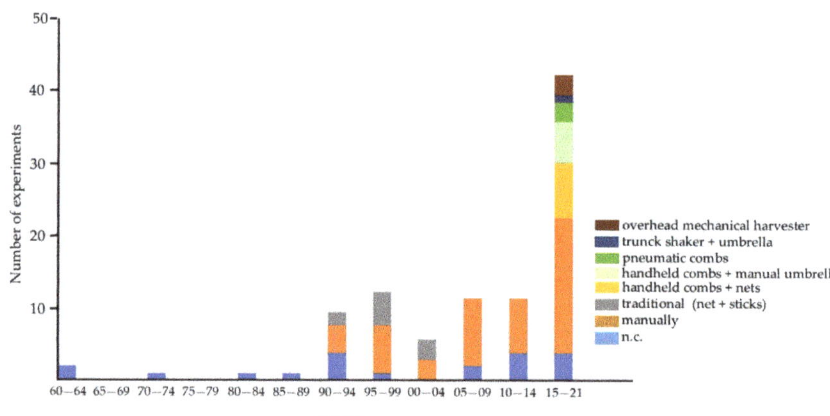

Figure 5. Harvesting methods used to harvest olive samples for the cold storage experiments.

The effects of a harvesting and conservation method, designed for small producers, on the fruit as well as the oil quality was presented by [45,46]. The studied harvest method consisted of hand-held pneumatic combs (branch shakers) together with a movable structure designed as an inverted umbrella that was placed around the tree, avoiding that the fallen fruit did not fall on the ground, but was recollected in perforated boxes, placed under the lowest part of the canvas [78]. After being harvested, the fruit was stored at the farm in a cold storage room at 5 °C for up to 14 days. Three different varieties ('Arbequina', 'Picual' and 'Verdial') were tested. Fruit evaluation, physicochemical analysis and panel analysis revealed a significant effect of variety, as well as the harvesting and conservation method. Calculation of the magnitude of the strength and panel preference discerned the explanatory weight of each factor and emphasized the need to consider the days of storage when evaluation their importance. When the storage time falls within the range of a few days, the quality is best guaranteed by an optimized harvesting method. When longer storage is expected, high day temperatures are expected or when dealing with sensitive varieties, cold storage turned out to be unavoidable when a high-quality end-product is the goal.

4. Effects on Fruit Physiology and Quality

Harvested olive fruit is a living tissue that respires, consuming oxygen and liberating carbon dioxide [72]. The respiratory process produces the energy required for metabolic activities and fruit maturation. Cooling fruit decreases the effectiveness of the fruit respiration and by doing so the commercial life is extended for the produce stored. By slowing down the metabolic processes, the weight and nutritional value losses are reduced, while pathogenic deterioration of the olive fruit is delayed [22].

4.1. Skin Color and Maturity Index

The progress of the ripening can be registered through changes in skin color, using the CIELAB color space. Alternatively, the maturity index (MI) was calculated following a method that allowed an appreciation of the overall ripening stage of the olive fruit, based on both the skin and flesh color [79].

Each of the studies that used a colorimeter observed a significantly better color for the cold stored olives as compared with the ones kept at room temperature [10,14,20, 28,38,39,45,47,48,55,68,69]. This effect became visible from the first week of storage and continued up to 30 days [10]. 'Picual' olive fruit kept at 5 °C maintained the same color during the first 14 days of storage independently of the atmosphere [14]. Similar results were obtained when measuring the color variance among 'Coratina', 'Manzanilla' and 'Picual' olives stored at 5 °C [48], and comparing the 'Arbequina', 'Picual' and 'Verdial' cultivars during 14 days at the same storage temperature [45]. A slight change in color after 7 days of storage of the 'Arbequina' variety was observed, independent of the harvesting method [39]. It was further reported that storage time and atmospheres had no significant effect on color [28,55]. In olive fruit stored at 5 °C under air, the red skin color was noticed after 45 days, while at 7.5 °C it was observable after 30 days [69]. A significant change in color in the 'Picual' cultivar after one week of storage at 5 °C was reported, while this took 3 weeks for the 'Manzanilla' and up to 5 weeks for the 'Verdial' cultivar [38].

The same tendencies are described when the MI is used to quantify the ripening process and a significant delay of the maturation due to low storage temperatures is observed [17,19,20]. A reddish-brown skin color, corresponding to a number 3, on the scale from 0 to 7, was attained after 15 days when stored at ambient temperature, 30 days when stored at 8 °C and 45 days at 5 °C. The obtained MI values in another study showed only a modest increase in the third and last week of storage at 4 ± 2 °C [29]. The same range of variation was noted with respect to the studied cultivars [38]. Concerning the Croatian varieties, no significant rise in the MI was reported after 1 week of storage at 4 °C [58].

4.2. Firmness

The majority of the published studies define the firmness as the resistance to a deformation of the skin measured either with a shoremeter or densimeter, equipped with a 5- [10,12,14,38,39] or 2-mm probe [55], expressed in Newton/cm^2, or with a momentum transfer generator (MTG), that reports the olive firmness as MTG units [28]. From the first studies on, it was demonstrated that the maintenance of flesh firmness depended on storage temperature [10]. At 5 °C, the initial values are maintained, while at ambient temperature they fell to almost 50% after 30 days of storage. At 8 °C, a moderate decrease was observed up to 60 days [17]. The fruit firmness of black-ripe olives declined by 24.4–45.9% in air and 34.2–38.0% in 2 kPa O_2 after 4 weeks of storage [28]. These observations were contrary to results on mature green olives which firmness was better when stored under 5 or 2 kPa O_2 [14]. However, 'Gordal' olives stored under 1 kPa O_2 did show a decrease in firmness [10]. The firmness of black-ripe olives at 20 °C decreased by 44% within the first 2 weeks, which was much lower than any of the other examined. Variations among cultivars were observed [26,38]. A clear difference was registered between the 'Mission' and 'Sevillano' varieties, which maintained their firmness up to 8 weeks, and the 'Acolano' and 'Manzanillo' cultivars, which, respectively, lost 9 and 29% of their firmness at 4 weeks and 16 and 53% at 8 weeks [26]. Significant differences occurred between 'Picual', showing a significant reduction of the firmness after 1 week, and the 'Manzanilla' and 'Verdial' varieties, only after 4 and 6 weeks, respectively [38]. Hand-harvested olives softened slower as compared to mechanic harvest. However, it was only after 10 days of storage at 5 °C that a significant difference was observed. A difference that was contributed to the internal partial breakage of the cellulose walls of the cells of the mesocarp during mechanical harvesting [39]. A similar observation was made in a recent research [45], in which the use of nets was the damaging factor.

4.3. Humidity

The amount of humidity lost during the storage is measured as the difference between fresh and dry mass and expressed as % [12]. Water loss increased significantly in studies that stored the fruit for more than 60 days [12,14]. Higher temperature and lower relative humidity do provoke higher losses in fruit stored at ambient temperature, while an advanced stage of fungal decomposition induces leakage of cell fluids [12,14]. Kept at 5 °C and up to 60 days, no differences were observed among the fruit stored under different atmospheric concentrations. No significant differences in water content in black-ripe olive were observed after storage of 2 weeks at 20 °C [26] and neither in the 'Gemlik' variety after 25 days of storage [55]. However, these results were not confirmed by other studies [48,57]. The former authors registered a significant decline in humidity after 1 week when stored at 20 °C, while at 5 and 10 °C this parameter did not change [48]. Nevertheless, after 2 weeks the fruit stored at the lowest temperature kept significantly higher moisture when compared with the one at 10 °C. The latte authors mentioned a significant decrease of water content after over 20 days for the fruit stored at 20 °C or 5 °C [57].

4.4. Weight Loss

Weight loss is one of the main postharvest changes in fruit and vegetables due to respiration and transpiration. It includes, beside the loss of water, also the loss of other substrates (sugars, organic acids ...) that are consumed by the fruit metabolism during the storage. To quantify the weight loss of the olives, the fruit is weighed with a balance before and after each treatment and storage duration. The difference is expressed as percentage loss. Weight losses increase with time over storage [26,27,39,45,48,68,69]. The highest losses were observed at 7.5 °C as compared to at 5 °C under air [27]. At a lower storage temperature and under controlled atmospheres, the weight loss decreased. Slight differences between the two studied varieties, 'Sevillano' and 'Manzanillo', were reported [26]. While the mass loss after 2 weeks increased in the former by 0.6%, the latter attained 1.5% when stored at 5 °C. At 20 °C the mass loss attained levels between 2.3 and 3%. Varietal variance was also observed among the 'Coratina', 'Manzanillo' and 'Picual' varieties with 16, 12 and 11% weight loss, respectively, when stored at 20 °C. When stored at 10 °C, the weight loss was more than 3 folds those at 5 °C. Prolonging the storage period led to a significant increase in weight loss for all the cultivars [48]. Similar varietal differences between 'Chondrolia' and 'Conservolea' were reported, although the authors pointed out that this might be due to high chilling injury of the epidermis of 'Chondrolia' [69]. Notwithstanding, the study underlined that the olive mass loss increased with storage temperature and duration for both cultivars. This process was accelerated when stored under a controlled atmosphere. The influence of CO_2 during the storage at 5 °C is reported as a higher value of average weight at 5 and 15 days with CA, however, the differences became insignificant after 25 days [55]. Significantly higher weight losses were observed in the mechanically harvested 'Arbequina' samples [39]. The difference was attributed to the possible internal breakage that led to a lower juice content. The recent experiments of Plasquy et al. showed only an effect of the conservation temperature and the storage time on the weight loss for the three varieties studied ('Arbequina', 'Picual' and 'Verdial') [45].

4.5. Oil Content and Oil Yield

The oil content was determined either by Soxhlet extraction [26,39,40,58] or the method described by Folch et al. [80] (cited in [48]). No significant changes depending on the fruit storage temperature were reported, indicating that accumulation of the oil did not continue during storage [39,40,58]. It was specified that although the oil content remained unchanged in four cultivars studied during storage at 5 °C, when stored at 20 °C, it was lower than the initial values for the 'Manzanillo' and 'Ascolano' varieties [26]. On the other hand, a significant decrease of the oil content was observed with an increasing storage temperature and the progress of the storage period up to seven weeks [48].

To estimate the oil yield of the olive under different storage treatments, the decanted volume obtained through extraction with an Abencor system was used, considering the value of 0.915 g mL^{-1} as the value for oil density to obtain the yield as expressed as the percentage of fresh weight [14,38–40,57,81,82]. An alternative method [83] takes the mass ratio of the mechanically extracted oil and the centrifuged olive paste and multiplies the result by 100 [58]. One study carried out on a pilot plant scale obtained the data from the producer [29]. The extractability index (EI) was calculated based on the mean value of the percentages of oil physically extracted from the total oil content.

All the published studies reported higher oil yields from olive stored at ambient temperature [11,38–40,57,58]. During the first 2 weeks of storage at 5 °C a significant reduction of the yield for the 'Picual' and 'Verdial' cultivars was observed, followed by a gradual increase up to levels significantly higher than the initial ones [38]. The 'Manzanilla' variety, on the other hand, had a low oil yield from day 0 (4.6% against 18.2% and 20.0% for the 'Picual' and 'Verdial' varieties, respectively) but did not show any significant change. This was related the higher values to the fact that the fruit could continue to ripen at room temperature, facilitating degradation of the walls of oil-bearing cells and consequently an improvement of the extraction process [38,39]. This hypothesis was supported by other authors [58] who also found the highest values of oil yield and EI in olives stored at room temperature. At the same time, it was underlined that the EI is highly dependent on the cultivar and fruit properties. An increase in oil yield in the first 2 weeks of cold storage was observed, followed by a decrease [29]. It is suggested that the increase is a consequence of cell-wall degradation due to low-temperature storage. After two weeks, most of the oil trapped in the cells was easily released. The decrease after 3 weeks was attributed to an advanced stage of hydrolysis.

4.6. Incidence of Decay

Fungal rot was determined visually when fungal mycelium was observed in randomly selected samples of 100 olives and expressed as a percentage of decay [12,14,17–20,26,39,43]. One study used an ordinal scale for 0 to 5 (0: not rot, 1, few rotting olives without sporulation; 2, 30% rotting but few sporulation; 3, 50% rotting and sporulation; 4, 75% rotting and sporulation; 5, 100% rotting and sporulation) [27].

Low temperatures do reduce fungal growth. Olive fruit from the 4 studied varieties ('Ascolano', 'Manzanillo', 'Mission' and 'Sevillano') only showed little or no rot visible on the fruit after 5 weeks of storage at 5 °C [4]. However, after 10 weeks of storage, considerable numbers of decayed fruit were recorded in all varieties. Stored at 12 °C, the decay of 'Picual' olives was more than 20% after only 7 days. At 5 °C, 45 days of storage were needed to attain the same level. An intermediate behavior was observed when stored at 8 °C [17,18]. The authors further observed that the entire batch of fruit, stored in air and ambient temperature, was affected at 30 days of storage. Similar results were obtained with the 'Villalonga' and 'Blanqueta' varieties [19]. Stored at 5 °C, only about 20% were rotten after 60 days, while this level was already attained before 20 days at ambient temperature. Under ambient temperatures, 'Blanqueta' was significantly more resistant when compared to 'Villalonga', but when refrigerated, no difference was found. A more pronounced intercultivar variance was observed between 'Manzanillo' and 'Ascolano' with a high decay incidence (12 and 7%, respectively, after 4 weeks, and 90 and 27% after

6 weeks at 5 °C) and the 'Sevillano' and 'Mission' olives, where the first had 13 and 11% decay after 6 and 8 weeks, respectively, and the second only 8% after 8 weeks. When kept at 20 °C, the 'Manzanillo' and 'Ascolano' olives showed 40 and 23% decay incidence after 2 weeks, the 'Sevillano' 6% and the 'Mission' cultivar no decay [26]. The fungal damage on 'Koroneiki' was manifest (score of 4) after 30 days of storage at 7.5 °C, independently whether CA was applied or not. At 0 °C in air and 5 °C under 2% of O_2 and 5% of CO_2, no rot (score 0) was observed after 30 days. At 5 °C in air, few rotting olives were present (score 1). At 60 days of storage the values for each treatment augmented 1 point [27]. The application of CA did not result in significant differences until 60 days of storage after which a higher degree of infection was observed. This shift was attributed to the lowered resistance to the molds, because of the combined effects for high CO_2 and/or low O_2 and low temperatures [14].

Fruit with physiological disorders or mechanical damage developed infection preferentially [19,20]. Mechanically harvested olives showed a significantly higher incidence of decay when compared to hand-picked olives, regardless of the storage temperature [39]. Independently the method of harvesting, the fruit stored at 3 °C showed a lower percentage of decay. Mechanically harvested 'Arbequina' olives attained almost 30% of decay after 4 days of storage at 18 °C, a level that was not attained when stored during 10 days at 3 °C. It was also remarkable that the harvesting system provoked mechanical damages that led to their rapid deterioration, even when the damage was not visually perceptible. A significantly higher decay incidence was reported between the fruit of the 'Manzanilla de Sevilla' and 'Manzanilla Cacereña' varieties in a hedgerow plantation, and hand-picked olives, although the difference was more pronounced in the first variety [43]. The authors referred to an earlier study in which they compared the damage caused by the used grape straddle harvester in the studied cultivars. The fact that a higher proportion of fruit with cuts was found in green 'Manzanilla de Sevilla' (18% versus 2%) may explain the intervarietal difference. Storage during 11 days at 2 °C was highly effective for controlling the fungal rot in the mechanically harvested 'Manzanilla de Cacereña', while the 'Manzanilla de Sevilla' attained an incidence of 30% under these conditions.

4.7. Microbiological Profile

To study the production of volatile phenols, related to the appearance of VOO sensory defects, the specific microbiological profile on the olive surface of fruit stored in plastic bags and open boxes at 5 ± 3 °C by night and 8 ± 3 °C by day up to 21 days was determined [41,42]. The viable-culturable cell number was determined on different substrates to evaluate the presence of fungi, lactic acid bacteria, enteric bacteria and *Pseudomonas*. The study revealed that only fungi were present at the initial time. Acetic bacteria attained levels above 104 cfu (colony forming unit) per gram after six days, while the presence of enteric bacteria reached this level at day 5 of storage in a bag and day 15 in an open box. The lactic bacteria became only present after 9 days and the *Pseudomonas* only after 15 days when stored in a bag. One study [33] focused on the presence of two enzymes, responsible for the oxidation process of the phenolic compounds. The research considered both the phenoloxidase and the peroxidase from the spontaneous microbiota as possible oxidoreductases enzymes involved in the polyphenol decay in the oily fraction of the fruit during storage at 15 °C and registered the total bacteria count, the total molds, and the total yeasts. The microbiological analysis showed a rapid increase of total bacteria and yeasts during the first 4 days of storage. Thereafter, they stabilized at 105 and 106 cfu/g. The molds increased rapidly but never attained a level above 102 cfu/g. The trials further showed that after 3 days of storage, only the total polyphenols presented in the oily fraction decreased rapidly. This was attributed to both the activity of the oxidoreductase's enzymes from the pulp and the microbiota on the carposphere and in the pulp of damaged fruit.

4.8. Respiration Rate

Low temperature has a direct effect on the respiration rate. Keeping the fruit at 5 °C reduces the metabolism of the fruit cells. When comparing the respiration rate of olive fruit ('Gordal' cultivar) over 10 h, a ten-fold higher CO_2 production was registered when stored at 18 °C instead of 5 °C [10]. Similar results were observed with 3 Italian cultivars ('Coratina', 'Olgiarola' and 'Leccino') [32]. When kept under 2% O_2 and 5 °C, the respiration rate was about three times lower than when stored in ambient air. In the same line, a significant effect on 'Manzanillo' olives stored at three temperatures between 5 and 10 °C of both the temperature and atmospheric composition was reported [28]. An increased respiration rate was described when the storage temperature rose, while black olives showed higher values than green at each temperature tested [68]. The values of the CO_2 production of 'Arbequina' olives that were systematically could be ordered according to the cold storage temperature [70].

The respiratory quotient (RQ), or the ratio of CO_2 produced to O_2 consumed, is assumed to be equal to 1.0 if the metabolic substrates are carbohydrates. The overall values were calculated to be a little above 1 for the green and black 'Conservolea' cultivar when stored at temperatures between 0 and 20 °C during the 24 days [68]. Similar values were obtained at harvest, but a pronounced decline was noticed parallel with the respiration rate for 2 days when stored at temperatures between 10 and 40 °C [81]. This decline was the most pronounced at 40 °C (RQ < 0.4) and the least at 10 °C (±0.6) for both studied cultivars ('Gordal' and 'Manzanilla'), indicating that above 10 °C sugars were not the only respiratory substrate consumed. The respiration rate increase for a 10 °C rise in temperature, or Q10, were similar in both studies, with values between 1.7 and 2.0 for the green and black 'Conservolea' [68] and a range between 1.2 and 2.5 for the 'Gordal' and 'Manzanilla' varieties [81] in the temperature intervals considered.

Once the fruit is detached, the concentration of substrates for respiration becomes limited. When stored at room temperature these substrates are consumed continuously, while when refrigerated this consumption is reduced. When brought at 20 °C, this fruit still disposes of a higher concentration and therefore, higher CO_2 production levels can be expected. This phenomenon is clearly illustrated in the study of Plasquy et al. in which the respiration rate is followed during 14 days of storage for three varieties ('Arbequina', 'Picual' and 'Verdial') at ambient temperature and 5 °C [45]. Meanwhile, higher respiration rates were observed when the fruit was damaged during the harvesting. The fruit of two studied varieties ('Manzanilla de Sevilla' y 'Manzanilla Cacereña'), harvested with a grape straddle harvester, showed a higher respiration rate than the hand-picked ones [43]. Similarly, higher CO_2 production was reported in naturally bruised 'Manzanilla' olives when compared to healthy [84].

4.9. Ethylene Production

Ethylene production during cold storage was significantly higher for mechanically harvested olives and equally attributed to the damage caused by the harvester [43]. A similar result was obtained by [45], who reported a significant effect of the harvesting method when comparing the traditional harvesting that implied the dragging of nets with olives over the ground, with a gentler picking using a manual inverted umbrella [45]. the ethylene production rate in green olives was almost undetectable for the duration and temperatures tested when manually picked olives were compared [68]. The black ones, on the contrary, showed a 10-fold increase in ethylene production between 0 and 10 °C, indicating that the critical temperature for its synthesis is below 10 °C. This result was in line with earlier published studies that demonstrated the significant effect of ripening on the biosynthesis of ethylene at ambient temperature [9,85,86].

5. Physico-Chemical Analysis of the Extracted Oil

Most of the published studies included a physicochemical analysis of the oil extracted from cold-stored fruits. As a minimum, this includes the official parameters as shared by the European Commission [87], the International Olive Council [1] and the US Department of Agriculture [88] comprising the level of free fatty acids, expressed as the % oleic acid; the peroxide value (PV), expressed in mEq O_2/kg oil; and the absorbance at 232 and 270 nm. Additional parameters such as the oxidative stability expressed as the oxidative induction time (h), the chlorophyll and carotenoid pigment profile, and the bitterness index are often included. A specific focus on the minor components of the oil implies the detection and quantification of the kind and amount of polyphenols, tocopherols, volatile compounds (VOC) and alkyl esters.

5.1. Level of Free Acidity

Free fatty acids (FFA) are produced by the hydrolysis of the oil-induced by lipolytic enzymes. These enzymes are normally present in the pulp and seed cells of the olive. When the integrity of the fruit is lost, the enzymes react with the oil that is contained in vacuoles. Unhealthy, damaged or bruised olives, together with detriment storage conditions are the main sources for high values of FFA. The amount of FFA is expressed as the percentage of gram of oleic acid per 100 g of oil. The maximum level for extra virgin olive oil (EVOO) is set at 0.8%. The level of acidity is always present as one of the key parameters to evaluate and compare the different storage conditions. Not only because it is one of the decisive elements in enhancing the quality of the oil, but also because of its sensitivity in detecting changes in the investigated oils. The published reports leave no doubt that the increase in the acidity level is positively related to the storage temperature as well as the period of time [10,11,17–20,26,27,30,31,34,39–44,46,47,49,51,55,57,89].

The combined effect of mechanized harvesting of olives for the mill and the conservation method was the subject of several studies [30,39,43,46]. The acidity of the oil produced from olives harvested with a combine harvester was, after 4 days and regardless of the storage temperature, higher when compared with the values of the oil obtained from fruits harvested by hand [39]. However, it was pointed out that the mechanization of the harvest does not compromise the quality of the oil produced from olives of the 'Picual' variety, although it is also not denied that an effect on certain other varieties such as 'Koroneiki' and 'Barnea' is possible [30]. Gently harvested olives maintained low FFA values over the storage period; however, hand-held combs, increased significantly after 1 week and hand-held machines even evolved above the 0.8% limit after 1 week. Fruit harvested with a straddle machine even attained 1.16% after 48 h [35]. The high incidence of decay is brought forward as the main reason why the acidity levels were significantly higher in oils extracted from mechanically harvested olives, although a varietal difference was noticed between the studied varieties [43].

When the time it takes to exceed the level of 0.80% is compared, the variety 'Picual', preserved at 5 °C needed between 23 [30], more than 30 [38] and up to 45 days at 1.00% [17,18]. The 'Coratina' variety was maintained for up to 30 days at 5 °C a FFA level below 0.8% [31]. Olives of the 'Arbequina' variety, stored at 3 °C and harvested with a harvester, exceed that threshold after 10 days, while when hand-picked, they reach up to 3 weeks [39]. The 'Verdial' variety was stored for more than 30 days at 5 °C before it exceeded 0.8% [38].

The intervarietal differences concerning the level of acidity were also noticed among three Portuguese varieties [47], among the 'Picual', 'Barnea' and 'Koroneiki' varieties [30], and between 'Arbequina' and 'Arbosana' [42]. Regarding the effect of the used harvesting and conservation method for the 'Arbequina' variety, conservation became prominent after four days of storage, while the 'Picual' cultivar showed to be more resistant to a rise in FFA. The 'Verdial' offered a more confusing profile. While in the first year an increased importance of the conservation method was measured, this effect was absent in the second year while the harvesting method was responsible for 30–60% of the variance [46].

Differences between the ripening stage were also observed. Comparing green mature, veraison and black olives of the 'Picual' cultivar during a storage period of 4 weeks at 5 °C, the green mature stayed below 0.8% up to 4 weeks, the veraison and black olives attained levels above 0.8% after, respectively, 3 and 2 weeks [44].

5.2. Peroxides

Peroxides are the primary products of the oxidation of olive oil. Peroxide value (PV) is a measure of total peroxides in olive oil expressed as mEq O_2 kg^{-1} oil and hence a major guide of quality. None of the published studies exceeded de maximum value of 20 mEq/kg oil when cold-stored. Notwithstanding, significant differences were observed due to the specific conditions of conservation.

After one week, a sharp rise in the peroxide value was observed up to 30 days at ambient temperature and 8 °C, followed by a decrease until 60 days. At 5 °C, the increase was lower and spread over 45 days [19]. Other authors reported an increase of the PV only for olives stored at 7.5 °C after 30 days [27]. Values higher than 20 mEq were registered in olives stored at 2 weeks after which a decline set in, probably due to the consumption of minor compounds as phenols and tocopherols that make the formation of peroxides difficult [47]. The initial increase followed by a decrease was attributed to the formation of secondary carbonyl compounds [89]. The storage conditions do play a significant role. A doubling of the PV was observed after 11 days when stored in plastic bags, while only after 25 days in plastic containers [51]. The effect of the ripeness of the fruit on the PV values was also reported [44]. A clear difference between on the one hand 'Arbequina' and 'Picual', and on the other 'Verdial' was noticed [46]. The latter attained the PV limit in one week when recollected with nets. The harvesting method stood out as the most influential factor for all varieties.

5.3. K270 and K232

UV-specific extinction determination permits an approximation of the oxidation process in unsaturated oils. At 232 nm, primary oxidation products show absorption (conjugated peroxides) which increase due to inappropriate storage of olive fruits or a defective extraction. At 270 nm secondary oxidation products, such as carbolynic compounds (aldehydes and ketones) are detected which indicates an advanced oxidation process. The maximum permitted values are 2.5 for K232 and 0.20 for K270. The results obtained in relation to the degree of oxidation as measured by these parameters are in line with the one observed with the peroxide values, with only a few studies reporting values that exceeded the set limits even when the fruit was cold-stored for more than a month.

The values of K232 and K270 remained practically constant when stored at 5 °C [10,26,27,31,34,40–42,55]. An increase in K232 and K270 at 15 days at ambient temperature was observed, surpassing the limits at 30 days [11]. 'Picual' olives, stored in plastic containers of 64 kg of fruit for two months, attained the maximum limit after 14 days when stored at ambient temperature, 33 days at 8 °C. At 5 °C the limit was not yet reached at the end of the experiment. similar values for K232 were obtained at different temperatures up to 30 days [19]. From then on, the samples differed significantly with the lowest values found in oils from olives stored at 5 °C. At 270 nm, and in contrast with the earlier published study [10], the storage temperature showed to significantly affect the presence of carbonylic compounds. [29] on the other hand, observed a significant effect of the K232 but not on the K270. In the same manner, [89] registered an increase of K232 after 7 days, regardless of the storage temperature, and all values surpassing the limit except when stored at 5% and kept a low O_2 concentration. The values of K270 showed a significant effect of storage temperature but all values were above maximum.

Mechanical harvesting exerted a significant effect on the K232 and K270 values of the extracted oils. The higher values observed led to an increase in the amount of conjugated fatty acids, meanwhile this process was accelerated at a storage temperature of 18 °C [39]. For the values of K270, no significant differences were found initially. However, in oils from mechanically harvested olives an increase took off after 4 days of storage at 3 °C, while no difference was observed in the oils extracted from manually harvested fruit. This was attributed to the internal rupture of the olive during the crushing in the course of the harvest. Similar changes were not reported in a recent research [46] as no significant effect for K232 was detected. At 270 nm, the effect of the harvesting method was significant in all the varieties. These results indicate that the damages provoked by a harvester are even more severe than those that occur when olives are picked from the ground. However, other authors did not observe the K232 and K270 values to be affected by the variety, type of harvesting and storage time [30,43]. Opposed results were reported, pointing to a varietal difference in the Portuguese cultivars for K232, but obtaining at 270 nm similar values [47]. Olives in open perforated boxes behaved better, as only after 17 days the limit value for K232 was attained [51]. In plastic bags, a drastic increase was observed, without of the limit values after 11 for K232 and 25 days for K270. Significant changes in the K270 due to the time of storage were showed [44]. After one week, green mature fruits exhibited higher values, while no changes were found in riper olives. The values of K232 on the other hand remained constant.

5.4. Oxidative Stability

The oxidative stability, measured with the Rancimat system, although it is not considered as one of the legally established parameters to evaluate the quality of the oils, provides an interesting estimate of the shelf life of the oil. A wide range of studies shows that the levels of oxidative stability are highly influenced by the genetics of the fruit. It is also accepted that not only the values within each variety are reduced during its conservation, but also that this process slows down when the fruits are subjected to low temperatures. Several studies have pointed out that up to at least 21 days, the values obtained with the Rancimat method did not change. Oils stored at 5 °C lost 35% of stability after 60 days of storage, while at 8 °C and ambient temperature, 70 and 93%, respectively [19]. The resistance of 'Koroneiki' oil to oxidation did not change at 5 °C in air during 30 days of storage [27]. It was reported that the stability was slightly reduced (<10) after 6 weeks of storage, which was seen as normal because fruit matured, and the stability decreases with olive maturation [38]. A reduction of 7% was also observed after 3 weeks of storage at 5 °C with the Tunisian cultivars 'Chemlali' and 'Chetoui' [49]. A slightly deviant result was obtained by Pereira et al., reporting a significant decrease during the first week of storage at 5 °C, although a difference between the varieties 'Cobrançosa' on the one hand and 'Madural' and 'Verdeal Transmontana' on the other was also noted [47]. Similar intervarietal differences were observed by Rinaldi et al. and Plasquy et al. [32,46]. In the latter study, the harvesting method showed to be of major importance for up to 4 days. Over 14 days, the strength of the used conservation method increased, although the maximum and rate varied along with the cultivars. In 'Arbequina' it attained almost 80% on day 14, in 'Verdial' 40% and 'Picual' 20%. In the same line, varietal differences with the lowest values for the mechanically harvested olives were reported [39,43].

A bad correlation between PV and stability was observed [11], suggesting the existence of other factors in addition to oxidation state that influence stability. The high concentration of free fatty acid oils extracted from fruit stored at ambient temperature underwent a remarkable decrease after 30 days while at 5 °C the stability was constant. The temperature-dependent loss of stability was attributed to thermocatalyzed hydraulic and oxidative processes, which act on the oil contained in the olives during their ripening, either as a consequence of their own metabolism or as a result of a pathogenic activity [19]. Meanwhile, the detected correlation between the total amount of phenols and the induction times points to the antioxidant activity of hydrophilic phenols [31,65].

5.5. Chlorophyll and Carotenoid Pigment Profile

The spectrometric detection of carotenoid and chlorophyll pigments at 460 and 670 nm, respectively, was not reported systematically. The main interest to study their evolution during cold storage consist of them being active as an antioxidant and their key role in the oil color, a factor that influences the choice of the consumer [90].

The interpretation of the results turned out to be very difficult given the complexity of factors that are involved and the varietal differences. A stable carotenoids content in the oils extracted from the 'Manzanilla' cultivar stored during 5 weeks at 5 °C was registered, while the 'Picual' and 'Verdial' cultivars presented a clear reduction of carotenoids during the first 3 weeks [38]. In the following 2 weeks, the level of the 'Picual' increased to attain the initial values, while the 'Verdial' increased only slightly. The chlorophyll content of the 'Picual' and 'Manzanilla' increased continuously during the first 6 weeks after which a steep decrease was observed for the 'Picual'. On the other hand, 'Verdial' showed more than 40% loss of chlorophyll content after the first week and from then on the same level was maintained. A decrease of chlorophyll in 'Koroneiki' olives was observed when stored below 7.5 °C, although did not register any change over 60 days when stored at 5 °C [27]. In the same line, only a slight decrease in carotenoids as well as chlorophyll content over 3-week storage at 5 °C was registered [49], while reported an increasing trend during the cold storage time [34]. The profile of the chlorophyll carotenoid content did not show significant differences in the Croatian varieties at ambient temperature and 4 °C [58]. Only those stored at −20 °C showed a significant change in both pigments. As the chlorophylls decreased, the level of carotenoids increased.

The erratic character of the results was also present in the studies that focused on the effect of the harvesting method. After 3 weeks of storage, the content of photosynthetic pigments became significantly higher in the manually harvested 'Arbequina' olives [39,70]. The phenomenon was attributed to a possible decrease in the consistency of the chloroplast wall, facilitating the release of the pigments in the oil. Similar results were found for the 'Manzanilla de Sevilla' but, at the same time, the 'Manzanilla Cacereña' showed a completely different profile [43]. The quantification of the effect of the harvesting and conservation method revealed significant differences among 'Arbequina', 'Picual' and 'Verdial' cultivars [46]. In 'Arbequina' the impressive importance of the storage method, explaining 80% of the variance from day 4 to 14, was placed against the results of the 'Picual' variety, where the importance of the conservation was only visible in the first year, showing a linear growth from the onset to the importance of 80%. In 'Verdial' the storage sensitivity was illustrated by a steep increase from day 4 to 8, explaining 90% of the variance for one year. In the other year, a shift toward the importance of the interaction of the factors harvesting and conservation (40%) was observed from day 8 up to day 14.

5.6. Bitterness Index (BI)

The significant correlation between the intensity of bitterness, as evaluated in a sensorial manner by a panel, and the measurement of the absorbance at 225 nm [82], inspired several research groups to follow the progress of this parameter throughout a cold storage period.

A sharp decrease of BI in the oils obtained from olives stored at ambient and 8 °C was found, while the ones stored at 5 °C only presented a slow decrease [19]. Similarly, 8-day storage at 20 °C led to a reduction in K225 of 58% in the monolayered olives and 4-fold in the case of layers up to 20 cm. At 10 °C the same results were obtained after 14 days. Notwithstanding this difference, the K225 index was reduced by 80% in all storage conditions at 20 days of storage [40]. The same gradual decrease was observed in 'Chemlali' stored for 4 weeks at ambient temperature. This progress was attributed to alternations or partial or total inhibition of the specific enzymes (glycosidases and estearases) which downplay the presence of secoiridoid derivatives of phenols and consequently the bitterness intensity [91]. A divergent result was also reported, indicating an initial increase of the K225 values for 'Chétoui' olives stored at 5 °C, in air or under CA. After two weeks, a decrease set in and a similar value was obtained for both conditions [89].

The magnitude of the effect of the harvesting and conservation methods used indicated varietal difference [46]. The 'Arbequina' cultivar showed to be the most vulnerable to the storage temperature from day 4 on, with conservation being the factor that explained up to 71% (year 1) and 83% (year 2) after 14 days. 'Picual' maintained a high explanatory power of the harvesting method for up to 14 days. 'Verdial' expressed a steady increase of the factor 'conservation', which attained a maximum above 60% at the end of the experiment.

5.7. Phenolic Content

The phenolic content of olive fruit depends greatly on the characteristics of the cultivation zone, the climatic conditions and the different techniques applied during the production, the harvest and the postharvest treatment [92]. During ripening, their amount increases to attain a maximum at veraison after which a decline starts. The phenolic composition of the olive fruit is overly complex, although three secoiridoid glucosides form the major part: oleuropein, ligustroside and demetyloleuropein. When the cell tissues are ruptured during grinding, these hydrophilic components contact with specific enzymes that catalyze and modulate the formation of hydrolyzed secoiridoid derivatives. These oil-soluble substances play not only an important role as antioxidants but also in the formation of the sensorial attributes of the oil. Once again, numerous phenolic compounds have been detected although the 4 most important are the aldehydes of the aglucones of oleuropein and ligustroside (3, 4-pHEA-EA and p-HPEA-EA and, through the mediation of specific β-glucosidases, dialdehydes of the same dimethylised aglucones (3, 4-pHPEA-EDA and p-HPEA-EDA) [93]. During extraction, the phenolic compounds undergo processes of oxidation, either by free radicals or through the enzymatic activity of polyphenol oxidase (PPO) and peroxidase (POX), that are equally liberated during the grinding. The final content of secoirridoids results of the equilibrium between the processes of hydrolysis, catalyzed by the β-glucosidases and the oxidation processes induced by the PPO and POX enzymes [93].

Their specific role as antioxidants and their critical contribution in the bitter and pungent taste of the oils attained escalating attention within the cold storage research. In addition to an interest in the evolution of the total phenol content under specific storage conditions, the focus was brought towards shifts in the polyphenol's profiles, the evolution in the activity of the different enzymes and the prospect to modify the bitterness of the extracted oil through a modification of the temperature and/or CA [61].

The reported total amount of phenols measured in the extracted oil was determined by a colorimetric method using a Folin-Cicalteau reagent and expressing the amount as ppm of caffeic acid; gallic acid equivalents; p-hydroxyphenolacetic acid and o-coumaric acid. The separation of olive oil phenolic compounds has been achieved by the application of all types of liquid chromatography and gas chromatography [94].

The majority of the studies observed a significant delay in the decrease of total polyphenols [16,27,29,31,33,34,39,40,49,57,58,62,95]. Nevertheless, varietal differences were equally reported and even contradicted the general trend [30,46,89]. The phenolic compounds of 'Manzanilla de Sevilla' and 'Manzanilla Cacereña', even stored at 2 °C, were strongly and negatively affected, especially those which were mechanically harvested [43]. Similarly, but less explicitly, a significant difference in the oils extracted from mechanically harvested fruit was registered [70].

Measurements of the phenols content in the fruit during cold storage elucidated the importance of the oxidation process of the phenols. More specifically, changes in oleuropein were reported that could only be explained by the effect of exogenous enzymes produced by microbial growth [29,40].

Monitoring POX, PPO and β-glucosidases in crude protein extracts during fruit ripening and storage revealed different trends of activities for different cultivars and storage conditions [53]. A significant reduction of the PPO during the first 2 days of storage was reported, and from then on, a gradual decrease [33]. POX showed a gradual decrease over the 7 days of storage at 15 °C. However, the fact that a multitude of factors (aerobic conditions, storage time and temperature, ripeness, initial quality and microbiological state, varietal resistance against microbiological attack) could influence the production of volatile phenols make it extremely difficult to isolate and describe their complex chemical nature [41,42,93].

5.8. Volatile Compounds

The volatile compounds of virgin olive oil are primarily generated during the extraction process. Most of them are aldehydes and alcohols of six and five straight-chain carbons (C5, C6) together with the corresponding esters, and are synthesized through the enzymatic lipoxygenase (LOX) pathway [95]. However, the handling of olive fruit and consequent storing can result in progressive cell disruption, increased enzymatic activity and ultimately a modification of the oil sensory quality. It is suggested that the premature activation of the LOX pathway alters not only the biochemical status of the fruit but also the amount of volatiles produced during the milling step [43].

While the total amount of volatiles decreases throughout the storage time it is equally true that the different C5 and C6 volatile compounds showed to evolve differently over the storage period, making a detailed interpretation difficult if not impossible [29,40]. Moreover, negative volatile compounds are formed from the metabolic action of yeast and molds [40].

Fruit damage caused by mechanical harvesting triggered the formation and emission of volatile compounds during storage, leading to a decrease in the amount of total volatiles found in the oils. After 24 h of storage, a significantly lower content in C5 and C6 compounds between the mechanically and hand-harvested olives was found, however, these differences were minimized after 11 days [43].

The evolution of the activity of LOX-derived volatiles under different storage conditions showed to be highly variety dependent. No significant differences were reported in the presence of C6 aldehydes and alcohols in the oil of 'Chétoui' variety when related to the olive ripening and storage [52]. Quite different results were obtained with 'Arbequina' olives, where the compounds were much higher in the oil from fruit with the same ripening but stored at 25 °C when compared with those at 5 °C. It is suggested that these high amounts are due to the rapid fruit degradation and cell disruption due to the higher storage temperature [52].

5.9. Tocopherols

Tocopherols (α-, β-, γ-, δ-) belongs, together with tocotrienols (α-, β-, γ-, δ-), to a group of plant soluble lipid compounds known as vitamin E. The antioxidant effect of tocopherols depends mainly on the temperature and lipid composition. α–Tocopherol is the most common form of vitamin E with the highest biological activity, representing 95% of the total tocopherol content in olive oil. The content of tocopherols is shown to be cultivar dependent and significantly affected by the harvesting system, fruit ripeness, crop year, the storage temperature and the storage time [39,43,46,96].

Fruit cold storage does delay the degradation of the tocopherols but cannot avoid it. A significant decrease in tocopherol when fruit was stored at 3 °C was reported as well as significant differences between three Portuguese cultivars [47]. Similarly, a significant effect of the cultivar was reported, comparing 'Ottobratica' and 'Carolea', as well as of the storage temperature on the total amount of tocopherols [34]. A focus on the effects of the harvesting methods revealed that the oil extracted from manually harvested fruit significantly contained higher values when compared with the mechanically harvested, regardless of the storage temperature [39]. At the same time, the fruit stored at 3 °C resulted in values that were significantly higher than when stored at 18 °C, regardless the system of harvesting. A significant effect of the harvesting method was reported but specified the presence of an extreme difference after one day of storage, although this discrepancy equalized after 6 days [43]. It was suggested that the internal damage caused in the fruit during mechanical harvesting, was responsible for the extremely rapid oxidation of tocopherols). A significant effect of the storage time and the harvesting method on the amount of α–tocopherols in three varieties ('Arbequina', 'Picual' and 'Verdial') was observed [46]. However, the effect of the conservation method was absent in both studied years in the 'Arbequina' and 'Verdial' cultivars and only in one year in the 'Picual'. The strength of the factor 'harvesting' varied between the two years and among the varieties.

5.10. Alkylic Esters

The extraction of oil from olives in which fermentative processes have started gives place to negative attributes, such as musty, winey-vinegary or muddy sediment. These sensory classifications were related to the content of free acid alcoholic ester's (FAAE) in the olive oil [97]. It was further observed that very high concentrations of FAAEs in the olive oil do reflect fermentative defects of the olive fruit, whereas other defects, such as oxidative ones or frozen olives, do not produce FAAEs [98]. FAAEs are formed by trans-esterification of free fatty acids with short-chain alcohols, mainly methanol and ethanol yielding methyl (FAME) and ethyl esters (FAEE), which are both extracted with the oil. However, while the methanol is also formed during the pectin degradation of the cell wall during the ripening of the olive fruit, the ethanol can only be produced by fermentation, except specific varieties that have been shown to produce ethanol during their ripening [99]. Since 2013, FAEE are considered as an important marker and as such included in the EU Commission Regulation [100].

Inappropriate practices during the harvest and storage of olives are important factors that do promote their hydrolytic and oxidative deterioration. The content of ethanol in ground-picked olives was significantly higher compared to fruit harvested from the tree [101]. However, the amount of ethyl esters in the extracted oil remained below the limit established for EVOO and was attributed to the short time between the recollection and the extraction which impeded the synthesis of higher concentrations [101]. The type of container used during post-harvest storage showed to be a relevant factor in explaining the high levels of FAEE when olives, stored in closed plastic bags, were compared with those kept in perforated plastic boxes [51].

Thus far, the presence of FAAE have been studied scarcely. A manifest effect of the olive storage temperature on the FAME and FAEE content was observed [34]. The limits of FAEE for EVOO (<30 mg/kg) was surpassed after 12 days of storage at 25 °C. When stored at 4 °C, one variety, 'Carolea', showed no increase of neither of the esters. The other variety, 'Ottobratica', showed a significant difference when two harvesting moments were compared. The latter harvested fruit attainted a level above the set maximum, indicating that the physical state of the drupes was declining and the aptitude to deterioration increasing. No significant differences were found in the treatments in the studied 'Rosignola' variety [58]. In the other variety, 'Istarska Bejelica', only an increase in the oils stored at room temperature was observed [58].

6. Fruit Temperature Management

6.1. Rationale for Controlling Postharvest Olive Fruit Temperature

The primary objective of cold storage of olive fruit was to spread the extraction process over a longer time to increase the processing capacity. The research was oriented to determine the maximum storage time of the fruit without jeopardizing the quality of the oils extracted therefrom. However, as the oil industry opted to amplify the production capacity, the proposed conservation method was introduced only sporadically. Subsequently, the fruit temperature was modified to induce precise characteristics of the extracted oil. Yet, the challenging technique came along with specific constraints and thwarted its incorporation at an industrial level. Currently, the olive oil industry is facing various new challenges that directly relate to the control and the modification of the fruit temperature. Climatological changes are provoking a shift in the phenology of the olive plant and are forwarding the optimum harvest time by several weeks [102,103]. Meanwhile, high-density orchards (HDO) do request specific varieties, such as 'Arbequina', 'Koroneiki' and 'Arbosana', which need to be harvested early in the season [104]. Finally, the success of fresh-flavored oils for the gourmet market, backed up by the growing importance of the international olive oil competitions, favors an early harvest [105]. Consequently, a harvesting campaign that envisions high-end quality olive oil is taking off in September, with day temperature that easily raises above 30 °C.

Forwarding the harvest leads unavoidably to a high field temperature of the fruit. This undesirable situation, when combined with a transport of a couple of hours in trailers of more than 10 t, during which the fruit temperature further increases, suffice to create conditions of asphyxia and induce anaerobic respiration. Consequently, processes of fermentation produce by-products that lead to detectable sensory defects and high levels of alcoholic esters in the extracted oil and ultimately to the rejection of the oil as EVOO [106]. This new situation has already alarmed the industry. Nocturnal harvesting was introduced as a possible solution but soon after prohibited for its disastrous side effects on the birdlife [107].

But even without devastating fermentation processes during the transport and storage, a high field temperature of the fruit impedes the extraction of high-quality olive oil. It is widely accepted that the ideal malaxation temperature is situated between 25 and 30 °C [108,109]. This becomes simply impossible with fruit at more than 25 °C, knowing that crushing itself rises the olive pasta at least 5 °C [110]. Under these conditions a full control of the malaxing parameters, essential to obtain the desired organoleptic characteristics of the oil, simply becomes impossible.

Up to now, controlling the temperature during the transport of the olives was hardly taken into consideration, similar to the necessity to adjust the temperature before the extraction process. Once the fruit was harvested, it was brought to the mill and extracted immediately if possible and stored for a short time when necessary. The mentioned environmental, agronomical and cultural factors are bringing the existing postharvest model under stress and urge for a more dynamic way to control the fruit temperature from the moment it is harvested up to its processing.

Developing specific cooling installations adapted to the different stages of the postharvest stages and production lines, do have in common the need to know the specific physiological and biothermal characteristics of the fruit. Several studies have been published in this respect and do confirm that olive fruit can be cooled to the desired temperature in a time of several minutes either by water or cooled air, without compromising its quality [111–113].

6.2. Biothermal Characteristics

Determining the cooling time needed to reach the desired temperatures is the first necessary step in the design and/or implementation of any cooling system. Existing physical models have low applicability at the industrial level since it is very difficult to take into account the physical and biochemical variability of the fruit with the parameters they contemplate. Therefore, an effective way to obtain information about the cooling process is by making physical measurements in a real process.

Specific data concerning the thermodynamic properties of olive fruit remains scarce. Ref. [111] published data on the thermal conductivity (k), specific heat (Cp) and thermal diffusivity (α) of olive fruit but did not take into account the geometrical variability that exists among the varieties and used simplified mathematical equations to calculate the constants. The calculations of the Cp were based on the chemical composition of 4 different varieties while the thermal conductivity was measured using the transient hot-wire method. The thermal diffusivity was calculated using the experimentally determined values of k and Cp. The results showed deviant results when comparing temperature-related models with empirical ones. A sophisticated mathematical model for heat conduction in stone fruits was developed to calculate the thermal properties of olive fruit [112]. The used empirical data stemmed in part from the precise measurement of the dimensions of one olive (c.v. 'Gordal') and the internal temperature change during its cooling in cold water. They calculated a Biot number of 4.43 using water as the transfer medium for an olive that weighed 10.44 g, which is by all standards an extreme weight for olive fruit. The Biot number (Bi) expresses the ratio of the internal resistance of conduction to the external resistance of convection (h). When $Bi < 0.1$, the internal resistance is considered to be negligible in comparison with the surface resistance and it is assumed that the material or produce heats or cools down uniformly [72,114]. As the values for h are medium-dependent, the obtained Biot-number is not valid when considering a pre-cooling treatment with cooled air [115]. Submerging picked olives in a cooled solution of diluted lye is a recognized and recommended treatment to avoid bruise damage for table olives [116,117]. However, when dealing with fruit intended for oil extraction, contact with water before storage is not recommended as it prompts deterioration and even fermentation processes [118].

Thermal imaging was used to measure the temperature change of individual olives of six different cultivars stored under room cooling conditions and related the obtained results to their physical and geometrical characteristics [113]. Thermodynamic parameters were further used to simulate the cooling process of the olive fruit by convection within a range of field temperatures and room cooling temperatures. The calculated cooling rates revealed a non-linear relation between the specific surface area and the cooling rate of an olive, suggesting that underlying factors need to be considered when calculating the specific heat of the fruit. To explain the observed flattening on both sides of the curve, further research is needed to clarify the influence of the stone when dealing with small olive fruit, as well as the increased Biot number for heavier olives. Nevertheless, the simulation clarified that a relatively short time is needed to cool the fruit to a desirable temperature when placed in a cold air environment. In the case of 'Arbequina' and 'Koroneiki' cultivars, fruit with a field temperature of 38 °C would attain a temperature of 22 °C when kept in an environment of −16 °C for less than 2 min. In the case of 'Hojiblanca' and 'Picual', the cooling time under these conditions increased to 3 min. Finally, the heavier 'Verdial' and 'Gordal' varieties needed up to 5 min to cool down to the desired temperature. Even though further research is needed to discard negative impacts on the fruit, the obtained results are useful for the development and the introduction of new techniques based on cooled air. A related experiment in which boxes with 10 kg of olives were brought to 5 °C by placing them at a room temperature of −18 °C, revealed no deterioration in the quality parameters of the oil [77].

The important role that forced air cooling can play in the field, at the farm and at the reception yard of the mill, does advocate for extensive studies on the application of this technique. More specifically, experiments at the laboratory level are needed to document the working of a cooling tunnel and to study the effect of the cooled forced air on the fruit characteristics and oil extracted therefrom. This implies the design of the laboratory scale cooling tunnel. An example of such a system is given in Figure 6. The showed system would imply the preliminary warming of the olive fruit in an incubator at a set temperature while the cooled air can easily be provided by a separate cooling group.

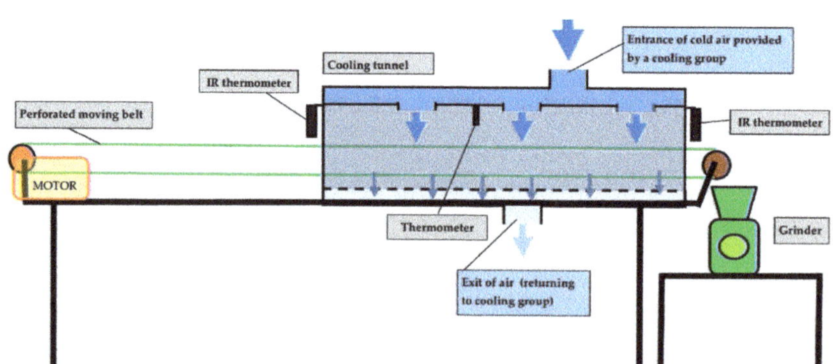

Figure 6. Experimental setup at laboratory scale to study the effect of cooled forced air on the quality of the fruit and the oil extracted therefrom.

6.3. Removing Field Heat

The effect of time and temperature on respiration has been studied on various intact fruit as well as fresh-cut produce [75,119–123]. In all of these studies, an Arrhenius-type equation showed to be very reliable to describe the respiration rate in function of the temperature. The studies on the respiration rate of cold-stored olives fruit have shown that the respiration rate increases at a higher temperature. In addition, damaged or bruised fruit do not only accelerate respiration [43,84] but also induces hydrolytic and oxidative deterioration. The presence of FAEE in the extracted oil does indicate the formation of ethanol during the transport or storage because of fruit damage as well as anaerobic respiration [51,101]. The rate of gas diffusion and respiration varies significantly when the fruit is piled up, as is the case when olives are collected and transported in large containers. Under such conditions, some parts risk becoming anaerobic while others remain aerobic.

Removing field heat as soon as possible stands as a compelling recommendation in the postharvest studies that focus on controlling the quality of the fruit [74,124]. The process of removing field heat is commonly referred to as precooling and implies the quick removal of heat, in a few minutes or a few hours, either in the field or at the processing plant. Prompt precooling is used for fruit and vegetables that need rapid cooling to maintain the high quality during the holding period. A delayed cooling, implying a lag between the harvest and the precooling treatment, generally reduced the quality and the storage life, although it is also reported that for specific fruits, such as cherry, peach, nectarine and apple, a cooling delay may be beneficial. Gradual cooling lowers the temperature of the fruit and vegetables to only certain degrees over hours or days to induce the cold tolerance in the produce.

Forced air cooling, hydro-cooling, liquid ice-cooling and vacuum cooling are the most commonly used precooling techniques [125–127]. All involve the fast transfer of heat from the produce to a cooling medium, such as air or water. However, the characteristics of the commodity and the container, the rate of cooling and the further storage conditions do define the most appropriate technique. Forced air-cooling is further differentiated in the circulation of cold air at a high velocity in refrigerated rooms, forcing cold air through bulk or stacked produce as they move through a cooling tunnel on a conveyer, and using the pressure differential technique that forces air through produce packed in containers.

When contemplating the possibilities to remove the field heat of harvested olives, at least three different circumstances need to be taken into account. Firstly, when there is a risk for fruit asphyxiation and fermentation processes during the transport and/or consequent storage. Secondly, when short-time storage before transport is necessary while the requested minimum amount of fruit for extraction is harvested. Thirdly, when the optimal extraction of the oil becomes impossible when the fruit arrives at the mill at a temperature already above the set maximum.

If the day temperature and transportation time are such that the respiration threshold will be attained before the processing of the fruit, prompt precooling come to the fore as the only solution to avoid deterioration in the batch of olive fruit. Water as a cooling medium can be excluded as wetting the olives before their transport will create even greater deterioration.

Forced air-cooling comes to the fore as a possibility and more specifically the use of a cooling tunnel in which the temperature of the fruit is lowered to at least 15 °C. The estimation of the cooling rates allowed to predict that this can be attained in the range of a couple of minutes [113]. However, the cooling mechanism needs to be capable to handle the considerable and continuous amount of olives that are harvested mechanically. Even though this looks not possible in the short term, it can be expected that either the next generation of straddle machines will include such a cooling extension or that the existing mobile pre-cooling facilities are adapted to the specific harvesting characteristics [125,128]. As the latter is designed to precool palletized produce, this would imply either the distribution of the harvested olives in containers or a rethinking of their design.

Cold storage for a short time at the farm can be provided using room cooling. The experience of a small-scale producer dealing with 5 to 6000 kg of olive fruit, harvested over a week, demonstrated already its feasibility [45,46]. However, natural convection air cooling is slow, especially when the room is very packed or when the olives are kept in 400 kg containers. In that case, it can take several days to lower the temperature to the desired 5 °C. Storage in smaller plastic boxes filled with 15–20 kg of olives and providing sufficient space around and below each box to allow air circulation, thus becomes unavoidable. An important drawback of this method is the logistic problem it creates when hundreds of boxes need to be transported to the mill. Filling 400 kg containers before the transport offers a solution but it is obvious that a precooling treatment before storing them at 5 °C would be a far better solution. The possibility to precool a container of 400 kg has shown to be possible at laboratory level [77]. However, it would imply a small-scale precooling tunnel at the farm or a portable forced-air cooling unit in the field, as described elsewhere [128], to implement the system.

When the olives arrive at the mill at a temperature that impedes a proper malaxation, two possible options are available, either the fruit is cooled at arrival or just before the grinding, or the temperature of the ground olive paste is lowered before entering the malaxer. Cooling of the olive paste through a tubular heat exchanger before the malaxation process is suggested when the thermal conditions are above the optimal temperature to extract a high-quality EVOO [129]. The solution is presented as an alternative to the use of cold, climatic chambers or the use of dry ice, practices they discard as being not easily adaptable to an industrial oil transformation process [129]. Yet, the experiment was performed with olive fruit at a moderate temperature of 27 °C that was cooled to 15 °C followed by a malaxation at 25 and 30 °C. Neither the cooling time nor details of the heat exchanger (capacity, energy) were reported. The option to cool the fruit has not attracted the same attention until recently. During the harvest campaign 2020–2021 an experiment in a mill in the province of Jaen, planned to inject cold air using a turbine (7.5 kW) into hoppers with a capacity of 45 and 60 tons to eliminate sensorial defects, attributed to anaerobic fermentation in the produced oils [130]. At the same time, a new study was announced by a private olive mill to implement and evaluate the use of hydro-cooling to lower the fruit temperature of the incoming fruit [131]. Hydro-cooling is considered as one of the fastest precooling techniques and consists of immersing or flooding products in chilled water or by spraying chilled water over the product. According to the press release, the study uses a shower cooling system to lower the fruit temperature. To the best of our knowledge, the results of both studies are not yet published. In the first experiment, the crucial question will be whether the adapted forced air-cooling system can penetrate homogeneously the bulk of fruit in the hopper. Flow field uniformity is regarded as an important factor that affects not only the effectiveness and the energy consumption of forced air cooling [126]. Concerning the second experiment, it can be expected that hydro-cooling will offer a suitable solution, however, this method can only be applied when the olives are processed immediately thereafter. The degree of olive damage at different points in the pre-processing steps in the mill increases from 28% at the reception to 59% after the washing and 78% during the storage of less than 12 h. [118]. When washed olives are stored, their ethanol content increases significantly due to the loss of integrity of the fruit and the availability of oxygen in the hopper [132].

A third alternative might consist of the cooling of the fruit immediately after the fruit is cleaned from branches and leaves. This could be accomplished by installing a forced-air cooling tunnel right after this cleaning or, when space impedes such a structure, by converting the existing ramps of the moving belts into air-cooled tunnels. Eventually, this solution can also be used for precooling the fruit for subsequent cold storage. In that case, the cooled fruit can be collected in containers of 400 kg, allowing traceability and easy portability towards the cooling room at the mill.

6.4. Adjusting to Desired Preprocessing Temperature

Even though removing the field heat or the cold storage of olives may be the ideal solution to maintain the fruit quality, it has one important drawback when the cold fruit needs to be processed. Similarly, to too warm olives, a too-cold temperature does impede proper malaxation and extraction [133].

The use of heat exchangers, microwave, ultrasound applications or combinations of them is presented as reliable solutions to adjust the temperature of the paste and to reduce the malaxing time [134–141]. Several models of heat exchangers are already on the market (Pieralisi Protoreattore, Alfa Laval EVO-line) while a sono-heat exchanger, being a combination of a heat exchanger and ultrasound technique, is expected to enter industrial production soon [142]. However, all these innovative flash heat treatments imply substantial investments, adjustments of the production lines and higher energy consumption [143].

The feasibility of warm-up olives during their immersion in the washing tub might be an economic alternative [144]. Experiments at the laboratory as well as pilot-plant level demonstrated that an immersion time between 15 and 20 s in water at 35–40 °C suffice to warm-up olives at 10 °C to the desired 27 °C. It was further demonstrated that the temperature of the paste when entering the malaxer, could be easily monitored with a thermostat and adjusted by turning the electric heaters in the washing tub on and off. These results look very promising. The used temperatures and immersion time are not expected to affect the biosynthesis of the olive aroma and neither will deteriorate the oxidative stability or lower the polyphenol content, as was the case in experiments that used water at 50 °C and 72 °C for 3 to 5 min to lower the bitterness of the fruit [60,63,145,146]. The energy cost to keep the water around 40 °C will depend on the characteristics of the washing tub (capacity, isolation) and the ambient temperature. Direct heating with electric resistances or an external heat exchanger is suggested as feasible economic solutions, that can be introduced to an existing washing installation. The lower investment costs and the easiness to integrate it into an existing production line makes it a valuable alternative for the costlier flash heat treatments.

6.5. Developing Innovating Plants

A technology maturity assessment system used by Clodoveo describes the progress in developing innovating plants to improve the VOO extraction process [142]. It offers a clear overview of the different steps that need to be taken from an initial idea to a fully commercial application. It comprises 10 steps of which the first four are related to the validation of an idea, and two to the developing of a prototype, the validation of the prototype system and finally two related to the production stage. Applied to the suggested forced air-cooling system, the initial idea is formulated, part of the basic research published and the concept and the application formulated. Once sufficient stakeholders are found, the building of a small-scale prototype can start. When the results of the laboratory experiments confirm the validity of the idea, the building of a large-scale prototype cooling tunnel or the adaptation of existing systems comes to the fore as the next step. Meanwhile, the specific environmental conditions in which the cooling tunnel is supposed to be integrated will bring the adaptation of the existing movable cooling systems or the incorporation of a cooling tunnel in a straddle harvester under the attention of the engineers with a focus on field precooling systems. The promising result in using the washing water to adjust the fruit temperature before grinding can be evaluated similarly. Once again, the different steps in the progress can be expected to come with specific challenges. The preliminary experiments already made clear that controlling the temperature of the water remains to be solved together with the necessity to continuously sanitize the used water to avoid bacterial growth.

7. Conclusions and Future Research Directions

The presented research made clear that the beneficial effects of cold storage are widely accepted. Overwhelming evidence, build up over the last 30 years, underscore the influence of cold storage on the quality of the olive fruit and the official and frequently used quality parameters of the oil extracted from cold-stored olives. As such, storage at 4–6 °C made it possible to delay the further processing of olive fruit by up to 4 weeks. The use of CA is not recommended as it did not show beneficial advantages. However, these general recommendations do imply healthy fruit, harvested with minimal damage and stored in optimal containers impeding asphyxiation and allowing the removal of respiration heat. Moreover, the observed intervarietal differences with respect to the vulnerability of the fruit and the importance of the degree of ripeness further caution overgeneralization. When the quality expectations rise and the production of premium oil becomes the main focus the conclusions become even less straightforward and urge for a further disclosure of the impact on the different minor components. The effect of cold storage on the amount of minor components such as polyphenols, tocopherols and volatile components as well as the formation of ethanol in the fruit during cold storage still needs further clarification. The mechanisms behind the observed and the reported contradictory shifts in composition and amount, continue to trigger new research questions. The role of ethanol in the synthesis of FAEE on the one hand and the importance of the factor temperature in attaining the threshold for aerobic respiration urge for a deeper understanding of the metabolic shifts during and within bulk olive transports. The effects of pressure, temperature and time on respiration and eventually the production of ethanol are still to be disclosed. Real-time measurement during bulk transport of the temperature and atmospheric gases could provide the necessary data to further construct and validate mathematical models.

The impact of increasing the mechanization of the harvest on the quality of the fruit calls for additional monitoring, especially when new machines and techniques are introduced. At the same time, broadening the scope of investigation towards not only the harvesting techniques at an industrial level but also at a small-scale level will document in a more precise way the mutual influence between the used harvesting and storage methods. The influence of the type and capacity of the storage unit has been left undiscussed for too long. Two recent publications of did address this subject [71,77], but the first reported only the effects on the official quality parameters, while the second restricted the storage time to less than 1 day. Whether cold storage will ever be a used technique in the postharvest handling of olive fruit, will largely depend on its ability to manage large quantities and workable solutions to precool the olives.

Cold storage is not just a method to prolong the quality of the harvested olives but can also be regarded as an important link in a more broadly defined fruit temperature management. From this perspective, new aspects come to the fore that have hardly been investigated in the handling of olive fruit. Recognizing that the removal of field heat is the first essential step in this process urge for specific knowledge on the biothermal parameters of olive fruit, the characteristics of the heat transfers in the different cooling mediums, and not in the least the evaluation of the different methods on the fruit quality and the oil extracted therefrom. Even though the preliminary results of Plasquy et al. do indicate that air at −18 °C can be used to precool olive fruit, it must be underlined that the experiments took place under room cooling conditions [77,113]. To the best of our knowledge, no data are available on the use of forced air for a controlled lowering of the olive fruit temperature.

Studying fruit temperature management demands a multidiscipline approach. The intertwining of agro-chemical and engineering facets needs input from each of these specialties. Thus far, the focus has been mostly on the physiological changes in the fruit and the chemical changes within the extracted oils. The further study of the biothermal characteristics of olive fruit as well as the development and adaptation of existing cooling devices count on specific engineering skills. The knowledge that has been build up in the field of fruit and vegetable conservation offers an impressive backing but does not suffice because of the idiosyncratic nature of the olive fruit harvest and fruit handling.

The final aim of fruit temperature management is a concise control of the temperature from the moment the olives fruit are harvested up to the moment the extraction process starts. This mission boils down to taking seriously the fact that olives are fruit and that virgin olive oil is nothing less than fresh fruit juice. Even though this phrase is often cited and partly substantiated the official definition, it is still too often overlooked and too easily taken for granted. However, the new climatological challenges, the increasing mechanization of the harvest and the demand for premium products are fueling a new awareness that the quality of the oil depends largely on that of the olive fruit. Keeping the quality of the fruit as optimal as possible thus becomes a logical conclusion, which acknowledges the key role that control of the fruit temperature plays in this process.

Author Contributions: E.P. is responsible for the conceptualization of this review paper. The authors state to have equally participated in the writing. All authors have read and agreed to the published version of the manuscript.

Funding: This research received no external funding.

Institutional Review Board Statement: Not applicable.

Informed Consent Statement: Not applicable.

Conflicts of Interest: The authors declare no conflict of interest.

References

1. IOC [International Olive Council]. *Trade Standard Applying to Olive Oils and Olive Pomace Oil*; COI/T.15/NC No 3/Rev. 16; IOC: Madrid, Spain, 2021. Available online: https://www.internationaloliveoil.org/wp-content/uploads/2021/07/COI-T15-NC3-REV-16-2021-_ENG.pdf (accessed on 28 June 2021).
2. Maxie, E.C. Storing olives under controlled temperature and atmospheres. *Calif. Olive Assoc. Annu. Tech. Rep.* **1993**, *42*, 34–40.
3. Maxie, E.C. Experiments on cold storage and controlled atmospheres. *Calif. Olive Assoc. Annu. Tech. Rep.* **1964**, *43*, 12–15.
4. Woskow, M.; Maxie, E.C. Cold storage studies with olives. *Calif. Olive Assoc. Annu. Tech. Rep.* **1965**, *44*, 6–11.
5. Cantarelli, C. Nuove soluzioni al problema della conservazione delle olive: Contributi sperimentali. *Riv. Ital. Sostanze Grasse* **1965**, *17*, 475–480.
6. Petruccioli, G.; Montedoro, G.; Cantarelli, C. Nuove prospettive nella tecnología della conservazione delle olive destinate all'estrazione. *Riv. Ital. Sostanze Grasse* **1970**, *47*, 475–480.
7. Montedoro, G.; Garofolo, L. Caratteristiche qualitative degli oli vergini di oliva. Influenza di alcune variabili: Varietà, ambiente, conservazione, estrazione, condizionamento del prodotto finito. *Riv. Ital. Sostanze Grasse* **1984**, *61*, 157–168.
8. Petruccioli, G.; Parlati, M.V. Studies on olives storage before oil extraction. III Effect of Storage conditions of olives on the chemical and organoleptic characteristics of oils. In *Third Subproject: Conservation and Processing of Foods—A Research Project (1982–1986)*; Abstract 20; National Council of Italy: Milan, Italy, 1987; pp. 83–84.
9. Kader, A.A.; Nanos, G.D.; Kerbel, E.L. Responses of 'Manzanillo' olives to controlled atmospheres storage. In Proceedings of the Fifth International Controlled Atmosphere Research Conference, Wenatchee, WA, USA, 14–16 June 1989; Volume 2, pp. 119–125.
10. García, J.M.; Streif, J. The effect of controlled atmosphere storage on fruit and oil quality of 'Gordal' olives. *Gartenbauwissenschaft* **1991**, *56*, 233–238.
11. Gutiérrez, F.; Perdiguero, S.; García, J.M.; Castellano, J.M. Quality of oils from olives stored under controlled atmosphere. *J. Am. Oil Chem. Soc.* **1992**, *69*, 1215–1218. [CrossRef]
12. García, J.M. Efecto del CO_2 en la atmósfera de almacenamiento sobre la fisiología de la aceituna de molino. *Grasas Aceites* **1993**, *44*, 81–84. [CrossRef]
13. García, J.M. Efecto del CO_2 en la atmósfera de almacenamiento sobre la calidad del aceite de oliva. *Grasas Aceites* **1993**, *44*, 169–174. [CrossRef]
14. Castellano, J.M.; García, J.M.; Morilla, A.; Perdiguero, S.; Gutierrez, F. Quality of Picual olive fruit stored under controlled atmospheres. *J. Agric. Food Chem.* **1993**, *41*, 537–539. [CrossRef]
15. Pérez-Camino, M.C.; García, J.M.; Castellano, J.M. Polar compound concentrations in virgin olive oils from stored cultivar Picual olive fruits. *J. Agric. Food Chem.* **1992**, *40*, 2260–2262. [CrossRef]
16. Maestro, R.; García, J.M.; Castellano, J.M. Changes in Polyphenol Content of olives stored in modified atmosoheres. *HortScience* **1993**, *28*, 749. [CrossRef]
17. García, J.M.; Castellano, J.M.; Gutiérrez, F.; Castellano, J.M.; Perdiguero, S.; Morilla, A.; Albi, M.A. Cold storage of mill olives. In *Contribution du Froid à la Preservation de la Qualité des Fruits, Legumes et Produits Halieutiques: Actes du Sumposium*; Bennani, A.L., Messaho, D., Eds.; Actes Éditions Rabat: Rabat, Maroc, 1994; pp. 77–86.
18. García, J.M.; Gutiérrez, F.; Castellano, J.M.; Perdiguero, S.; Morilla, A.; Albi, M.A. Storage of olives destined for oil extraction. *Acta Hortic.* **1994**, *368*, 673–681. [CrossRef]
19. García, J.M.; Gutiérrez, F.; Barrera, M.J.; Albi, M.A. Storage of Mill Olives on an Industrial Scale. *J. Agric. Food Chem.* **1996**, *44*, 590–593. [CrossRef]
20. García, J.M.; Gutiérrez, F.; Castellano, J.M.; Perdiguero, S.; Morilla, A.; Albi, M.A. Influence of storage temperature on fruit ripening and olive oil quality. *J. Agric. Food Chem.* **1996**, *44*, 264–267. [CrossRef]
21. Canet, M.; García, J.M. Repercusión de la frigoconservación de la aceituna de molino en el proceso de producción de aceite de oliva virgen. *Grasas Aceites* **1999**, *50*, 181–184. [CrossRef]

22. García, J.M.; Yousfi, K. The postharvest of mill olives. *Grasas Aceites* **2006**, *57*, 16–24. [CrossRef]
23. Monteleone, E.; Caporale, G.; Lencioni, L.; Favati, F.; Bertuccioli, M. Optimizing of virgin olive oil quality in relation to fruit ripening and storage. In *Food Flavors: Generation, Analysis and Process Influence*; Charalambous, G., Ed.; Elsevier: Amsterdam, The Netherlands, 1995; pp. 397–418.
24. Gardiman, M.; Tonutti, P.; Pizzale, L.; Conte, L.; Carazzolo, A. The effect of hypoxic and CO_2-enriched atmospheres on olive ripening and oil quality. *Acta Hortic.* **1997**, *474*, 525–530. [CrossRef]
25. Koprivnjak, O.; Procida, G.; Benčić, D.; Zelinotti, T. Effect of olive fruit storage in sea water on oil quality. *Food Technol. Biotechnol.* **1999**, *37*, 209–214.
26. Agar, I.T.; Hess-Pierce, B.; Sourour, M.M.; Kader, A.A. Quality of fruit and oil of black-ripe olives is influenced by cultivar and storage period. *J. Agric. Food Chem.* **1998**, *46*, 3415–3421. [CrossRef]
27. Kiritsakis, A.; Nanos, G.D.; Polymenopoulos, Z.; Thomai, T.; Sfakiotakis, E.M. Effect of fruit storage conditions on olive oil quality. *J. Am. Oil Chem. Soc.* **1998**, *79*, 721–724. [CrossRef]
28. Agar, I.T.; Hess-Pierce, B.; Sourour, M.M.; Kader, A.A. Identification of optimum preprocessing storage conditions to maintain quality of black ripe 'Manzanillo' olives. *Postharvest Biol. Technol.* **1999**, *15*, 53–64. [CrossRef]
29. Kalua, C.M.; Bedgood, D.R., Jr.; Bishop, A.G.; Prenzler, P.D. Changes in Virgin Olive Oil Quality during Low-Temperature Fruit Storage. *J. Agric. Food Chem.* **2008**, *56*, 2415–2422. [CrossRef] [PubMed]
30. Dag, A.; Boim, S.; Sobetin, Y.; Zipori, I. Effect of mechanically harvested olive storage temperature and duration on oil quality. *Horttechnology* **2012**, *22*, 528–533. [CrossRef]
31. Clodoveo, M.L.; Delcuratolo, D.; Gomes, T.; Colelli, G. Effect of different temperatures and storage atmospheres on Coratina olive oil quality. *Food Chem.* **2007**, *102*, 571–576. [CrossRef]
32. Rinaldi, R.; Amodio, M.L.; Colelli, G.; Clodoveo, M.L. Controlled atmosphere storage of 3 Italian cultivars of olives for oil production. *Acta Hort.* **2010**, *857*, 97–105. [CrossRef]
33. Zullo, B.A.; Di Stefano, M.G.; Cioccia, G.; Ciafardini, G. Evaluation of polyphenols decay in the oily fraction of olive fruit during storage using a mild sample handling method. *Eur. J. Lipid Sci. Technol.* **2014**, *116*, 160–168. [CrossRef]
34. Piscopo, A.; De Bruno, A.; Zappia, A.; Gioffrè, G.; Grillone, N.; Mafrica, R.; Poiana, M. Effect of olive storage temperature on the quality of Carolea and Ottobratica oils. *Emir. J. Food Agric.* **2018**, *30*, 563–572. [CrossRef]
35. Famiani, F.; Farinelli, D.; Urbani, S.; Al Hariri, R.; Paoletti, A.; Rosati, A.; Esposto, S.; Selvaggini, R.; Taticchi, A.; Servili, M. Harvesting system and fruit storage affect basic quality parameters and phenolic and volatile compounds of oils from intensive and super-intensive olive orchards. *Sci. Hortic.* **2020**, *263*, 109045. [CrossRef]
36. Gutiérrez, F.; Varona, I.; Albi, M.A. Relation of acidity and sensory quality with sterol content of olive oil from stored fruit. *J. Agric. Food Chem.* **2000**, *48*, 1106–1110. [CrossRef] [PubMed]
37. Luaces, P.; Pérez, A.G.; Sanz, C. Effect of cold storage of olive fruits on the lipoxygenase pathway and volatile composition of virgin olive oil. *Acta Hort.* **2005**, *682*, 993–998. [CrossRef]
38. Yousfi, K.; Cayuela, J.A.; García, J.M. Reduction of virgen olive oil bitterness by fruit cold storage. *J. Agric. Food Chem.* **2008**, *56*, 10085–10091. [CrossRef] [PubMed]
39. Yousfi, K.; Weiland, C.M.; García, J.M. Effect of harvesting system and fruit cold storage on virgin olive oil chemical composition and quality of superintensive cultivated 'Arbequina' olives. *J. Agric. Food Chem.* **2012**, *60*, 4743–4750. [CrossRef] [PubMed]
40. Inarejos-García, A.M.; Gómez-Rico, A.; Salvador, M.D.; Fregapane, G. Effect of Preprocessing olive storage conditions on virgin olive oil quality and composition. *J. Agric. Food Chem.* **2010**, *58*, 4858–4865. [CrossRef] [PubMed]
41. Vichi, S.; Romero, A.; Gallardo-Chacón, J.; Tous, J.; López-Tamames, E.; Buxaderas, S. Influence of olives' storage conditions on the formation of volatile phenols and their role in off-odor formation in the oil. *J. Agric. Food Chem.* **2009**, *57*, 1449–1455. [CrossRef]
42. Vichi, S.; Romero, A.; Gallardo-Chacón, J.; Tous, J.; López-Tamames, E.; Buxaderas, S. Volatile phenols in virgin olive oil: Influence of olive variety on their formation during fruit storage. *Food Chem.* **2009**, *116*, 651–656. [CrossRef]
43. Morales-Sillero, A.; Pérez, A.G.; Casanova, L.; García, J.M. Cold storage of 'Manzanilla de Sevilla' and 'Manzanilla Cacereña' mil olives from super-high density orchards. *Food Chem.* **2017**, *237*, 1216–1225. [CrossRef]
44. Petrón, M.J.; Timón, M.L.; Carrapiso, A.I.; Andrés, A.I. Effect of ripening and olive cold storage on oil yield and some olive oil characteristics. *Riv. Ital. Sostanze Grasse* **2018**, *95*, 255–259.
45. Plasquy, E.; Blanco-Roldán, G.; Florido, M.C.; García, J.M. Effects of an integrated harvest system for small producers on the quality of the recollected olive fruit. *Grasas Aceites* **2021**, *72*, e436. [CrossRef]
46. Plasquy, E.; Florido, M.C.; Sola-Guirado, R.R.; García, J.M. Effect of a harvesting and conservation method for small producers on the quality of the produced olive oil. *Agriculture* **2021**, *11*, 417. [CrossRef]
47. Pereira, J.A.; Casal, S.; Bento, A.; Oliveira, M.B.P.P. Influence of olive storage period on oil quality of three Portuguese cultivars of Olea Europea, Cobrançosa, Madural and Verdial Transmontana. *J. Agric. Food Chem.* **2002**, *50*, 6335–6340. [CrossRef]
48. El-Nagaar, N.I.; El-Saedy, R.M. Storage temperature and storage duration affect fruit and oil quality of Coratina, Manzanillo and Picual olives. *Alex. Sci. Exch. J.* **2010**, *31*, 137–154. [CrossRef]
49. Ouni, Y.; Nabil, B.Y.; Zarrouk, M.; Flamini, G. Influence of olive storage period on volatile compounds and oil quality of two Tunisian cultivars of Olea europea, Chemlali and Chetoui. *Int. J. Food Sci. Technol.* **2011**, *46*, 1245–1252. [CrossRef]

50. Nabil, Y.; Ouni, Y.; Dabbech, N.; Baccouri, B.; Abdelly, C.; Zarrouk, M. Effect of olive storage period at two different temperatures on oil quality of two Tunisian cultivars of Olea europea, Chemlali and Chétoui. *Afr. J. Biotechnol.* **2012**, *11*, 888–895. [CrossRef]
51. Jabeur, H.; Zribi, A.; Abdelhedi, R.; Bouaziz, M. Effect of olive storage conditions on Chemlali olive oil and the effective role of fatty acids alkyl esters in checking olive oil authenticity. *Food Chem.* **2015**, *169*, 289–296. [CrossRef]
52. Hbaieb, R.H.; Kotti, F.; Gargouri, M.; Msallem, M.; Vichi, S. Ripening and storage conditions of Chétoui and Arbequina olives: Part I. Effect on olive oils volatiles profile. *Food Chem.* **2016**, *203*, 548–558. [CrossRef]
53. Hbaieb, R.H.; Kotti, F.; Cortes-Francisco, N.; Caixach, J.; Gargouri, M.; Msallem, M.; Vichi, S. Ripening and storage conditions of Chétoui and Arbequina olives: Part II. Effect on olive endogenous enzymes and virgin olive oil secoiridoid profile determined by high resolution mass spectrometry. *Food Chem.* **2016**, *210*, 631–639. [CrossRef] [PubMed]
54. Oueslati, I.; Zarrouk, M.; Flamini, G. Storage of olives (*olea europaea*): Effect on the quality parameters and the radical scavenging activity of the total fraction, lipidic and methanolic fractions of the VOO. *Acta Hort.* **2013**, *997*, 129–136. [CrossRef]
55. Bozdogan, A.; Eker, T.; Konuskan, D.; Oz, A.T.; Kafkas, E. Effects of the storage of Turkish Gemlik olives under CO_2 and N_2 on the phenolic compounds and fatty acid compositions of olive oils. *J. Food Meas. Charact.* **2019**, *13*, 187–195. [CrossRef]
56. Vali, R.; Ghorbani, S.; Hajnajari, H. Effect of temperature and storage period on olive (*Olea europaea* cv. Zard) fruit in olive oil quality. *J. Food Agric. Environ.* **2011**, *9*, 74–77.
57. Taluri, S.S.; Jafari, S.M.; Bahrami, A. Evaluation of changes in the quality of extracted oil from olive fruits stored under different tempertures and time intervals. *Sci. Rep.* **2019**, *9*, 19688. [CrossRef]
58. Bubola, K.B.; Lucić, M.; Novoselić, A.; Krapac, M.; Lukić, I. Olive fruit refrigeration during prolonged storage preserves the quality of virgin olive oil extracted therefrom. *Foods* **2020**, *9*, 1445. [CrossRef] [PubMed]
59. García, J.M.; Yousfi, K.; Mateos, R.; Olmo, M.; Cert, A. Reduction of oil bitterness by heating of olive (*Olea europaea*) fruits. *J. Agric. Food Chem.* **2001**, *49*, 4231–4235. [CrossRef]
60. García, J.M.; Yousfi, K.; Oliva, J.; García-Diaz, M.T.; Pérez-Camino, M.C. Hot Water Dipping of Olives (*Olea europaea*) for Virgen Oil Debittering. *J. Agric. Food Chem.* **2005**, *53*, 8248–8252. [CrossRef] [PubMed]
61. García, J.M.; Plasquy, E.; Jiménez-Romero, A.R. Efecto de los tratamientos postcosecha en el contenido fenólico del aceite de olive virgen. *Actas Hortic.* **2019**, *84*, 26–30.
62. Yousfi, K.; Cayuela, J.A.; García, J.M. Effect of temperature, modified atmosphere and Ethylene during olive storage on quality and bitterness level of the oil. *J. Am. Oil Chem. Soc.* **2009**, *86*, 291–296. [CrossRef]
63. Yousfi, K.; Moyano, M.J.; Martinez, F.; Cayuela, J.A.; García, J.M. Postharvest Heat Treatment for Olive Oil Debittering at the Industrial Scale. *J. Am. Oil Soc.* **2010**, *87*, 1053–1061. [CrossRef]
64. Asheri, M.; Sharifani, M.M.; Kiani, G. An examination into the effects of frozen storage of olive fruit on extracted olive oils. *Adv. Hort. Sci.* **2017**, *31*, 191–198. [CrossRef]
65. Poerio, A.; Bendini, A.; Cerretani, L.; Bonoli-Carbognin, M.; Lercker, G. Effect of olive fruit freezing on oxidative stability of virgin olive oil. *Eur. J. Lipid Sci. Technol.* **2008**, *110*, 368–372. [CrossRef]
66. García-Vico, L.; García-Rodriguez, R.; Sanz, C.; Pérez, A.G. Biochemical aspects of olive freezing-damage: Impact on the phenolic and volatile profiles of virgin olive oil. *Food Sci. Technol.* **2017**, *86*, 240–246. [CrossRef]
67. Masella, P.; Guerrini, L.; Angeloni, G.; Spadi, A.; Baldi, F.; Parenti, A. Freezing/storing olives, consequences for extra virgin olive oil quality. *Int. J. Refrig.* **2019**, *106*, 24–32. [CrossRef]
68. Nanos, G.D.; Agtsidou, E.; Sfakiotakis, E.M. Temperature and propylene effects on ripening of green and black 'Conservolea' olives. *HortScience* **2002**, *37*, 1079–1081. [CrossRef]
69. Nanos, G.D.; Kiritsakis, A.K.; Sfakiotakis, E.M. Preprocessing storage conditions for green 'Conservolea' and 'Chondrolia' table olives. *Postharvest Biol. Technol.* **2002**, *25*, 109–115. [CrossRef]
70. Yousfi, K.; Weiland, C.M.; García, J.M. Responses of fruit physiology and virgin oil quality to cold storage of mechanically harvested 'Arbequina' olives cultivated in hedgerow. *Grasas Aceites* **2013**, *64*, 572–582. [CrossRef]
71. Guerrini, L.; Corti, F.; Cecchi, L.; Mulinacci, N.; Calamai, L.; Masella, P.; Angeloni, G.; Spadi, A.; Parenti, A. Use of refrigerated cells for olive cooling and short-term storage: Qualitative effects on extra virgin olive oil. *Int. J. Refrig.* **2021**, *127*, 59–68. [CrossRef]
72. Ramaswamy, H.S. *Post-Harvest Technologies of Fruit & Vegetables*; DEStech Publilcations: Lancaster, PA, USA, 2015; ISBN 978-1-932078-27-5.
73. Becker, B.R.; Misra, A.; Fricke, B.A. Bulk refrigeration of fruits and vegetables. Part 1: Theoretical considerations of heat and mass transfer. *HVACR Res.* **1996**, *2*, 122–134. [CrossRef]
74. Brosnan, T.; Sun, D.W. Precooling techniques and applications for horticultural products. A review. *Int. J. Refrig.* **2001**, *24*, 154–170. [CrossRef]
75. Redding, G.P.; Yang, A.; Shim, Y.M.; Olatunji, J.; East, A. A review of the use and design of produce simulators for horticultural forced-air cooling studies. *J. Food Eng.* **2016**, *190*, 80–93. [CrossRef]
76. Mercier, S.; Villeneuve, S.; Mondor, M.; Uysal, I. Time-Temperature Management Along the Food Cold Chain: A review of Recent Developments. *Compr. Rev. Food Sci. Food Saf.* **2017**, *16*, 647–667. [CrossRef]
77. Plasquy, E.; Florido, M.C.; Sola-Guirado, R.B.; García-Martos, J.M. Precooling and cold storage of olives (cv Picual) in containers with a capacity of 400 kg. *Grasas Aceites* **2022**, *73*, e467. [CrossRef]
78. Plasquy, E.; Sola-Guirado, R.R.; Florido, M.C.; García, J.M.; Blanco Roldán, G. Evaluation of a manual olive fruit harvester for small producers. *Res. Agric. Eng.* **2019**, *65*, 105–111. [CrossRef]

79. Uceda, M.; Frías, L. Épocas de recolección. Evolución del contenido graso del fruto y de la composición y calidad del aceite. In *Proceedings of II Seminario Oleícola*; IOOC, Ed.; International: Córdoba, Spain, 1975.
80. Folch, J.; Lees, M.; Stanley, G.H.S. A simple method for the isolation and purification of total lipides from animal tissues. *J. Biol. Chem.* **1957**, *226*, 497–509. [CrossRef]
81. García, P.; Brenes, M.; Romero, C.; Garrido, A. Respiration and psysicochemical changes in harvested olive fruits. *J. Hortic. Sci.* **1995**, *70*, 925–933. [CrossRef]
82. Gutiérrez Rosales, F.; Perdiguero, S.; Gutiérrez, R.; Olias, J.M. Evaluation of the bitter taste in virgin olive oil. *J. Am. Oil Chem. Soc.* **1992**, *69*, 394–395. [CrossRef]
83. Koprivnjak, O.; Brkić Bubola, K.; Kosić, U. Sodium chloride compared to talc as processing aid has similar impact on volatile compounds but more favorable on ortho-diphenols in virgin olive oil. *Eur. J. Lipid Sci. Technol.* **2016**, *118*, 318–324. [CrossRef]
84. Segovia-Bravo, K.; García-García, P.; Lopéz-Lopéz, A.; Garrido-Fernández, A. Effect of bruising on respiration, superficial colour and phenolic changes in fresh Manzanilla olives (Olea europea pomiformis): Development of treatments to mitigate browning. *J. Agric. Food Chem.* **2011**, *59*, 5456–5465. [CrossRef] [PubMed]
85. Maxie, E.C.; Catlin, P.B.; Hartman, H.T. Respiration and ripening of olive fruits. *Proc. Am. Soc. Hort. Sci.* **1960**, *75*, 275–291.
86. Fernández-Bolaños, J.; Heredia, A.; Vioque, B.; Castellano, J.M.; Guillén, R. Changes in cell-wall-degrading enzyme activities in stored olives in relation to respiration and ethylene production. *Z. Lebensm. Unters. Forsch. A* **1997**, *204*, 293–299. [CrossRef]
87. EC 2568/91. Characteristics of olive and olive pomance oils and their analytical methods. *Off. J.* **1991**, *L248*, 1–82.
88. USDA [United States Dept of Agriculture]. *United States Standards for Grades of Olive Oil and Olive-Pomace Oil*; United States Department of Agriculture: Washington, DC, USA, 2010. Available online: https://www.ams.usda.gov/sites/default/files/media/Olive_Oil_and_Olive-Pomace_Oil_Standard%5B1%5D.pdf (accessed on 25 June 2021).
89. Ben-Yahia, L.; Baccouri, B.; Ouni, Y.; Hamdi, S. Quality, stability and radical scavenging activity of olive oils after Chétoui olives (*Olea europaea* L.) storage under modified atmospheres. *Food Sci. Technol. Int.* **2012**, *18*, 353–365. [CrossRef]
90. Beltrán, G.; Aguilera, M.P.; Del Rio, C.; Sanchez, S.; Martínez, L. Influence of fruit ripening process on the natural antioxidant content of Hojiblanca virgin olive oils. *Food Chem.* **2005**, *89*, 207–215. [CrossRef]
91. Oueslati, I.; Krichene, D.; Manaï, H.; Taamalli, W.; Zarrouk, M.; Flamini, G. Monitoring the volatile and hydrophilic bioactive compounds status of fresh and oxidized Chemlali virgin olive oils ovr olive storage times. *Food Res. Int.* **2018**, *112*, 425–433. [CrossRef] [PubMed]
92. Morales-Silero, A. Influencia de la recolección en la composición fenólica. *Actas Hortic.* **2019**, *84*, 22–25.
93. Pérez, A.P.; Sanz, C. Composición fenólico del aceite de oliva: Compuestos y enzimas implicados. *Actas Hortic.* **2019**, *84*, 7–9.
94. Tsimidou, M.Z. Analytical Methodologies: Phenolic compounds related to olive oil taste issues. In *Handbook of Olive Oil, Analysis and Properties*, 2nd ed.; Aparicio, R., Harwood, J., Eds.; Springer: New York, NY, USA, 2013; pp. 311–334.
95. Hbaieb, R.H.; Kotti, F.; García-Rodriguez, R.; Gargouri, M.; Sanz, C.; Pérez, A.G. Monitoring endogenous enzymes during olive fruit ripening and storage: Correlation with virgin olive oil phenolic profiles. *Food Chem.* **2015**, *174*, 240–247. [CrossRef] [PubMed]
96. Beltrán, G.; Jiménez, A.; del Rio, C.; Sánchez, S.; Martínez, L.; Uceda, M.; Aguilera, M.P. Variability of vitamine E in virgin olive oil by agronomical and genetic factors. *J. Food Compost. Anal.* **2010**, *23*, 633–639. [CrossRef]
97. Bendini, A.; Valli, E.; Cerretani, L.; Chiavaro, E.; Lercker, G. Study on the effects of heating of virgin olive oil blended with mildly deodorized olive oil: Focus on the hydrolytic and oxidative state. *J. Agric. Food. Chem.* **2009**, *57*, 10055–10062. [CrossRef] [PubMed]
98. Gómez-Coca, R.B.; Cruz-Hidalgo, R.; Fernades, G.D.; Pérez-Camino, M.C.; Moreda, W. Analysis of methanol and ethanol in virgin olive oil. *MethodsX* **2014**, *1*, 207–211. [CrossRef]
99. García-Vico, L.; Belaj, A.; León, L.; de la Rosa, R.; Sanz, C.; Pérez, A.G. A survey of ethanol content in virgin olive oil. *Food Control* **2018**, *91*, 248–253. [CrossRef]
100. EC 1348/2013. Regulation No 2568/91/EEC on the characteristics of olive oil and olive-residue oil and on the relevant methods of analysis. *Off. J.* **2013**, *L338*, 31–67.
101. Beltrán, G.; Sánchez, R.; Sánchez-Ortiz, A.; Aguilera, M.P.; Bejaoui, M.A.; Jiménez, A. How 'ground-picked' olive fruits affect virgin olive oil ethanol content, ethyl esters and quality. *J. Sci. Food Agric.* **2016**, *96*, 3801–3806. [CrossRef] [PubMed]
102. Lorite, I.J.; Gabaldón-Leal, C.; Ruiz-Ramos, M.; Belaj, A.; de la Rosa, R.; León, L.; Santos, C. Evaluation of olive response and adaptation strategies to climate change under semi-arid conditions. *Agric. Water Manag.* **2018**, *204*, 247–261. [CrossRef]
103. Benlloch-González, M.; Sánchez-Lucas, R.; Bejaoui, M.A.; Benlloch, M.; Fernández-Escobar, R. Global warming effects on yield and fruit maturation of olive trees growing under field conditions. *Sci. Hortic.* **2019**, *249*, 162–167. [CrossRef]
104. Rallo, L.; Díez, C.M.; Morales-Silero, A.; Miho, H.; Priego-Capote, F.; Rallo, P. Quality of olives: A focus on agricultural preharvest factors. *Sci. Hortic.* **2018**, *233*, 491–509. [CrossRef]
105. Rébufa, C.; Pinatel, C.; Artaud, J.; Girard, F. A comparative study of the main international extra virgin olive oil competitions: Their impact on producers and consumers. *Trends Food Sci. Technol.* **2021**, *107*, 445–454. [CrossRef]
106. Di Serio, M.G.; Giansante, L.; Di Loreto, G.; Faberi, A.; Ricchetti, L.; Di Giacinto, L. Ethyl esters versus fermentative organoleptic defects in virgin olive oil. *Food Chem.* **2017**, *219*, 33–39. [CrossRef]
107. Ministerio de Agricultura, Pesca y Alimentación. Real decreto 41/2021. *BOE* **2021**, *23*, 7955–7977.
108. Inarejos-García, A.M.; Gómez-Rico, A.; Salvador, M.D.; Fregapane, G. Influence of malaxation condition on virgin olive oil yield, overall quality and composition. *Eur. Food Res. Technol.* **2009**, *228*, 671–677. [CrossRef]

109. Peri, C. Quality Excellence in Extra Virgin Olive Oils. In *Olive Oil Sensory Science*; Monteleone, E., Langstaff, S., Eds.; John Wiley: Chichester, UK, 2014; pp. 3–32.
110. Caponio, F.; Gomes, T.; Summo, C.; Pasqualone, A. Influence of the type of olive-crusher used on the quality of extra virgin olive oils. *Eur. J. Lipid Sci. Technol.* **2003**, *105*, 201–206. [CrossRef]
111. Al-Widyan, M.I.; Rababah, T.M.; Mayyas, A.; Al-Shbool, M.; Yang, W. Geometrical, thermal and mechanical properties of olive fruits. *J. Food Process. Eng.* **2010**, *33*, 257–271. [CrossRef]
112. Cuesta, F.; Alvarez, M. Mathematical modeling for heat conduction in stone fruits. *Int. J. Refrig.* **2017**, *80*, 120–129. [CrossRef]
113. Plasquy, E.; García, J.M.; Florido, M.C.; Sola-Guirado, R.R. Estimation of the cooling rate of six olive cultivars using thermal imaging. *Agriculture* **2021**, *11*, 164. [CrossRef]
114. Çengel, Y.N.; Ghajar, A.J. *Heat and Mass Transfer Fundamentals & Applications*, 5th ed.; McGraw-Hill Education: New York, NY, USA, 2015.
115. Wang, G.; Zhang, X. Evalutation and optimization of air-based pre-cooling for higher postharvest quality: Literature review and interdisciplinary perspective. *Food Qual. Saf.* **2020**, *4*, 59–68. [CrossRef]
116. Ben-Shalom, N.; Harel, E.; Mayer, A.M. Enzymatic browning in green olives and its prevention. *J. Sci. Food Agric.* **1978**, *29*, 398–402. [CrossRef]
117. Rejano-Navarro, L.; Sánchez-Gomez, A.H.; Vega Macías, V. Nuevas tendencias en el tratamiento alcalino "cocido" de las aceitunas verdes aderezadas al estilo español o sevillano. *Grasas Aceites* **2008**, *59*, 197–204.
118. Vichi, S.; Boynuegri, P.; Caixach, J.; Romero, A. Quality losses in virgin olive oil due to washing and short-term storage before olive milling. *Eur. J. Lipid Sci. Technol.* **2015**, *177*, 2015–2022. [CrossRef]
119. Bower, J.H.; Jobling, J.J.; Patterson, B.D.; Ryan, D.J. A method for measuring the respiration rate and respiratory quotient of detached plant tissues. *Postharvest Biol. Technol.* **1998**, *13*, 263–270. [CrossRef]
120. Fonseca, S.; Oliveira, F.; Brecht, J.K. Modelling respiration rate of fresh fruits and vegetables for modified atmosphere packages: A review. *J. Food Eng.* **2002**, *52*, 99–119. [CrossRef]
121. Caleb, O.J.; Mahajan, P.V.; Opara, U.L.; Witthuhn, C.R. Modelling the respiration rates of pomegranate fruit and arils. *Postharvest Biol. Technol.* **2012**, *64*, 49–54. [CrossRef]
122. Iqbal, T.; Rodrigues, F.A.S.; Mahajan, P.V.; Kerry, J.P. Effect of time, temperature, and slicing on respiration rate of mushrooms. *J. Food Sci.* **2009**, *74*, 6. [CrossRef]
123. Waghmare, R.B.; Mahajan, P.V.; Annapure, U.S. Modelling the effect of time and temperature on respiration rate of selected fresh-cut produce. *Postharvest Biol. Technol.* **2013**, *80*, 25–30. [CrossRef]
124. Elansari, A.M. Design aspects in the precooling process of fresh produce. *Fresh Prod.* **2009**, *3*, 49–57.
125. Elansari, A.M.; Fenton, D.L.; Callahan, C.W. *Postharvest Technology of Perishable Horticultural Commodities*; Yahia, E.M., Ed.; Woodhead Publishing: Duxford, UK, 2019; pp. 161–207. [CrossRef]
126. Duan, Y.; Wang, G.; Fawole, A.; Verboven, P.; Zhang, X.; Wu, D.; Opara, U.L.; Nicolaï, B.; Chen, K. Postharvest precooling of fruit and vegetables: A review. *Trends Food Sci. Technol.* **2020**, *100*, 278–291. [CrossRef]
127. Wang, S.; Tang, J.; Cavalieri, R.P. Modeling fruit internal heating rates for hot air and hot water treatments. *Postharvest Biol. Technol.* **2001**, *22*, 257–270. [CrossRef]
128. Kitinoja, L.; Thompson, J.F. Pre-cooling systems for small-scale producers. *Stewart Postharvest Rev.* **2010**, *2*, 2–14. [CrossRef]
129. Veneziani, G.; Esposto, S.; Taticchi, A.; Urbani, S.; Selvaggini, R.; Di Maio, I.; Sordini, B.; Servili, M. Cooling treatment of olive paste during the oil processing: Impact on the yield and extra virgin olive oil quality. *Food Chem.* **2017**, *221*, 107–113. [CrossRef]
130. Mercacei. El lado oscuro: Tolvas y aceite de oliva virgen extra. *Mercacei* **2020**. Available online: https://www.mercacei.com/imprimir-noticia.asp?noti=53283 (accessed on 25 June 2021).
131. Olimerca. Túnel Criogénico Para Mejorar la Calidad del Aceite Virgen Extra. Olimerca, 03/11/2020. Available online: https://www.olimerca.com/noticiadet/tunel-criogenico-para-mejorar-la-calidad-del-aceite-virgen-extra/f75ce756489ab60d321ea7120bffa889 (accessed on 25 June 2021).
132. Beltrán, G.; Hueso, A.; Bejaoui, M.A.; Gila, A.M.; Costales, R.; Sánchez-Ortiz, A.; Aguilera, M.P.; Jiménez, A. How olive washing and storage affect fruit ethanol and virgin olive oil ethanol, ethyl esters and composition. *J. Sci. Food Agric.* **2020**. epub ahead of print online. [CrossRef]
133. Dourou, A.M.; Brizzolara, S.; Famiani, F.; Tonutti, P. Effects of pre-processing low temperature conditioning of olives on volatile organic compounds (VOC) profiles of fruit paste and oil. *Acta Hortic.* **2019**, *1256*, 53–57. [CrossRef]
134. Amirante, P.; Clodoveo, M.L.; Dugo, G.; Leone, A.; Tamborrino, A. Advance technology in vigin olive oil production from traditional and de-stoned pastes: Influence of the introduction of a heat exchanger on oil quality. *Food Chem.* **2006**, *98*, 797–805. [CrossRef]
135. Clodoveo, M.L. Malaxation: Influence on virgin olive oil quality. Past, present and future. An overview. *Trends Food Sci. Technol.* **2012**, *25*, 13–23. [CrossRef]
136. Esposto, S.; Veneziani, G.; Taticchi, A.; Selvaggini, R.; Urbani, S.; Di Maio, I.; Sordini, B.; Minnocci, A.; Sebastiani, L.; Servili, M. Flash Thermal Conditiong of Olive Pastes during the Olive Oil Mechanical Extraction Process: Impact on the Structural Modifications of Pastes and Oil Quality. *J. Agric. Food Chem.* **2013**, *61*, 4953–4960. [CrossRef]
137. Fiori, F.; Di Lecce, G.; Boselli, E.; Pieralisi, G.; Frega, N.G. Effects of olive paste fast preheating on the quality of extra virgin olive oil during storage. *LWT—Food Sci. Technol.* **2014**, *58*, 511–518. [CrossRef]

138. Leone, A.; Esposto, S.; Tamborrino, A.; Romaniello, R.; Taticchi, A.; Urbani, S.; Servilli, M. Using a tubular heat exchanger to improve the conditioning process of the olive paste: Evaluation of yield and olive oil quality. *Eur. J. Lipid Technol.* **2016**, *118*, 308–317. [CrossRef]
139. Bejaoui, M.A.; Sánchez-Ortiz, A.; Sánchez, S.; Jiménez, A.; Beltrán, G. The high-power ultrasound frequency: Effect on the virgin olive oil yield and quality. *J. Food Eng.* **2017**, *207*, 10–17. [CrossRef]
140. Cecchi, L.; Bellumori, M.; Corbo, F.; Milani, G.; Clodoveo, M.L.; Mulinacci, N. Implementation of the Sono-Heat-Exchanger in the Extra Virgen Olivo Oil Extraction Process: End-User validation and Analytical Evaluation. *Molecules* **2019**, *24*, 2379. [CrossRef]
141. Tamborrino, A.; Romaniello, R.; Caponio, F.; Squeo, G.; Leone, A. Combined industrial olive oil extraction plant using ultrasounds, microwave, and heat exchange: Impact on olive oil quality and yield. *J. Food Eng.* **2019**, *245*, 124–130. [CrossRef]
142. Clodoveo, M.L. Industrial Ultrasound Applications in the Extra-Virgin Olive Extraction Proces: History, Approaches, and Key Questions. *Foods* **2019**, *8*, 121. [CrossRef]
143. Stillitano, T.; Falcone, G.; De Luca, A.I.; Piga, A.; Conte, P.; Strano, A.; Gulisano, G. A life cycle perspective to assess the environ-mental and economic impacts of innovative technologies in extra virgin olive oil extraction. *Foods* **2019**, *8*, 209. [CrossRef]
144. Plasquy, E.; García, J.M.; Florido, M.C.; Sola-Guirado, R.R.; García-Martín, J.F. Adjustment of olive fruit temperature before grinding for olive oil extraction; Experimental study and pilot plant trials. *Processes* **2021**, *9*, 586. [CrossRef]
145. Luaces, P.; Pérez, A.G.; Sanz, C. Effect of the blanching process and olive fruit temperature at milling on the biosynthesis of olive oil aroma. *Eur. Food Res. Technol.* **2006**, *224*, 11–17. [CrossRef]
146. Cruz, S.; Yousfi, K.; Oliva, J.; García, J.M. Heat Treatment Improves Olive Oil Extraction. *J. Am. Oil Soc.* **2007**, *84*, 1063–1068. [CrossRef]

processes

MDPI

Article

Adjustment of Olive Fruit Temperature before Grinding for Olive Oil Extraction. Experimental Study and Pilot Plant Trials

Eddy Plasquy [1,*], José María García Martos [1,*], María del Carmen Florido Fernández [2], Rafael Rubén Sola-Guirado [3] and Juan Francisco García Martín [4]

1 Department of Biochemistry and Molecular Biology of Plant Products (CSIC), Instituto de la Grasa, 41013 Seville, Spain
2 Department of Crystallography, Mineralogy and Agricultural Chemistry, ETSIA, University of Seville, 41013 Seville, Spain; florido@us.es
3 Department of Mechanics, University of Cordoba, 14014 Cordoba, Spain; ir2sogur@uco.es
4 Departamento de Ingeniería Química, Universidad de Sevilla, 41012 Seville, Spain; jfgarmar@us.es
* Correspondence: eddy.plasquy@telenet.be (E.P.); jmgarcia@cica.es (J.M.G.M.)

Abstract: Harvesting at high temperatures and bulk transport can negatively influence the quality of olives and lead to undesirable alterations in the extracted oil. Cooling the fruit in the field would be the most logical solution, but it means that the olives arrive too cold at the mill for immediate processing. In this work, the use of warm water in the washing tub to warm up the fruit before grinding instead of flash heat treatment on the paste was assessed in two experiments. In the first one, at the laboratory level, the temperature after milling was determined in three olive cultivars, previously stored at 5 or 10 °C, and then submerged at different water temperatures (25, 30, and 35 °C) for 15, 30, 45, and 60 s. In the second one, two batches of olives were cooled in the field at 5 °C and then conditioned with washing water to obtain a paste at the entrance of the pilot plant malaxer at 27 °C. The temperature of the olives was measured at five points from the discharging up to their entering, as paste, into the malaxer. The results demonstrated the feasibility of the method as the temperature of the ground olives was kept at the desired temperature (28 ± 1 °C). The trials highlight the potential for automating an even more precise adjustment of the temperature of the olives before milling once the washing tub is equipped with a safe heating system.

Keywords: malaxation; olive fruit; olive oil; postharvest; thermal treatment; washing

Citation: Plasquy, E.; García Martos, J.M.; Florido Fernández, M.d.C.; Sola-Guirado, R.R.; García Martín, J.F. Adjustment of Olive Fruit Temperature before Grinding for Olive Oil Extraction. Experimental Study and Pilot Plant Trials. *Processes* **2021**, *9*, 586. https://doi.org/10.3390/pr9040586

Academic Editor: Carl Schaschke

Received: 3 March 2021
Accepted: 24 March 2021
Published: 27 March 2021

1. Introduction

The quality of virgin olive oil and its organoleptic characteristics are directly related to the integrity of the harvested fruit as well as the conditions under which the extraction is carried out. In this respect, multiple studies have underlined the importance of the used harvesting method and the postharvest conditions of the fruit before its processing [1,2]. Available methods to harvest olives [3] influence the quality parameters, and the detrimental effects of a mechanized harvest can be aggravated during their transport and storage [4,5].

When storage is needed and mechanical harvesting is employed, a low temperature is recommended [6,7]. Climatological changes are provoking a shift in the phenology of the olive tree and are bringing forward the harvest time by several weeks to warmer seasons [8,9]. The high temperatures of harvested fruit along with a transport to the olive oil mill of a couple of hours in trailers may create conditions of asphyxia and induce anaerobic respiration [10]. Consequently, fermentation processes produce undesirable by-products that lead to detectable organoleptic defects and high levels of alcoholic esters in the extracted oil and ultimately to the rejection of the oil as extra virgin olive oil [11].

The conditions under which the malaxing phase takes place (time and temperature) are known to play a crucial role in the olive oil yield, the phenolic content, and the formation of volatile components [12–14]. During the slow and continuous kneading of the ground olives (olive paste), the temperature influences the physicochemical quality parameters and the enzymatic activity. Temperatures above 35 °C increase the oil yield but induce lipolytic, hydrolytic, and oxidative degradation, affecting both phenols and volatile compounds [15]. Malaxing temperatures below 14 °C negatively affect the oil extraction performance and induce compositional changes in the resulting olive oil [12–16]. Extending the malaxation time up to 60 min increases the oil yield but leads to a decrease in its phenolic content and drastically reduces the shelf life of the olive oil [17]. Shortening of the kneading reduces the enzymatic activity necessary for the development of sensorily interesting volatile components. Nowadays, it is widely accepted that the ideal malaxing temperature ranges between 25 and 30 °C for a malaxing time between 30 and 45 min [12–14].

Grinding the olives raises the temperature by at least 5 °C [18]. Consequently, obtaining the ideal malaxation temperature becomes exceedingly difficult with olive fruits that arrive at the olive oil mill at a temperature above 25 °C. Cooling the olive fruit with cold air or hydro-cooling, or passing the paste through a heat exchanger, have already been suggested as potential alternatives [16,19,20]. However, cooling the fruit right after the harvest seems more logical, as it does have the additional advantages of reducing the fruit metabolism and avoiding the action of enzymatic agents involved in deterioration processes [21,22]. Although cooling the olives in the field still has to be developed on an industrial scale, a small-scale work demonstrated the feasibility of an on-farm cooling room [23].

The aforementioned small-scale work at the farm level also revealed important issues that are easily overlooked when theorizing the possibility of cold storage of olives [24]. Supplying olives to olive oil mills at 5 °C does not imply that the olives can be easily processed. Even after being warmed up during its transport to the mill, the cleaning, the washing, and finally the grinding made the temperature of the olive paste remain far below the desired temperature in the malaxer and could not be warmed up to the desired temperature by the inbuilt heating system. Therefore, a heating phase is required before entering the malaxation machine.

During the last decade, flash heat treatment has been suggested to overcome the limitations of the traditional malaxation process. It is argued that malaxers are inefficient not only in terms of working continuity but also for the great amount of thermal energy that is required to warm the olive paste after crushing because malaxers are poor heat exchangers [13]. Ultrasound, microwave, heat exchangers, and combinations of them have been proposed as alternatives for malaxers [25–31]. Currently, several models of heat exchangers (Pieralisi Protoreattore, Pieralisi, Jesi, Italy; Alfa Laval EVO-line, Alfa Laval, Lund, Sweden) are available on the market, while the combination of a heat exchanger and ultrasound (sono-heat exchanger) is in its final stage before industrial production [32]. However, these innovative technologies imply serious investments, modifications of the production lines, and higher energy consumption [33].

Using warm washing water to bring the olive fruit to the ideal temperature before grinding has not been taken into consideration to date. Currently, the washing tubs are filled with cold water from the tap (10–12 °C). The water then further warms or cools during the day, depending on the air temperature. The washing phase is solely regarded as a step to remove impurities and pesticide residues [34]. This work aimed to assess whether the use of warm washing water could solve the specific problem that arises when olives arrive too cold at the mill, as in the on-the-farm cooling case, and as such can be considered as a potential alternative to optimize the malaxation process.

2. Materials and Methods

Two experiments were carried out at the Instituto de la Grasa (Spanish National Research Council) in Seville (Spain) between September and December 2019. At laboratory scale, the heating of different varieties of olives with water at different temperatures and times was simulated to determine the temperature of the paste after milling. At the available pilot plant of the Instituto de la Grasa, the temperature of two batches of olives, previously stored at 5 °C at the farm, was controlled by keeping the washing water at a chosen temperature using heaters. The used olive fruits were harvested in the Del Cetino farm in Bollullos par del Condado (Huelva, Spain).

2.1. Laboratory-Scale Experiment

Fifty kilograms of three different cultivars, 'Arbequina', 'Cobrançosa', and 'Gordal' (*Olea europaea* L.), were hand-picked, immediately brought to the research center, and stored at 5 °C or 10 °C in perforated plastic boxes inside two cooling rooms. The varieties were chosen for their distinct mean weight. All the fruits were healthy, free of diseases, and without signs of deterioration. The maturity index was determined according to the Jaen method, and their weight measured (Table 1) [35].

Table 1. Maturity index (M.I.), weight, and proportionality constant (k) empirically calculated for different varieties, at two different initial temperatures (T_i) and after being submerged in water at different temperatures.

Variety	M.I.	Weight (g)	T_i (°C)	Value of k T_w = 25 °C	 T_w = 30 °C	 T_w = 35 °C
Arbequina	2.1 ± 0.2	0.84 ± 0.17	7.1	−0.056 (0.019; 1.43 °C) *	−0.043 (0.017; 2.87 °C)	−0.032 (0.012; 1.79 °C)
			12.2	−0.045 (0.015; 1.96 °C)	−0.043 (0.017; 1.29 °C)	−0.031 (0.010; 1.85 °C)
Cobrançosa	1.9 ± 0.3	2.75 ± 0.17	7.1	−0.055 (0.019; 2.29 °C)	−0.037 (0.017; 3.45 °C)	−0.028 (0.011; 1.88 °C)
			12.2	−0.043 (0.020; 2.67 °C)	−0.042 (0.014; 1.09 °C)	−0.031 (0.010; 1.94 °C)
Gordal	1.7 ± 0.4	8.72 ± 1.73	7.1	−0.021 (0.010; 2.34 °C)	−0.017 (0.007; 2.79 °C)	−0.015 (0.003; 0.95 °C)
			12.2	−0.020 (0.010; 2.85 °C)	−0.015 (0.005; 1.04 °C)	−0.012 (0.004; 2.46 °C)

* mean values of k (calculated at 15, 30, 45, and 60 s); values within brackets (SD; RMSE).

Throughout the experiment, a cooled pot was used to transport the olives to the laboratory (10 m distance), while an isolating plate was placed on top of it to reduce the warming of the fruit during the transfer. The experiment consisted of immersing roughly 0.5 kg of each batch in a stirring water bath at three different temperatures: 25 (T25), 30 (T30), and 35 (T35) °C, for 4 different time intervals: 15 (t15), 30 (t30), 45 (t45), and 60 (t60) s.

A Unitronic 320 OR stirring water bath (P-Selecta, Barcelona, Spain) was filled with approximately 20 dm^3 of water. A plasticized iron wire basket equipped with a lid (25 × 15 × 7 cm) was placed on the moving plate to keep the olives together and allow the water to flow. The recipient could contain 0.5 kg. Keeping the olives moving was crucial to simulate the passing of the olives in the washing tub and to attain a maximum heat transfer during the immersion time. When the set temperature was attained, a pot of olives was taken out of the cooling room, brought to the laboratory, and poured into the basket as fast as possible. The lid was closed, the moving system activated, and a stopwatch pressed. At the set time, the basket was taken out of the water, quickly stirred to remove as much water as possible, and transferred to the close-by grinder. The olives were ground in the MM-100 hammer mill which forms part of the Abencor (MC2 Ingeniería y Sistemas S.L., Seville, Spain) laboratory sets for olive analysis, present in the laboratory and used for the extraction of small samples of olive oil. The paste was collected in a flat plastic bowl and quickly shaped into a rectangular block with the aid of a spatula once the grinding was finished. Immediately afterward, the internal temperature was measured with a digital probe thermometer. The control trial consisted of fruit that was ground and measured without being immersed. Each trial was performed in triplicate. During the whole process, either an IR thermometer or a digital probe thermometer was used to properly control the temperature.

The grinding of the olive fruit in the hammer mill increased the temperature of the resulting paste. This temperature increase (ΔT_g) was determined experimentally for each variety and each storage temperature. The temperature of the intact fruit on arrival in the lab (T_i) and that of the paste obtained from this fruit (T_{ig}) allowed for calculating ΔT_g. The values of ΔT_g were used to calculate the fruit temperature during the immersion:

$$T_{ft}(t) = T_{pt}(t) - \Delta T_g \quad (1)$$

(for t = 15, 30, 45, and 60 s)

T_{ft} = Temperature of the fruit, after being immersed in water for a given time t.

T_{pt} = Temperature of the paste, obtained from fruit immersed in water for a given time t.

ΔT_g = Temperature increase as a consequence of the grinding.

Based on Newton's cooling law, the temperature difference between the fruit and its surroundings (water) allows for calculating the rate of heat loss as a function of time when assuming a low Biot number and a specific heat capacity (k) independent of temperature. The solution to that equation describes an exponential decrease in the temperature difference over time:

$$T(t) = T_a + (T_i - T_a)e^{kt} \quad (2)$$

where k can be expressed as:

$$k = \frac{\ln\left(\frac{T(t) - T_a}{T_i - T_a}\right)}{t} \quad (3)$$

Knowing the initial temperature of the fruit (T_i), the calculated values of T_{ft} for the immersion time (t) and the given water temperature (T_a) allowed for calculating the value of k for each batch of olives at different times, thus obtaining the average k value for each variety. When the proportionality constant is known, the temperature at a set time can be calculated, given the temperature of the object and the temperature of the surroundings.

2.2. Pilot Plant Experiment

The experiment was performed with two batches of olives ('Arbequina' and 'Picual'). Both were harvested, stored, and transported similarly to those of the laboratory-scale experiment. The harvested fruit was cleaned to remove leaves and twigs, weighed, temporarily stored in a hopper, and finally washed before being ground as the first step of the extraction process (Figure 1). The experimental mill extracted only one batch a day and the production line was cleaned between tests.

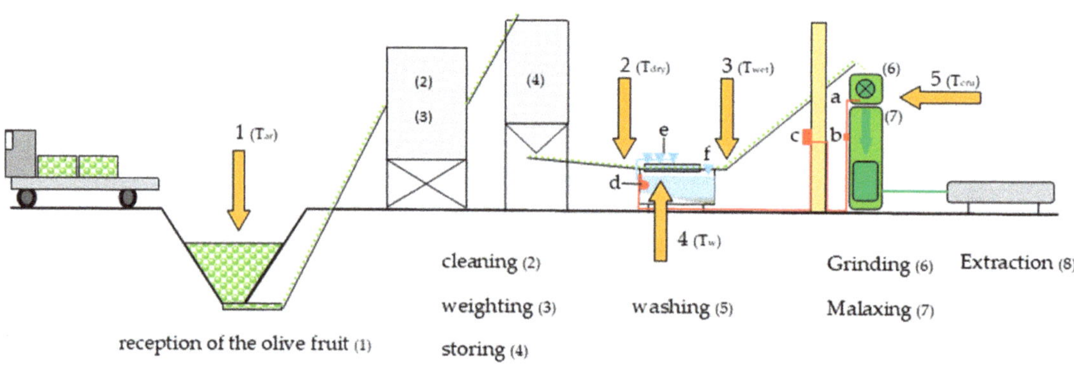

Figure 1. Processing steps of the olive fruit from their discharge up to malaxation of the paste. The different measurement points are indicated: At the reception of the olive fruit (**1**); before entering the washing tub (**2**); right after leaving the washing tub (**3**); and after being crushed (**5**). The water temperature is also continuously measured (**4**). The automized control system (red line) consisted of a temperature gauge at the exit of the grinder (**a**), a thermostat (**b**), a relay (**c**), and 4 heaters placed in the water tub (**d**). The water circulated in the washing tub from the lower to the upper part (**e**) and back (**f**).

The Lav Air/R-5000 washing tub (JAR, Mancha Real, Spain) has a capacity of 1 m^3 and is divided into two parts. In the upper part, water flows through an open stainless box of 50 × 100 cm, in which injected air provokes fierce turbulence of the water. The olives in the water are pushed forward by the ones entering and end up on a vibrating perforated plate and finally an inclined plate from where they fall on a moving belt that transports them to the indoor part of the milling facility. The lower part of the tub is an open reservoir for the circulating water and is designed to facilitate sedimentation. A pump continuously moves water from the lower to the upper part where it falls back into the reservoir through the perforated plate. It was estimated that between 15 to 20 kg were washed at a time. The average immersion time was estimated as the relation between the fruit quantity (Cw, kg) in the washing tub and the washing capacity (kg/h), being 18 s (Cw = 15 kg) and 24 s (Cw = 20 kg) when processing 3000 kg/h and between 11 s (Cw = 15 kg) and 15 s (Cw = 20 kg) when processing 5000 kg/h.

The olives were directed toward an FP HP 40 hammer mill (Pieralisi España, S.L.U., Mengíbar, Spain). The grinder was mounted on top of the malaxer so that the crushed olives (or paste) fell straight into it. A malaxer model 1250 2C E (Pieralisi España, S.L.U., Mengíbar, Spain) was used, which has a capacity of 6 m^3 and is composed of two connected chambers, one on top of the other. Within each chamber, two slow-turning propellers move the paste continuously. The chambers can be heated by hot water, circulating between the double walls, to warm the olive paste if necessary. The whole malaxing process was fully automized and continuously monitored through sensors that registered the fill level and the temperature of the paste as well as the circulating water and the malaxing time. After 2700 s at 28 °C in the malaxer, the paste was injected into the decanter, the next step being to extract the oil by gravitational forces. Once the extraction started, the process continued until the malaxer was empty.

Based on the results of the laboratory-scale experiment, it was decided to bring the temperature of the washing water to 40 °C. During the discharge of the olives, an operator filled the tub with hot water while the temperature was controlled with a digital thermometer of which the probe was fixed in the water. At the same time, four electric heaters were hung in the water at the edge of the bath to keep the water temperature at 40 °C during the whole process.

The temperature of the olives was measured at 4 different points from their discharge at the reception yard up to their entering in the malaxer, while the temperature of the water in the washing tub was continuously measured (Figure 1). The first measurement was done on the discharged pile of olives (arrival temperature, T_{ar}). The second took place after the olives had been cleaned, weighed, and temporarily stored in a hopper. The temperature of the olive fruit was measured during the transport from the hopper to the washing tub (T_{dry}). The third measurement took place right after the washing, on a sample of roughly 500 g olive fruit, collected in a plastic jar before falling on the transporting belt (T_{wet}). All these measurements were performed with an IR thermometer. Finally, the last measurement was realized on the crushed olives before entering the malaxer (T_{cru}). Measurement of the temperature of the ground olives before they entered the malaxer was hindered by the safety prescriptions that made it impossible to directly access the grinder. Therefore, a 15 cm stainless steel shaft was mounted in the front lid of the grinder and placed right below the exit. A PTC S6-S temperature probe (Osaka Solutions, S.L., Barcelona, Spain) was placed in the shaft together with thermal silicon grease to facilitate the heat transfer. The digital display/thermostat of the Mundocontrol FN-42 probe (Salvador Escoda S.A., Barcelona, Spain, 0.1 °C precision) was fixed at one side of the malaxer and made it possible to register the temperature of the olive paste. The temperature of the washing water (T_w) was measured by a probe that was fixed in the lower water compartment of the washing tub. The digital thermometer was fixed on the side to allow easy reading. Three operators recorded the temperatures T_{dry}, T_{wet}, T_w, and T_{cru} every 3 to 4 min until all olives were completely ground. The air temperature (T_{air}) was measured at the start of the experiment.

For the second trial, an electric circuit (Figure 1a–d) was mounted to facilitate the automized activation of the heaters in the washing tub. The digital display was replaced by a thermostat that was set at 27 °C (±1 °C). The thermostat was connected to a relay that in turn switched the heaters on and off.

2.3. Statistical Analysis

The data of the experiments were organized in Microsoft Excel 2010 (version 2101) (Microsoft, Redmond, WA, USA) sheets and further analyzed and visualized with IBM SPSS Statistics (version 24) for Windows (IBM, Armonk, NY, USA). ANOVA was applied for assessing, firstly, the effect of the variety, initial fruit temperature, water temperature, and time of immersion on the fruit temperature at time t (four-way), and for each variety separately (three-way), secondly, the effect of the variety on the value of the k (one-way) within each variety, and, finally, the effect of the initial temperature and the water temperature (two-way). When a significant effect was detected, a Tukey test was applied for differentiating mean values. The root mean square error (RMSE) was calculated to evaluate the average deviation of the predicted values, obtained from a theoretical exponential curve, from the actual values.

3. Results

3.1. Laboratory-Scale Trials

The measurement of the temperature during the transfer of the olive pots revealed that there were no significant differences between the three cultivars at any of the two storage temperatures. The olives cooled at 5° C attained 7.1 ± 0.2 °C (Ti7) and the ones stored at 10.0 °C reached 12.2 ± 0.2 °C (Ti12). These values were taken as the initial temperatures at time 0, i.e., at the moment of immersion.

The average temperature of the paste from olives at 7.1 °C was 12.0 ± 0.2 °C (Tg7), while the olives at 12.2 °C reached 17.6 ± 0.4 °C (Tg12). No significant differences were observed among varieties. The mean values of Tg7 and Tg12 were considered in further calculations. The mean values of the temperature of the fruit at the distinct immersion times are represented in Figure 2.

The temperature of 'Arbequina' at T_i = 7.1 °C increased similarly in the three assayed water temperatures. A steep increase in the temperature was noticeable from the start, after which a flattening of the curves set in. At t15, fruit immersed in water at 35 °C was already 5 °C warmer than that immersed in water at 25 °C, increasing almost 7 °C at t60. A similar profile was observed at T_i = 12.2 °C, although with smaller differences between T25 and T35 at t15 and t60. The 'Cobrançosa' variety showed a similar temperature increase to that of 'Arbequina', characterized by a steep increase during the first 15 s, after which the temperature of the fruit flattened. However, the gap between the obtained fruit temperatures at T25 and T35 was less pronounced when compared to 'Arbequina' for both initial temperatures. The profiles of the 'Gordal' samples did not present a similar steep increase at the start, and, especially, the fruit with T_i = 12.2 °C. The spread between T25 and T35 reached only a few degrees at t15 while at t60 it widened to almost 5 °C.

The levels of significance (three-way ANOVA) of the different factors on the fruit temperature showed similar results for the three varieties (Table 2).

The initial fruit temperature (T_f), the temperature of the water (T_w), and the time of immersion (t) presented a highly significant effect on the fruit temperature at t. For the interaction of $T_f \times T_w$, no significant effect was observed for the 'Arbequina' and the 'Gordal' varieties. In the first case, the temperature increase during the first 15 s of immersion led to almost similar values at t15 for the trials at different T_f. From then on, and up to 60 s, the effect of T_f was hardly noticeable. In the second case, this correction is less pronounced at t15, and also attained a lower temperature when compared to the other varieties. During t15 and t60, the fruit temperature of the 'Gordal' variety evolved similarly although with a greater range between the different Tw than with the 'Arbequina' variety. Concerning the interactions between T_f and t, and $T_w \times$ t, a highly significant effect was observed, indicating that the effect of the initial fruit temperature, as well as the water temperature, on the fruit temperature at each t, varied for each t. Finally, when the interaction between the three factors was taken into account, no significant effect was observed in the case of 'Arbequina', due to highly similar progression at both T_f from t15 on.

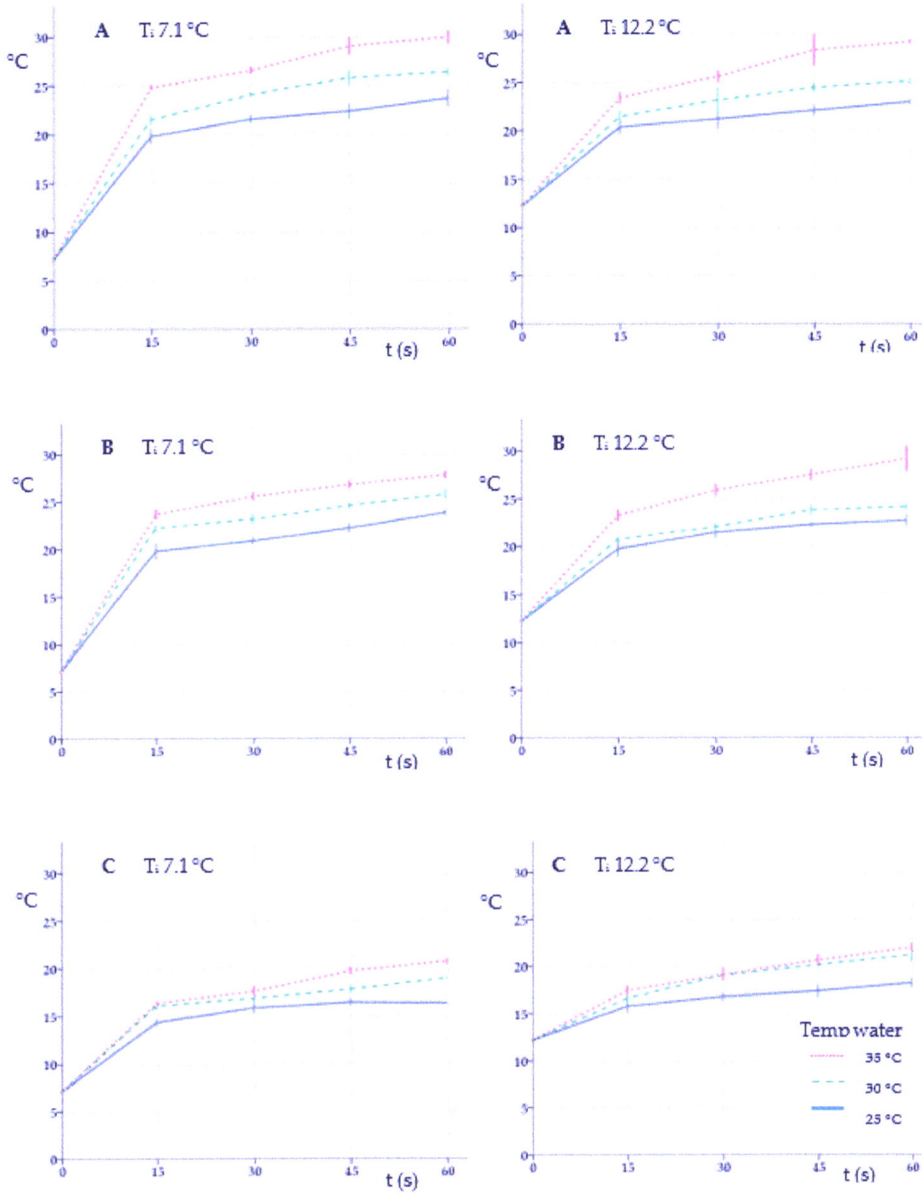

Figure 2. Evolution of the fruit temperature of 3 olive varieties (A: Arbequina; B: Cobrançosa; C: Gordal) at two initial temperatures (T_i) and immersed in moving water at different temperatures. Vertical bars express mean ± SD.

Table 2. Level of significance of the effect of the factors initial fruit temperature (T_f), temperature of the water (T_w), the time of immersion (t), and their interactions, on the fruit temperature for the three varieties ('Arbequina', 'Cobrançosa', and 'Gordal').

Variety	T_f	T_w	t	$T_f \times T_w$	$T_f \times t$	$T_w \times t$	$T_f \times T_w \times t$
Arbequina	0.001	0.000	0.000	0.254	0.000	0.000	0.405
Cobrançosa	0.000	0.000	0.000	0.000	0.000	0.000	0.000
Gordal	0.000	0.000	0.000	0.059	0.000	0.000	0.000
All varieties	0.000	0.000	0.000	0.073	0.000	0.000	0.158

These observations were confirmed by the calculated mean of the different k values, for each variety and temperature (Table 1).

A one-way ANOVA revealed significant differences ($p < 0.001$) among varieties, while the post hoc Tukey test ($p < 0.05$) detected a significant difference between 'Arbequina' and 'Cobrançosa' varieties on the one hand, and the 'Gordal' variety on the other. A two-way ANOVA within each variety showed no significant effect of the initial temperature, the water temperature, or their interaction on the value of k, for none of the varieties. The RMSE ranged between 0.95 °C ('Gordal'; T_i = 7.1; T_w = 35 °C) and 2.85 °C ('Gordal'; T_i = 12.2 °C; T_w = 25 °C). The results of the different trials varied within each variety and initial fruit temperature (Figure 2). The deviations were largely due to the difference between the calculated and the measured values at t15. The higher values of the latter at t15 (data not shown) were a consequence of the applied measurement technique which implied a correction of the temperature of the paste, subtracting the temperature generated by the grinding (Equation (1)). During the time between taking out the fruit and the actual grinding, a certain amount of time went by, during which the temperature of the fruit could further evolve, leading unavoidably to a variation in the exact time. This time was relatively much higher at t15.

The results made clear that to raise the temperature of olives from 7 and 12 °C to 20–25 °C, respectively, it sufficed to submerge them into water at a temperature between 30 and 35 °C, respectively, for 10–20 s. However, this estimation did not take into account the specific industrial processing conditions. For example, the time needed to transport the olives from the washing tub to the grinder, during which they cool down. To take this unavoidable cooling into account, it was estimated that raising the water temperature by an additional 5–10 °C would be enough. At the same time, the results showed a possible solution, knowing that the average immersion time was calculated to be in the range of 15–20 s, equal to the calculated average washing time in an industrial washing tub. Finally, these water temperatures neither damaged the fruit tissues nor jeopardized the quality of the olives. Therefore, the water temperature in the washing tub was set to 40 °C for the industrial-scale trials.

3.2. Industrial-Scale Trials

The measurements during the two trials allowed us to monitor the fruit temperature during its processing, from the discharge at the reception yard of the olive oil mill up to entering the malaxer (Figure 1). The first trial was carried out with a batch of 2134 kg 'Arbequina' olives. On arrival, the fruit temperature was 12.8 ± 0.3 °C (T_{ar}). The outdoor temperature was 13.2 °C at 9:00 a.m. The processing of the fruit started at 9:45 a.m. and took 63 min. For the second trial, a batch of 4721 kg 'Picual' olives was used. On arrival, their temperature was 11.1 ± 0.2 °C (T_{ar}). The outdoor temperature at 9 a.m. was 11.3 °C. The cleaning process started at 9:30 a.m. All the fruit was processed and ground within 118 min. The processing times were slightly below the calculated capacity of the washing tub. The rate at which the olives left the hopper was regulated by lifting manually a small slider and controlled by an operator of the olive oil mill. The mean immersion time was adjusted accordingly to 26 ± 5 s.

During the first trial, the fruit temperature at the exit of the hopper (T_{dry}) increased from 13.2 °C at the first measurement at t3 to 16.8 at t55 (Figure 3). From t9 to t41, the temperature remained constant around 15 °C, from then on it rose consistently to attain its maximum at t55. The increase was a logical consequence of the natural warming up during the day.

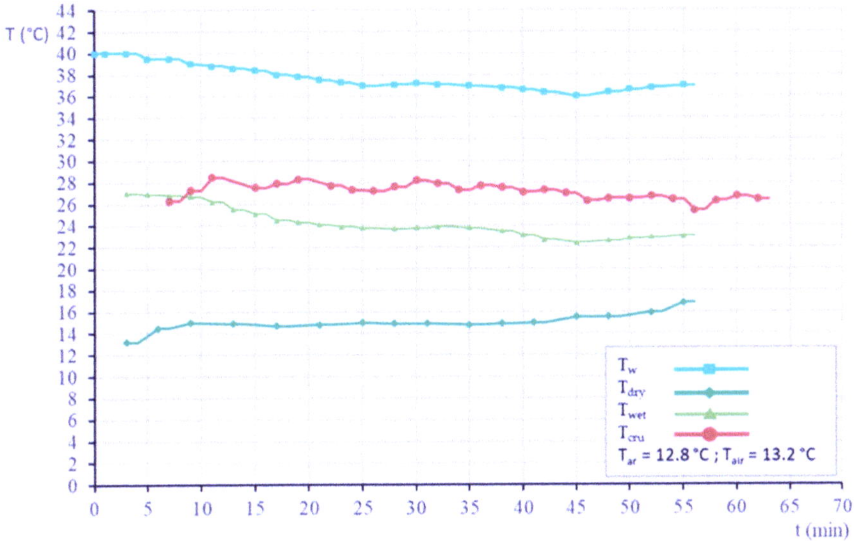

Figure 3. Evolution of the temperature during the processing of a batch of olives ('Arbequina' cv., 2134 kg). Each point represents a registration on one of the 4 different sites: the fruit temperature at the exit of the hopper (T_{dry}), the water temperature in the washing tub (T_W), the temperature exiting the washing tub (T_{wet}), and the temperature of the crushed olives (paste) before entering the malaxer (T_{cru}). On arrival, the fruit temperature was 12.8 ± 0.3 °C.

The washing tub was filled with warm water at 40 °C at the start of the processing. The different heaters were activated from the start to compensate for the cooling of the water. Notwithstanding, after a few minutes, the water started to cool down steadily to 37 °C at t25 and further to 36 °C at t45. At that moment it was decided to inject warm water into the tub, resulting in the temperature rising to 37 °C at t55. It was clear that the five heaters were not enough to keep the water at the desired temperature. In addition, it became clear that the water strains, produced within the tub, made it more difficult to warm the water from above.

The temperature of the olives measured after the washing (T_{wet}) followed a similar pattern of T_W. Starting at 27 °C at t3, the temperature declined from t9 and attained values just below 24 °C from t25 up to t35. From then on, a new pronounced decline set in, with the lowest value found at t45 (22.4 °C). Subsequently, the temperature rose again to 23 °C at t55. The fluctuations in the water temperature did have an immediate effect on the temperature of the olives and confirmed the efficacy of the used method. The data were also in agreement with the results of the laboratory experiments as they fell within the expected values.

Finally, the temperature of the olive paste just before entering the malaxer (T_{cru}) showed an equally consistent profile. The first olives were ground at t7, thus the measured temperature of 26.3 °C rose quickly to 28.5 °C at t11. From then on, the temperature maintained values between 28.3 and 27 °C up to t44, after which it declined to 25.4 °C at t56. In the final phase, the temperature rapidly attained values above 26 °C. The olives cooled down during their transport to the grinder since the ambient temperature was at least 10 °C lower than the fruit temperature. Meanwhile, the grinding generated heat, hence the olive paste attained the desired temperature for the malaxing (27 °C). Therefore, it sufficed to keep this temperature constant during the process without the need to heat-up the circulating water around the malaxer. The lower temperatures at the start of the process can be attributed to the heating of the metal grinder which absorbed the first generated heat produced by the grinding. Once stabilized, more heat was transferred to the paste. It was also observed that the major temperature shifts occurred with a delay of several minutes, being the time needed to transport the wet olives to the grinder. Consequently, it is shown that there exists a stable relationship between the temperature of the paste and the wet olives and, given the evenly strict relation between the fruit and the water temperature, also with the temperature of the latter.

During the second trial, the temperature of the fruit exiting the hopper (T_{dry}) remained constant at around 11 °C throughout the processing (Figure 4). The day temperature hardly rose during the forenoon and this meant that the fruit was not warmed up. Considering that the day temperature, as well as the temperature of the fruit on arrival, was several degrees lower than in the first trial, the water temperature was set to 45 °C instead of 40 °C. The heaters were controlled by a thermostat, connected to the thermometer probe in the grinder, but were activated from t0 to compensate for the cooling of the water. The supply of warm water was secured to correct this if necessary. The water temperature remained constant throughout the processing, except at t88, when it decreased to 44 °C, and t112, when temperature further declined to 43 °C. The temperature of the wet olives (T_{wet}) fell to 23 and 22 °C from t0 to t88. From then on, the temperature declined to 20 °C at t112. Similar to what was pointed out in the first trial, a strict relation between the T_w and T_{wet} was observed. The temperature of the paste during the whole grinding process reached around 28 °C, except for the first 5 min, where it was a few degrees lower due to the warming of the grinder. From t100, a minor decrease was observed, although without being below 27.2 °C at t110. This change was produced shortly after the temperature of the wet olives presented a small decline.

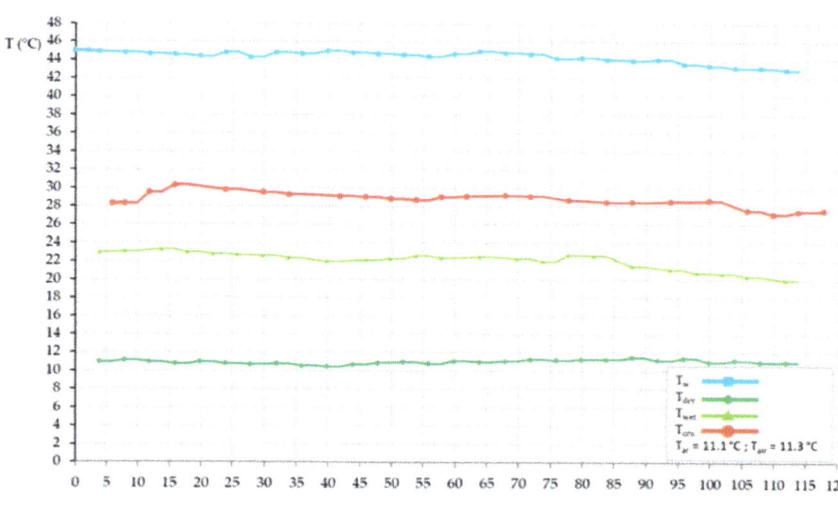

Figure 4. Evolution of the temperature during the processing of a batch of 'Picual' olives. Each point represents a registration on one of the 4 different temperature measurement points: fruit temperature at the exit of the hopper (T_{dry}), water temperature in the washing tub (T_w), temperature of the fruit exiting the washing tub (T_{wet}), and temperature of the crushed olives (paste) before entering the malaxer (T_{cru}). On arrival, the fruit temperature was 12.8 ± 0.3 °C. The outdoor temperature was 11.3 °C at 9 a.m.

4. Discussion

This work intended to investigate whether it was feasible to warm up olives at 10 °C or less during their immersion in the washing tub to obtain ground fruit at a temperature close to 27 °C before entering the malaxer. The laboratory experiment showed that for various varieties and water temperatures, this goal can be reached with an immersion time around 15 to 20 s and temperatures in a range of 35–40 °C, depending on fruit characteristics. The trials at the pilot plant did confirm this assumption on a pilot scale.

The temperature rise during the first 15 s stood out as the most significant observation in the first experiment, as well as the consistent separation of the curves at the different water temperatures. The spread between T25 and T35 varied according to the varieties. This indicated that the effect of the water temperature varied across the cultivars as well as across the initial fruit temperatures. The smaller the fruit and the greater the temperature difference, the greater the effect of the water temperature on the measured fruit temperature at t15. It was also observed that the warming up of the different fruit batches of each variety followed an exponential curve that flattened when the initial temperature of the fruit was higher or the water temperature lower. This shift was the most pronounced in 'Arbequina' as compared to 'Gordal'. 'Cobrançosa', on the contrary, showed only slight differences with the former. The ANOVA confirmed the highly significant effect of all the factors studied (variety, fruit temperature, water temperature, and time), and their interactions. The fact that no effect in the interaction between T_f and T_w was detected for 'Arbequina' can be explained by the more pronounced flattening of the curve from t15 (Figure 2). In the case of the 'Gordal' variety, this lack of interaction effect can be attributed to the greater variability that is observed within the samples at t60 (Figure 2).

The ANOVA applied to the values of the proportionality constant did confirm that they vary among the cultivars, given their different weights. However, they are not expected to present deviant values within each cultivar, as the constant k is specific for a given body and its surroundings, which was the case. The fact that there were no significant differences between 'Arbequina' and 'Cobrançosa', despite their difference in weight, can be explained using the results of an empirical study that determined the cooling rate of six different cultivars using thermal imaging [20].

It was shown that the relation between the cooling rate of olive fruit and the specific surface area can be characterized by a sigmoid (Boltzmann) function and suggested that the flattening of the curve can be attributed to the higher Biot number of the heaviest cultivar, 'Gordal', on one side, while the stone/flesh relation of the lightest fruit stabilized the cooling rate on the other side. The curve also revealed that other varieties with a heavier weight, such as 'Verdial', which had a mean weight of 2.95 ± 0.42 g, did present values that were close to the one observed for 'Koroneiki'. The values for k obtained for 'Arbequina' and 'Cobrançosa' varieties (mean weight of 0.84 ± 0.17 g and 2.75 ± 0.17 g, respectively) are thus comparable with the one obtained in the cited study, although in this case, the fruit was warmed up instead of cooled down. As thermodynamics works in the same way in both directions, these results do provide a reasonable explanation. However, the obtained results deviate strongly from the one published by other authors [36], who predicted a cooling time of 296 s to bring 'Gordal' olives from 17.7 °C to 4 °C in water at 0.46 °C. The mathematical calculations made by these authors, based on the biometrical and thermal characteristics of one olive, did not coincide with the empirically measured values obtained in the present two experiments.

The second experiment made clear that a large number of continuously changing variables influenced the heat exchange between the fruit and the different surroundings (air and water). As a result, accurate estimations of the required water temperature were not feasible. However, when the temperature in the grinder is constantly monitored with a thermostat, the temperature of the water can be perfectly adjusted to provide the necessary heat to the fruit. Overall, the data of the second trial do confirm the results of the first one. The lower temperature of the olive fruit was correctly predicted and corrected using a higher water temperature. The experience of the first trial meant that the water temperature could be better kept under control, even when dealing with olives at a lower initial temperature. The thermostat worked correctly but the heaters were not able to keep the temperature at the set level. On several accounts, warm water was added to the washing tub.

The used temperatures and the immersion time remained far below the conditions (3 min in 60 °C) that other authors [37] applied to study the blanching effect on the biosynthesis of olive oil aroma, as well as experiments in which heat treatment was applied to olive fruits to lower their bitterness, using water at 50 and 72 °C for 3 to 5 min [38–40]. As the fruit temperature measured right after the immersion (T_{wet}) was always lower than 25 °C after an average immersion time of 25 s, it can be assumed that modifications of the aroma profile, as well as the decrease in oxidative stability and polyphenol content, as pointed out in these studies, were not present.

The results of the flash heat treatment experiments demonstrated that bringing the temperature to the ideal malaxation temperature with a heat exchanger leads to a reduction in the malaxing time without significant modifications of the organoleptic characteristics [26,28,31]. Providing paste at the desired temperature further guarantees a constant temperature from the moment the malaxation starts, which is equivalent to extending the malaxation time by an interval equal to the time that the mixer employs to heat all the olive paste mass [25]. Whether this is also the case when the fruit is warmed before the crushing needs to be experimentally confirmed. Nevertheless, given the similar conditions of the paste entering the malaxer when compared with a heat exchanger, a strong argument can be made that this will be the case. Similarly, a significant impact on the plant's working capacity, oil yield, and oil quality can be expected to be the same as those achieved with a heat exchanger [26]. Bringing the olives to an adequate temperature significantly reduces the need to warm up the wall of the malaxer, and thus implies a lower energy cost during the malaxation process. At the same time, it avoids the risk of overheating the fraction that is in direct contact with the heated wall which, once attaining 45 °C, produces irreversible and detrimental changes to the chemical composition of the oil [41].

The energetic cost to keep a washing tub at a temperature between 35 and 45 °C will largely depend on its capacity, isolation, and ambient temperature. The potential sources and equipment for this necessary heat need further study but direct heating with electric resistances or a heating system based on an external heat exchanger may be a feasible, economic solution that can even be introduced to existing washing tubs. The latter solution can easily be combined with a filtering and UV system to keep the circulating water clean and free of bacterial contamination.

5. Conclusions

To optimize the malaxing process, flash heat treatments are proposed to control the temperature of the paste before entering the malaxer. While the commercially available and experimental systems are intended to control the temperature of the ground olives or paste, the present work studied the possibility to adjust the temperature of the intact fruit before grinding. At the laboratory level, the possibilities to warm up olive fruit were explored under different conditions. Based on these results, it was concluded that it is feasible to bring the fruit to the ideal temperature for malaxing without applying excessive heat and within a time frame that is adjusted to the characteristics of an industrial washing tub. Two trials at a pilot plant showed a satisfactory result and confirmed the estimated values. Meanwhile, the trials made clear that the washing tub needed to be equipped with a warming system that is safe and efficient to guarantee a constant water temperature during the processing. Once the equipment is installed, a quick and precise adjustment of the paste can be attained by controlling the temperature of the washing water through a thermostat that monitors the entering paste. The presented solution has the potential to be easily integrated into an existing production line with lower investment costs than those of the commercially available and experimental flash heat treatment systems.

Author Contributions: Conceptualization, E.P.; methodology, E.P.; software, E.P.; validation, J.M.G.M., M.d.C.F.F. and R.R.S.-G.; formal analysis, E.P. and R.R.S.-G.; investigation, E.P.; resources, J.M.G.M.; data curation, E.P.; writing—original draft preparation, E.P. and J.F.G.M.; writing—review and editing, J.M.G.M., M.d.C.F.F., R.R.S.-G. and J.F.G.M.; visualization, E.P.; supervision, J.M.G.M., M.d.C.F.F. and R.R.S.-G.; project administration, E.P. All authors have read and agreed to the published version of the manuscript.

Funding: This research received no external funding.

Institutional Review Board Statement: Not applicable.

Informed Consent Statement: Not applicable.

Data Availability Statement: Not applicable.

Acknowledgments: We thank the Del Cetino Olive Farm and their personnel for their collaboration, as well as Fernando Martínez, Paco Hidalgo, and Cristobal Hidalgo and the staff of the experimental plant of the Instituto de la Grasa for the technical assistance.

Conflicts of Interest: The authors declare no conflict of interest.

References

1. Rallo, L.; Díez, C.M.; Morales-Silero, A.; Miho, H.; Priego-Capote, F.; Rallo, P. Quality of olives: A focus on agricultural preharvest factors. *Sci. Hortic.* **2018**, *233*, 491–509. [CrossRef]
2. Famiani, F.; Arinelli, D.; Urbani, S.; Al Hariri, R.; Paoletti, A.; Rosati, A.; Esposto, S.; Selvaggini, R.; Taticchi, A.; Servilli, M. Harvesting system and fruit storage affect basic quality parameters and phenolic and volatile compounds of oils from intensive and super-intensive olive orchards. *Sci. Hortic.* **2020**, *263*, 109045. [CrossRef]
3. Sola-Guirado, R.R.; Castro-García, S.; Blanco-Roldán, G.L.; Jiménez-Jiménez, F.; Castillo-Ruiz, F.J.; Gil-Ribes, J.A. Traditional olive tree response to oil olive harvesting technologies. *Biosyst. Eng.* **2014**, *118*, 186–193. [CrossRef]
4. D'Imperio, M.; Gobbino, M.; Picanza, A.; Costanzo, S.; Della Corte, A.; Mannina, L. Influence of harvest and period on olive oil composition: An NMR and statistical study. *J. Agric. Food Chem.* **2010**, *58*, 11043–11051. [CrossRef]
5. Morales-Silero, A.; García, J.M. Impact assessment of mechanical harvest on fruit physiology and consequences on oil physicochemical and sensory quality from 'Manzanilla de Sevilla' and 'Manzanilla Cacereña' super-high-density hedgerows. A preliminary study. *J. Sci. Food Agric.* **2015**, *95*, 2445–2453. [CrossRef]
6. García, J.M.; Yousfi, K. The postharvest of mill olives. *Grasas Aceites* **2006**, *57*, 16–24. [CrossRef]
7. Brkić, K.; Lukić, M.; Novoselić, A.; Krapac, M.; Lukić, I. Olive Fruit Refrigeration during Prolonged Storage Preserves the Quality of Virgin Olive Oil Extracted Therefrom. *Foods* **2020**, *9*, 1445. [CrossRef] [PubMed]
8. Lorite, I.J.; Gabaldón-Leal, C.; Ruiz-Ramos, M.; Belaj, A.; de la Rosa, R.; León, L.; Santos, C. Evaluation of olive response and adaptation strategies to climate change under semi-arid conditions. *Agric. Water Manag.* **2018**, *204*, 247–261. [CrossRef]
9. Benlloch-González, M.; Sánchez-Lucas, R.; Bejaoui, M.A.; Benlloch, M.; Fernández-Escobar, R. Global warming effects on yield and fruit maturation of olive trees growing under field conditions. *Sci. Hortic.* **2019**, *249*, 162–167. [CrossRef]
10. Fernández-Bolaños, J.; Heredía, A.; Vioque, B.; Castellano, J.M.; Guillén, R. Changes in cell-wall-degrading ezyme activities in stored olives in relation to respiration and ethylene production. *Z Lebesm. Unters. Forsch. A.* **1997**, *204*, 293–299.
11. Di Serio, M.G.; Giansante, L.; Di Loreto, G.; Faberi, A.; Ricchetti, L.; Di Giacinto, L. Ethyl esters versus fermentative organoleptic defects in virgin olive oil. *Food Chem.* **2017**, *219*, 33–39. [CrossRef]
12. Inarejos-García, A.M.; Gómez-Rico, A.; Salvador, M.D.; Fregapane, G. Influence of malaxation condition on virgin olive oil yield, overall quality and composition. *Eur. Food Res. Technol.* **2009**, *228*, 671–677. [CrossRef]
13. Clodoveo, M.L. Malaxation: Influence on virgin olive oil quality. Past, present and future. An overview. *Trends Food Sci. Technol.* **2012**, *25*, 13–23. [CrossRef]
14. Peri, C. Quality Excellence in Extra Virgin Olive Oils. In *Olive Oil Sensory Science*; Monteleone, E., Langstaff, S., Eds.; John Wiley: Chichester, UK, 2014; pp. 3–32.
15. Selvaggini, R.; Esposto, S.; Taticchi, A.; Urbani, S.; Veneziani, G.; Di Maio, I.; Sordini, B.; Servili, M. Optimization of the temperature and oxygen concentration conditions in the malaxation during the oil mechanical extraction process of four Italian olive cultivars. *J. Agric. Food Chem.* **2014**, *62*, 3813–3822. [CrossRef] [PubMed]
16. Dourou, A.M.; Brizzolara, S.; Meoni, G.; Tenori, L.; Famiani, F.; Luchinat, C.; Tonutti, P. The inner temperature of the olives (cv Leccino) before processing affects the volatile profile and the composition of the oil. *Food Res. Int.* **2020**, *129*, 108861. [CrossRef] [PubMed]
17. Amirante, P.; Cini, E.; Montel, E.; Pasqualone, A. Influence of mixing and extraction parameters on virgin olive oil quality. *Grasas Aceites* **2001**, *52*, 198–201. [CrossRef]
18. Caponio, F.; Gomes, T.; Summo, C.; Pasqualone, A. Influence of the type of olive-crusher used on the quality of extra virgin olive oils. *Eur. J. Lipid Sci. Technol.* **2003**, *105*, 201–206. [CrossRef]

19. Veneziani, G.; Esposto, S.; Taticchi, A.; Urbani, S.; Selvaggini, R.; Di Maio, I.; Sordini, B.; Servili, M. Cooling treatment of olive paste during the oil processing: Impact on the yield and extra virgin olive oil quality. *Food Chem.* **2017**, *221*, 107–113. [CrossRef] [PubMed]
20. Plasquy, E.; García, J.M.; Florido, M.C.; Sola-Guirado, R.R. Cooling rate estimation of six olive cultivars using thermal imaging. *Agriculture* **2021**, *11*, 164. [CrossRef]
21. Brosnan, T.; Sun, D. Precooling techniques and applications for horticultural products. A Review. *Int. J. Refrig.* **2001**, *24*, 154–170. [CrossRef]
22. Elansari, A.M. Design aspects in the precooling process of fresh produce. *Fresh Prod.* **2009**, *3*, 49–57.
23. Plasquy, E.; Blanco-Roldán, G.; Florido, M.C.; García, J.M. Effects of an integrated harvest system for small producers on the quality of the recollected olive fruit. *Grasas Aceites* **2021**, *72*, e436. [CrossRef]
24. García, J.M.; Gutiérrez, F.; Barrera, M.; Albi, M. Storage of mill olives on an industrial ccale. *J. Agric. Food Chem.* **1996**, *44*, 590–593. [CrossRef]
25. Amirante, P.; Clodoveo, M.L.; Dugo, G.; Leone, A.; Tamborrino, A. Advance technology in vigin olive oil production from traditional and de-stoned pastes: Influence of the introduction of a heat exchanger on oil quality. *Food Chem.* **2006**, *98*, 797–805. [CrossRef]
26. Esposto, S.; Veneziani, G.; Taticchi, A.; Selvaggini, R.; Urbani, S.; Di Maio, I.; Sordini, B.; Minnocci, A.; Sebastiani, L.; Servili, M. Flash Thermal Conditiong of Olive Pastes during the Olive Oil Mechanical Extraction Process: Impact on the structural modifications of pastes and oil quality. *J. Agric. Food Chem.* **2013**, *61*, 4953–4960. [CrossRef]
27. Fiori, F.; Di Lecce, G.; Boselli, E.; Pieralisi, G.; Frega, N.G. Effects of olive paste fast preheating on the quality of extra virgin olive oil during storage. *LWT Food Sci. Technol.* **2014**, *58*, 511–518. [CrossRef]
28. Leone, A.; Esposto, S.; Tamborrino, A.; Romaniello, R.; Taticchi, A.; Urbani, S.; Servilli, M. Using a tubular heat exchanger to improve the conditioning process of the olive paste: Evaluation of yield and olive oil quality. *Eur. J. Lipid Technol.* **2016**, *118*, 308–317. [CrossRef]
29. Bejaoui, M.A.; Sánchez-Ortiz, A.; Sánchez, S.; Jiménez, A.; Beltrán, G. The high-power ultrasound frequency: Effect on the virgin olive oil yield and quality. *J. Food Eng.* **2017**, *207*, 10–17. [CrossRef]
30. Cecchi, L.; Bellumori, M.; Corbo, F.; Milani, G.; Clodoveo, M.L.; Mulinacci, N. Implementation of the sono-heat-exchanger in the extra virgen olive oil extraction process: End-user validation and analytical evaluation. *Molecules* **2019**, *24*, 2379. [CrossRef]
31. Tamborrino, A.; Romaniello, R.; Caponio, F.; Squeo, G.; Leone, A. Combined industrial olive oil extraction plant using ultrasounds, microwave, and heat exchange: Impact on olive oil quality and yield. *J. Food Eng.* **2019**, *245*, 124–130. [CrossRef]
32. Clodoveo, M.L. Industrial ultrasound applications in the extra-virgin olive extraction proces: History, approaches, and key questions. *Foods* **2019**, *8*, 121. [CrossRef]
33. Stillitano, T.; Falcone, G.; De Luca, A.I.; Piga, A.; Conte, P.; Strano, A.; Gulisano, G. A life cycle perspective to assess the environmental and economic impacts of innovative technologies in extra virgin olive oil extraction. *Foods* **2019**, *8*, 209. [CrossRef] [PubMed]
34. Kalogianni, E.P.; Georgiou, D.; Hasanov, J.H. Olive oil processing: Current knowledge, literature gaps, and future perspectives. *J. Am. Oil Chem. Soc.* **2019**, *96*, 481–507. [CrossRef]
35. Uceda, M.; Frías, L. Harvest dates: Evolution of the fruit oil content, oil composition and oil quality. In Proceedings of the 2nd International Olive Oil Seminary, Cordoba, Spain, 6–17 October 1975; pp. 125–130.
36. Cuesta, F.J.; Alvarez, M.D. Mathematical modelling for heat conduction in stone fruits. *Int. J. Refrig.* **2017**, *80*, 120–129. [CrossRef]
37. Luaces, P.; Pérez, A.G.; Sanz, C. Effect of the blanching process and olive fruit temperature at milling on the biosynthesis of olive oil aroma. *Eur. Food Res. Technol.* **2006**, *224*, 11–17. [CrossRef]
38. García, J.M.; Yousfi, K.; Oliva, J.; García-Diaz, M.T.; Pérez-Camino, M.C. Hot water dipping of olives (*Olea europaea*) for virgen oil debittering. *J. Agric. Food Chem.* **2005**, *53*, 8248–8252. [CrossRef] [PubMed]
39. Cruz, S.; Yousfi, K.; Oliva, J.; García, J.M. Heat Treatment Improves Olive Oil Extraction. *J. Am. Oil Soc.* **2007**, *84*, 1063–1068. [CrossRef]
40. Yousfi, K.; Moyano, M.J.; Martinez, F.; Cayuela, J.A.; García, J.M. Postharvest heat treatment for olive oil debittering at the industrial scale. *J. Am. Oil Soc.* **2010**, *87*, 1053–1061. [CrossRef]
41. Boselli, E.; Di Lecce, G.; Strabbioli, R.; Pieralisi, G.; Frega, N. Are virgin olive oils obtained below 27 C better than those produced at higher temperatures? *LWT Food Sci. Technol.* **2009**, *42*, 748–757. [CrossRef]

processes

Article

Use of Natural Microtalcs during the Virgin Olive Oil Production Process to Increase Its Content in Antioxidant Compounds

Sebastián Sánchez [1,2,*], Inmaculada Olivares [1,2], Juan G. Puentes [1,2], Rafael Órpez [1,2], M. Dolores La Rubia [1,2], Rafael Pacheco [1,2] and Juan F. García Martín [1,3,*]

1 University Institute of Research on Olive Groves and Olive Oils, GEOLIT Science and Technology Park, University of Jaen, 23620 Mengibar, Spain; momerino@ujaen.es (I.O.); jpuentes@ujaen.es (J.G.P.); rorpez@ujaen.es (R.Ó.); mdrubia@ujaen.es (M.D.L.R.); rpacheco@ujaen.es (R.P.)
2 Department of Chemical, Environmental and Materials Engineering, University of Jaen, 23071 Jaen, Spain
3 Departamento de Ingeniería Química, Facultad de Química, Universidad de Sevilla, 41012 Seville, Spain
* Correspondence: ssanchez@ujaen.es (S.S.); jfgarmar@us.es (J.F.G.M.)

Abstract: During the olive oil production process, certain olive varieties, such as 'Hojiblanca' and 'Picual', create pastes from which it is difficult to separate the oil, resulting in low extraction yields. To improve oil extraction, one alternative is the addition of natural microtalcs (NMT). In the present study, a NMT of great purity ($CaCO_3$ concentration less than 6 wt.%) and small average particle size ($\phi \leq 2.1$ μm) was added in the malaxation stage on an industrial scale at two olive mills. In one of them and using 'Hojiblanca' olives, the performance of the high-purity NMT was compared with that of a traditional NMT, while in the other, the effect of its dosage in the quality of 'Picual' oils was assessed. The performance of the high-purity NMT was evaluated in terms of industrial oil yield, extractability index, quality parameters and oxidative stability of the resulting oils. The addition of the high-purity NMT not only increased the extraction yields but also improved the quality of the virgin olive oils, especially in relation to antioxidant compounds (tocopherols and phenolic compounds). Increases of 10.4% in phenolic compounds and of 21.5% in the tocopherols were found, thus enhancing the oxidative stability of the oils.

Keywords: extractability index; industrial oil yield; natural microtalc; oxidative stability; phenolic compounds; tocopherols; virgin olive oils

Citation: Sánchez, S.; Olivares, I.; Puentes, J.G.; Órpez, R.; La Rubia, M.D.; Pacheco, R.; García Martín, J.F. Use of Natural Microtalcs during the Virgin Olive Oil Production Process to Increase Its Content in Antioxidant Compounds. *Processes* **2022**, *10*, 950. https://doi.org/10.3390/pr10050950

Academic Editor: Carla Silva

Received: 2 April 2022
Accepted: 6 May 2022
Published: 10 May 2022

1. Introduction

Virgin olive oils are defined by the International Olive Council as "the oils obtained from the fruit of the olive tree (*Olea europaea* L.) solely by mechanical or other physical means under conditions, particularly thermal conditions, that do not lead to alterations in the oil, and which have not undergone any treatment other than washing, decantation, centrifugation, and filtration" [1].

In certain olive growing areas, there are serious difficulties in the olive oil production process with certain varieties of olives. In these cases, after the grinding stage, pastes are produced from which it is difficult to separate the fat matter (called "strong or difficult pastes"). At the macroscopic level, retention and occlusion of the liquid phases in the pastes, oil-in-water emulsions, and very fine solids in the form of a slurry are observed, resulting in lower extraction yields at the industrial scale.

An increase in the temperature used in the malaxation step usually improves the oil extraction yield because of the reduction of the oily phase viscosity. However, an excess of temperature during malaxation can jeopardize the nutritional and health values of extra virgin olive oil (EVOO), leading to undesirable effects, such as accelerating the oxidative process and the loss of volatile compounds responsible for the aroma of the oil [2].

In order to improve the management of these difficult pastes in the production of olive oils without increasing the temperature and to comply with the definition of virgin olive oil, one of the widely accepted alternatives is the addition of a technological adjuvant to these pastes during the malaxation step. Among these adjuvants, the most used, proposed by the International Olive Council and accepted by the European Union, are natural microtalcs (NMT) [3,4].

The use of natural microtalcs (hydrated magnesium silicate) has been authorized by European regulation since 1986, solely for food purposes, provided that no changes in the physical and organoleptic properties of olive oil are produced. The current European Union Regulation allows a maximum concentration of 6 wt.% of calcium carbonate in NMT.

Initially and for three decades, the aim of adding microtalcs was only to increase the extraction yields without affecting the natural properties of the pastes or modifying the physical–chemical and sensory characteristics of the virgin oils obtained [5–7], keeping intact the nutritional quality of the oils obtained.

The addition of microtalcs promotes the breakdown of oil-in-water emulsions, thus facilitating the extraction of oil from mesocarp cells, as the oil does not remain in the pomace and consequently the oil yield increases [2,8,9]. This breakage of oil-in-water emulsions can be explained by the fact that the use of microtalcs during the malaxation step greatly lessens the action of pectic substrates and proteins, which are the main factors responsible for the formation of oil-in-water emulsions during the malaxation step [8]. Notwithstanding, excessive talc addition leads to a negative effect (taste loss) [8].

The main objective of the present work was to increase the industrial oil yield and extractability index and to improve the physicochemical composition of olive oils for olive varieties prone to forming difficult pastes by the addition of a new, commercially available, high-purity natural microtalc. A comparative study on the use of this high-purity NMT with a small average particle size ($\phi \leq 2.1$ μm) in the industrial-scale production of virgin olive oils was carried out. Two varieties were assayed, namely 'Hojiblanca' and 'Picual', in which pastes are difficult to separate the fat matter (the olive oil). With the former, the performance of the high-purity NMT in the malaxation stage was compared with that of a traditional NMT. With the later, the efficiency of the process with the addition or without the addition of the high-purity NMT was assessed. The influence of the addition of the high-purity NMT was evaluated not only on extraction yields, but also on the improving in the quality of virgin olive oils produced, especially in relation to antioxidant compounds (tocopherols and phenolic compounds) and thus on the oxidative stability of the olive oils.

2. Materials and Methods

2.1. Raw Materials

Olives from the variety 'Hojiblanca' were supplied by the olive oil mill 'Oleoalgaidas S.C.S.' (Villanueva de Algaidas, Malaga, Spain) during the 2011/2012 harvesting, while 'Picual' olives were supplied by the olive oil mill 'Aceites San Antonio S.L.U.' (Escañuela, Jaen, Spain) during the 2016/2017 harvesting.

Two natural microtalcs were used, both authorized for food use. The first NMT was the FC8KN natural microtalc, supplied by Mondo Minerals BV (Amsterdam, The Netherlands), which was used for the production of virgin olive oils in both harvestings. This NMT is characterized by its high purity (97 wt.%, calcium carbonate concentration less than 6 wt.%) and small particle size (with a D98% top cut of 8.6 μm, and an average particle size for D50% of 2.1 μm). Its BET specific surface area (ISO 4652) is 7 m^2/g, as indicated by the supplier. This means that it can be added in a lower percentage to olive pastes compared to traditional microtalcs. For comparison, a traditional NMT (17.5 μm, D90% < 40.0 μm, calcium carbonate concentration not specified) was also used in the production of virgin olive oils with 'Hojiblanca' olives during the 2011/12 harvesting.

2.2. Oil Olive Production at Industrial Scale

The sampling was carried out at different times in the olive ripening process during the 2011/12 and 2016/17 harvest seasons. This research was carried out on an industrial scale in the production lines of two olive oil mills that use the continuous centrifugation process with two-outlet decanters. One olive oil mill was 'Oleoalgaidas S.C.A.' (Villanueva de Algaidas, Malaga, Spain) from 'Grupo DCOOP' and the other was 'Aceites San Antonio S.L.U.' (Escañuela, Jaen, Spain).

Three- and four-body malaxers were used in the malaxation step process, 90 min being the malaxation time and 5000 kg h^{-1} the mass flow rate of the paste for the 'Oleoalgaidas S.C.A.' oil mill (Figure 1) and 60 min the malaxation time and 3000 kg h^{-1} the mass flow rate of the paste for the 'Aceites San Antonio S.L.U.' oil mill (Figure 2). The natural microtalcs were continuously added through dosage equipment during the malaxation step, before the centrifugation process of the olive pastes, at a percentage of 0.3 wt.% in relation to the mass flow rate of the paste in the production line. The NMT dispenser was placed at the beginning of the last body of the malaxer, because it was found in preliminary assays that this position was the one that led to a greater increase in industrial yields (data not shown). Two samples of each produced oil were collected at the outlet of the horizontal centrifuge in a time interval of less than 1 h.

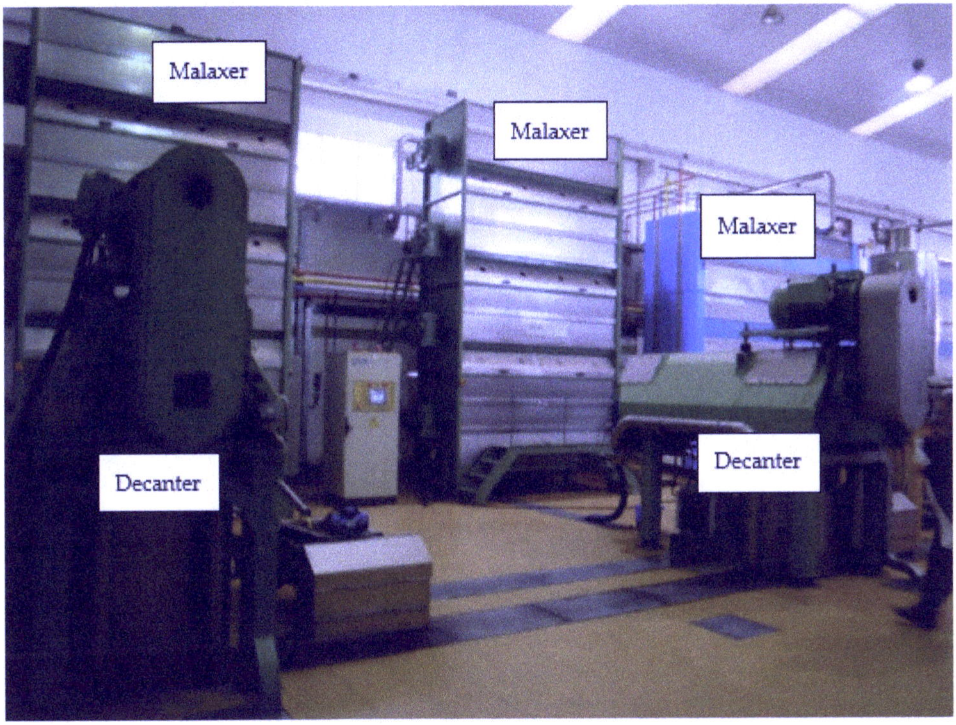

Figure 1. Malaxers and decanters in two different lines of the olive oil extraction process in the 'Oleoalgaidas S.C.A.' olive mill.

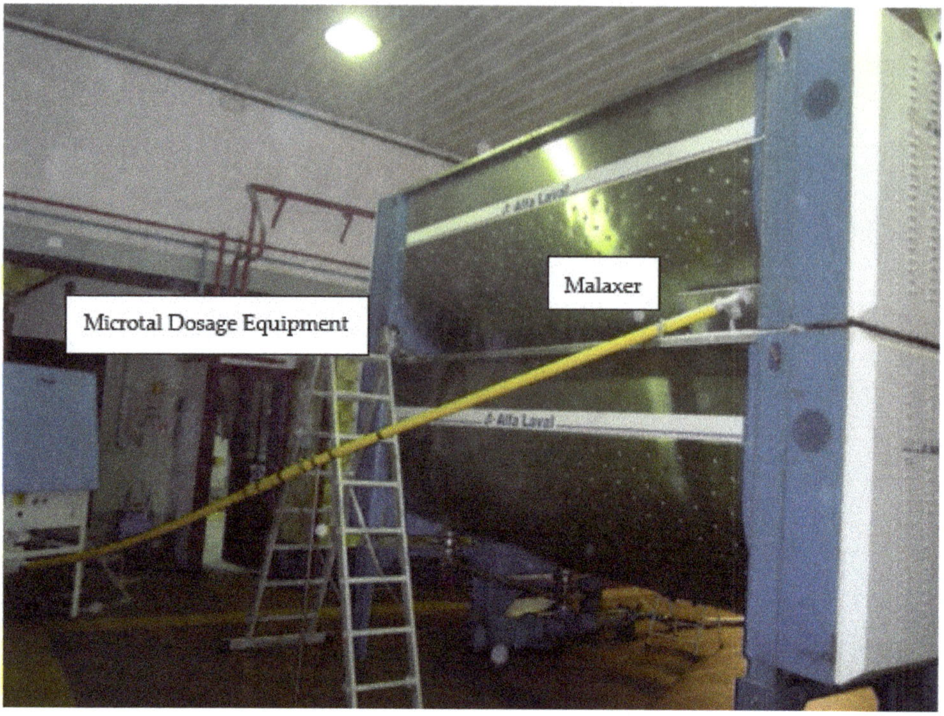

Figure 2. Malaxer and NMT dosage equipment at the 'Aceites San Antonio S.L.U.' olive mill.

2.3. Industrial Oil Yield and Extractability Index

About 750 g olives from each sample batch were taken to the laboratory and ground in an MM-100 hammer mill of an Abencor ('MC2 Ingeniería y Sistemas S.L.', Seville, Spain) mini-plant system for olive oil production, as described elsewhere [10], which is widely accepted to obtain the actual values of oil yield at the industrial scale. The extractability index is defined as the ratio between the industrial yield and the content of total fat matter on a wet basis, and is usually expressed as a percentage [11]. The total fat matter on a wet basis was determined by extraction in Soxhlet equipment with technical hexane.

2.4. Analytical Methods

The olive oils obtained were characterized according to the following quality parameters: free acidity, peroxide value, and ultraviolet absorption (K_{270}, K_{232} and ΔK). These parameters were determined following the analytical methods described in the CEE/2568/91, CEE/1429/92 regulations and subsequent modifications of the European Commission [12,13]. A brief description of the procedures can be found elsewhere [14]. The free acidity, expressed as percentage of oleic acid, was determined by placing the olive oil in an ethanol/ethyl ether solution (1:1 v/v) along with a few drops of phenolphthalein and then neutralized with 0.1 M KOH.

Regarding the peroxide value, expressed in $mEqO_2$ per kg of oil, it was determined by letting a mixture of oil and acetic acid chloroform react in the dark with a solution of potassium iodide; then, free iodine was titrated with a sodium thiosulfate solution.

The ultraviolet absorption parameters, K_{232}, K_{270}, and ΔK (absorption of a 1% solution of oil in cyclohexane at 232 and 270 nm), were measured in a UV-VIS spectrophotometer Spectronic Helios γ (Thermo Fisher Scientific, Waltham, MA, USA).

Similarly, the content of total phenolic compounds was determined as well as the oxidative stability of the oils produced. Regarding the evaluation of oxidative stability, the oxidation induction time was measured by Rancimat equipment, Mod. 743 (Metrohm Hispania, Madrid, Spain). Briefly, 3 g of oil were weighed and, after heating to 98 °C, an airflow was bubbled through it with a volumetric flow rate of 10 $dm^3 \cdot h^{-1}$. The results obtained referring to the oxidative stability were expressed as induction time in hours [15]. On the other hand, the content of total phenolic compounds was determined by extraction with methanol:water (60:40 v/v) solution and measurement of the absorbance of the complex formed between phenolic compounds and the Folin-Ciocalteu reagent at λ = 725 nm in a UV-VIS spectrophotometer Spectronic Helios γ (Thermo Fisher Scientific, Waltham, MA, USA), as described elsewhere [16].

The contents of α-, β-, γ-, δ- and total tocopherols were determined by high-performance liquid chromatography following a standardized method [17]. For this purpose, a Prominence modular HPLC system (Shimadzu Corporation, Kyoto, Japan), equipped with a fluorescence detector and a LiChrospher 5 µm Sil 60A analytical column (Phenomenex, Torrance, CA, USA) were used. Propan-2-ol in n-hexane (0.5:99.5 v/v) was used as mobile phase with a flow of 0.5 cm^{-3} min^{-1} and a temperature of 24 °C.

3. Results and Discussion

3.1. Industrial Oil Yield

Higher industrial oil yields were achieved with the 'Picual' variety than with 'Hojiblanca' variety (data not shown). Initially, this fact is logical, since 'Picual' olives have higher fat matter content. Based on the experimental results, an increase in the industrial oil yield in oil was observed when the FC8KN natural microtalc was added during the malaxation step at different times of the elaboration process (Figure 3).

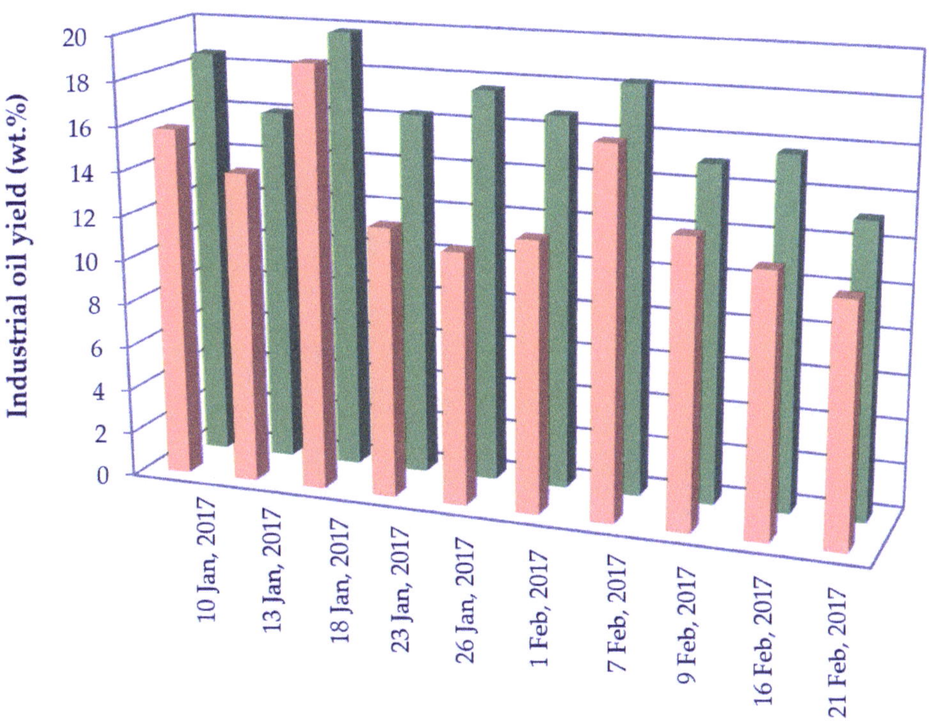

Figure 3. Industrial oil yield for the 'Picual' variety at different times during the 2016/2017 harvesting without (■) and with (■) FC8KN natural microtalc addition during the malaxation step.

In general, the addition of the FC8KN natural microtalc improved the industrial oil yield in all cases compared to the control without the addition of NMT. The average increase obtained for the samples shown in Figure 3 was 25%, which is similar to the increases found by other authors on the use of NMT but with higher dosages [18,19] and other olive varieties [19–21]. Of note is the increase in industrial oil yield obtained for the sample from 26 January 2017. The extraction process without NMT rendered an industrial oil yield of 11.34 wt.%, while it rose to 17.65 wt.% when using the FC8KN natural microtalc, which represents an increase of 56%. This increase is much higher than the highest increase in industrial oil yield (32% using 1 wt.% NMT and 'Hojiblanca' variety) that can be found in the literature [18]. Since the NMT dosage used in this work (0.3 wt.%) is even lower than those recommended in the available literature (0.5–2.0 wt.%) [22,23], the high industrial oil yields obtained can be attributed to the physicochemical characteristics of the FC8KN natural microtalc [18,23,24].

Physicochemical characteristics such as particle size (D50) and microtalc purity play an important role in breaking oil-in-water emulsions and improve the oil extraction process. The ability of microtalc to increase oil yields could be explained by its physical action, since it adsorbs the oil droplets retained in the cell walls, forming larger droplets and therefore facilitating oil extraction [25]. In olive paste, oil and water can be trapped in agglomerating microgels. Additionally, both paste and microtalc have lipophilic properties, so they tend to retain oil. The microtalc facilitates the release of the oil in an extractable form, breaking the reticular structure formed by the links among microgels [4,26,27].

Regarding the degree of depletion of the pomaces produced and with varieties very prone to forming difficult pastes ('Hojiblanca' and 'Picual'), values very similar to those reached at the laboratory level were obtained at the industrial level in both oil mills (2011/12 and 2016/17 harvestings, respectively). These results agree with those found by other authors using the 'Hojiblanca' variety in a process with a two-outlet decanter [22]. Additionally, it should be noted that the difference in consumption between both microtalcs was significant at each of the sampling times assayed (Figure 4). Thus, a reduction of more than 40% in used NMT was achieved in the experience of 12 January 2012 when using the FC8KN natural microtalc.

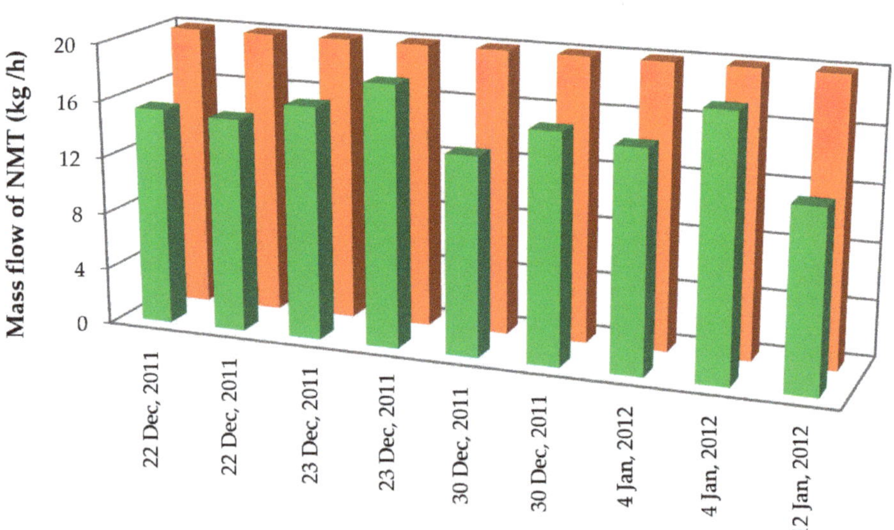

Figure 4. Mass flow of traditional (■) and FC8KN (■) natural microtalcs in processes with olives of the 'Hojiblanca' variety.

Traditional microtalcs, large in size, are usually added at the beginning of the malaxation stage together with the paste. However, the FC8KN microtalc was added at the beginning of the last body of the malaxer (corresponding to the last third of the malaxation total time), reducing the contact time between NMT and pastes and increasing oil yields and extractability indexes. Although the microtalc is a natural product, it is advisable to reduce at minimum the contact time between the olive paste and the technological adjuvant, since the microtalc is not a component of the olives.

In relation to the extractability index, extractability indexes over 80% were achieved with FC8KN microtalc for both varieties. To be specific, an extractability index value of 88.9% was obtained from olives of the 'Picual' variety with a maturity index of 5.62, an average fruit weight of 2.47 g, and a pulp/stone ratio of 2.17.

In general, it was demonstrated that when an olive paste has a water percentage between 48 and 51 wt.%, the addition of a NMT leads to an increase in the extractability index (data not shown), mainly when the FC8KN microtalc was used.

A visual aspect observed in the olive oils at the exit of the decanter was that they were cleaner (free of solid particles) and had higher pigmentation (greener). A similar fact occurred at the oultlet of the horizontal centrifuge. This fact could be attributed to a higher content of chlorophyll pigments and a lower content of water microdroplets and solid microparticles.

3.2. *Effect of the Additon of FC8KN Natural Microtalc on Olive Oil Quality Parameters*

From the results obtained in the characterization of the olive oils of both varieties ('Picual' and 'Hojiblanca') according to the olive oil quality parameters, namely acidity, peroxide index, and UV absorption, it is observed that all of them comply with the standards of the European Commission for edible oils [1]. In fact, 'Hojiblanca' oils produced before December 30 are within the normal ranges for EVOO, while oils obtained after December 30 are considered virgin olive oils (Table 1). As for 'Picual' oils, all are in accordance with the limits to be labelled as EVOO (Table 2).

Table 1. Quality parameters of olive oils obtained by adding FC8KN natural microtalc to olive pastes of the 'Hojiblanca' variety.

Haversting Date	Acidity (%)	Peroxide Value (mEq O_2/kg)	K_{232}	K_{270}	ΔK
22 Dec, 2011	0.20 ± 0.00	16.94 ± 2.96	1.53 ± 0.01	0.13 ± 0.01	0.01 ± 0.01
23 Dec, 2011	0.24 ± 0.02	15.08 ± 0.77	1.51 ± 0.12	0.12 ± 0.02	0.00 ± 0.00
30 Dec, 2011	0.23 ± 0.04	16.61 ± 0.01	1.48 ± 0.05	0.11 ± 0.00	0.00 ± 0.00
4 Jan, 2012	0.20 ± 0.03	22.84 ± 1.61	1.58 ± 0.06	0.10 ± 0.02	0.00 ± 0.00
12 Jan, 2012	0.20 ± 0.00	17.53 ± 0.00	1.62 ± 0.00	0.12 ± 0.00	0.00 ± 0.00
20 Jan, 2012	0.21 ± 0.02	20.52 ± 1.75	1.55 ± 0.04	0.12 ± 0.02	0.00 ± 0.00
Limits fixed by the IOC [1]	≤0.80	≤20.0	≤2.50	≤0.22	≤0.01

Table 2. Quality parameters of olive oils obtained by adding FC8KN natural microtalc to olive pastes of the 'Picual' variety.

Haversting Date	Acidity (%)	Peroxide Value (mEq O_2/kg)	K_{232}	K_{270}	ΔK
10 Jan, 2017	0.17 ± 0.00	2.93 ± 0.04	1.56 ± 0.04	0.18 ± 0.02	0.00 ± 0.00
18 Jan, 2017	0.17 ± 0.01	3.65 ± 0.06	1.64 ± 0.01	0.13 ± 0.00	0.00 ± 0.00
23 Jan, 2017	0.19 ± 0.02	4.77 ± 0.29	1.74 ± 0.01	0.15 ± 0.00	0.00 ± 0.00
26 Jan, 2017	0.17 ± 0.00	2.67 ± 0.28	1.73 ± 0.01	0.16 ± 0.00	0.00 ± 0.00
1 Feb, 2017	0.18 ± 0.01	3.83 ± 0.18	1.55 ± 0.03	0.12 ± 0.00	0.00 ± 0.00
23 Feb, 2017	0.28 ± 0.00	4.76 ± 0.23	1.59 ± 0.03	0.11 ± 0.01	0.00 ± 0.00
Limits fixed by the IOC [1]	≤0.80	≤20.0	≤2.50	≤0.22	≤0.01

On the other hand, it should be noted that the addition of the FC8KN natural microtalc has resulted, with respect to the characterization of the olive oils obtained, in lower values in these quality parameters when compared with those obtained without the addition of this natural adjuvant. On the other hand, if these quality parameters are compared with the addition of a different microtalcs (traditional NMT vs. FC8KN NMT), it was observed that the values of acidity, peroxide value, and UV absorption were very similar for the 11/12 harvesting with 'Hojiblanca' [28]. However, the peroxide value was slightly lower in oils produced using the FC8KN natural microtalc, results that can be found in that previous study. These results confirm the findings of other authors using different olive varieties, who have reported that microtalc addition protects the oil from oxidation [19,21,29,30].

From the results obtained in the determination of phenolic compounds, it is highlighted that there is a higher concentration in the olive oils obtained in production processes with the addition of NMT. This effect is clearly reflected in the 2016/2017 campaign and for the 'Picual' variety (Figure 5). In this research, increases of 10.4% in the fraction of total phenolic compounds were achieved when the FC8KN natural microtalc was added in the malaxation stage of the olive oil production process.

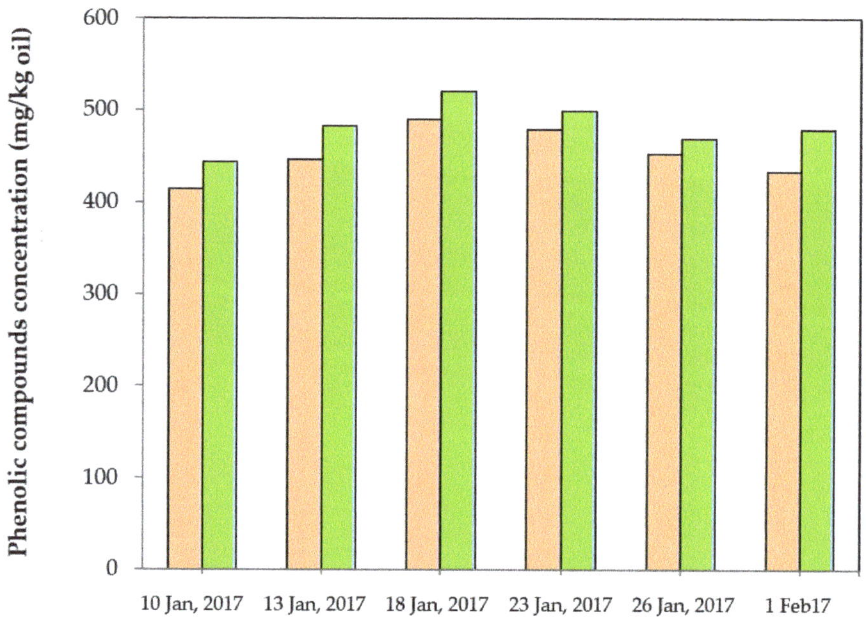

Figure 5. Concentration of total phenolic compounds in oils of the 'Picual' variety at different times of the 2016/2017 harvesting, produced without (■) and with the addition (■) of the FC8KN natural microtalc during the malaxation stage in the oil extraction process.

3.3. *Effect of the Additon of FC8KN Natural Microtalc on Olive Oil Tocopherol Content*

In this research, the total tocopherol content of the olive oils obtained from 'Picual' variety (2016/2017 harvesting) was close to or greater than 300 mg/kg oil when using the FC8KN NMT, reaching 443.4 mg/kg oil in one of the periods (Figure 6). These values are in agreement with the work of Cayuela and García [31], who reported a range of 55.2–466.4 mg/kg oil for the concentration of total tocopherols.

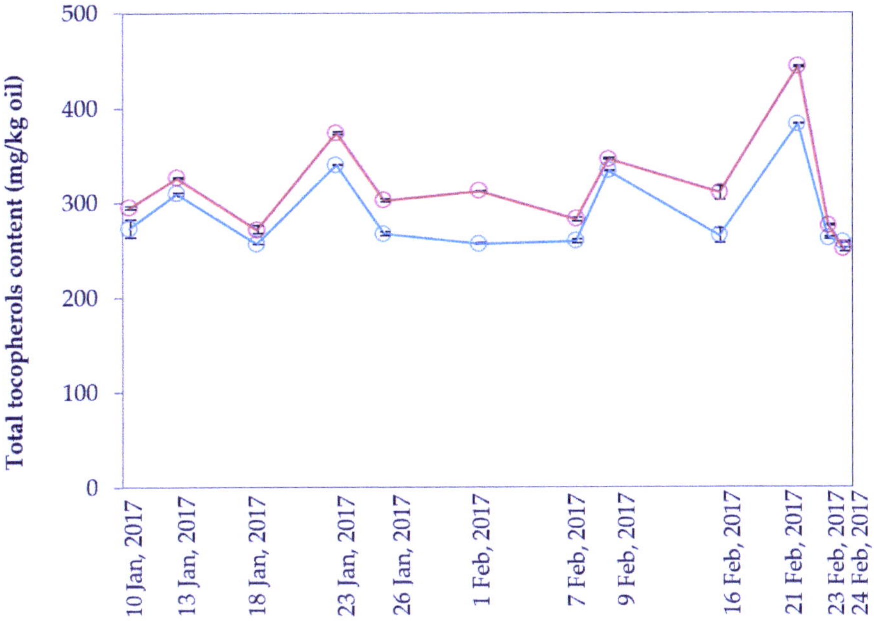

Figure 6. Concentration of total tocopherols in oils of the 'Picual' variety at different times of the 2016/17 harvesting, produced without (•) and with addition (•) of the FC8KN natural microtalc during the malaxation stage.

The effect of microtalc addition on the tocopherol content of olive oils has not been extensively studied to date, and most studies found a significant decrease in the tocopherol content when adding microtalc in the malaxation stage [19,21]. By contrast, increases of 21.5% in the fraction of total tocopherols and of 22.1 and 25.6% in α- and β-tocopherol, respectively, were obtained when the FC8KN NMT was used in the malaxation stage of the extraction process of the 'Picual' variety (Table 3) when comparing with olive oils from the same olive batches produced without the addition of NMT. This result could be attributed to the lower particle size and higher purity of the FC8KN microtalc.

Table 3. Average increase in the concentration of tocopherols in olive oils from the 'Picual' variety, produced at different times of the 2016/2017 harvesting, when adding the FC8KN natural microtalc during the malaxation step.

Harvesting Date	%Δ α-Tocopherol	%Δ β-Tocopherol	%Δ γ-Tocopherol	%Δ Total Tocopherols
10 Jan, 2017	9.5	10.3	−12.9	7.8
13 Jan, 2017	5.3	5.5	9.8	5.5
18 Jan, 2017	4.6	25.6	10.2	6.0
23 Jan, 2017	10.0	7.8	11.2	10.0
26 Jan, 2017	13.1	10.5	15.4	13.1
1 Feb, 2017	22.1	25.3	7.1	21.5
7 Feb, 2017	8.6	6.6	16.9	9.0
9 Feb, 2017	4.1	1.0	1.6	3.7
16 Feb, 2017	17.6	9.0	13.6	17.0
21 Feb, 2017	16.9	6.4	5.8	15.9
23 Feb, 2017	4.7	5.2	15.6	5.1

Tocopherols have an important activity as natural antioxidant agents and confer stability to the oils that contain them. Consequently, they are important constituents in virgin olive oils, since they give it stability and they have a beneficial biological role as antioxidants. The α-tocopherol (90–95% of the total tocopherol content) is the main homologue of the forms of vitamin E that exerts an important antioxidant action, since it reacts with the peroxide radicals of the fatty acids, which are the primary products of fat auto-oxidation, thus stopping spoilage in the early stages. In general, the contents of the β and γ forms are below 10% [32]. δ-tocopherol is usually present in such low concentrations that some authors even state that this form is not present in olive oils [33,34]. In this sense, in the trials carried out in this work, δ-tocopherol has not been detected in any of the sampling periods.

3.4. Effect of the Additon of Natural Microtalc on Olive Oil Oxidative Stability

Regarding the evaluation of oxidative stability, the Rancimat method is an accelerated stability test that provides very useful information about the resistance of the oil to the oxidation process. In general, a higher stability can be observed in the oils produced with smaller particle size microtalc (Figure 7). When traditional NMT was used in the process, the average oxidative stability for the oils produced from the 'Hojiblanca' variety was 50.6 h, with a range between 21.1 and 83.1 h, while using the FC8KN natural microtalc the average oxidative stability in the oils produced was 67.7 h, with a range between 34 and 107.9 h. These last results are very close to those obtained by Beltrán et al. with the same variety ('Picual'), in a process at the mini-plant level (Abencor system), during three consecutive years of harvesting, and using a natural microtalc [35].

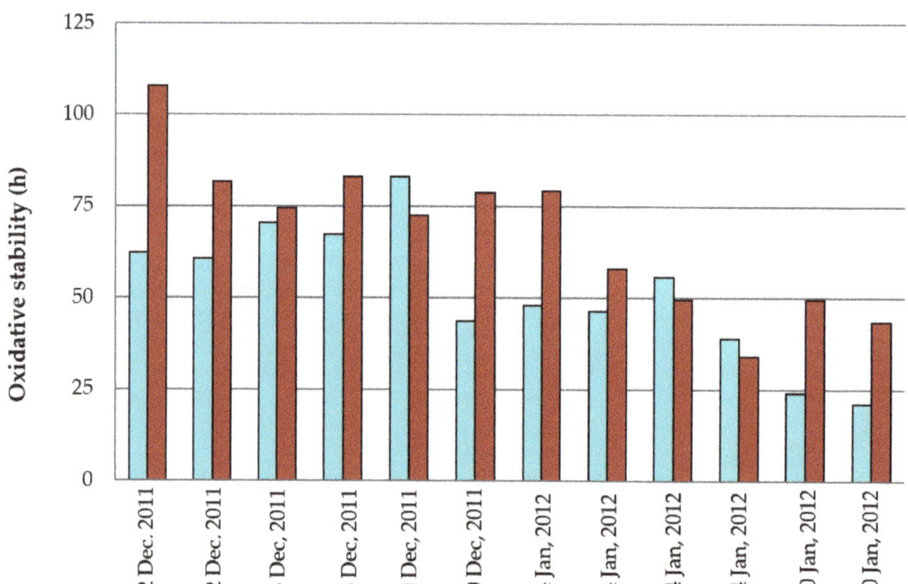

Figure 7. Oxidative stability of virgin olive oils of the 'Hojiblanca' variety produced using the FC8KN (■) and a traditional (■) NMT.

4. Conclusions

The FC8KN natural microtalc assayed in this work is characterized by its small particle size and extraordinary purity. These characteristics lead to a lower percentage added to the olive pastes compared to a traditional NMT; the FC8KN natural microtalc dosage (0.3 wt.%

in relation to the mass flow rate of the paste in the production line) was lower than those reported in the available literature (between 0.5 and 2.0 wt.%).

In the production of olive oils from both the 'Picual' and 'Hojiblanca' varieties, the industrial yield and extractability index increased when a natural microtalc was added during the malaxation stage. The olive oil yield increased by 25% on average when using the NMT with a smaller particle size (FC8KN). A maximum of 56% increasing in industrial oil yield was achieved with FC8KN when compared with the same process without the addition of NMT. This increase in industrial oil yield is much higher than those reported by other authors on the use of other microtalcs and at higher dosages.

Taking into account the quality parameters, the olive oils produced adding the FC8KN natural microtalc showed lower values (therefore higher quality) in these parameters when compared to those obtained without the addition of NMT. It is worth noting that improved pigmentation (higher concentration of chlorophylls) was visually observed for both the 'Picual' and 'Hojiblanca' varieties when adding FC8KN microtalc.

Finally, olive oils produced with the FC8KN natural microtalc showed a higher content of phenolic compounds and thus higher oxidative stability, both in the 'Picual' and 'Hojiblanca' varieties. To be specific, increases of up to 10.4% in total phenolic compounds, 21.5% in total tocopherols, and 22.1% in α-tocopherol were achieved when the FC8KN natural microtalc was added in the malaxation stage of the olive oil production process.

All in all, the FC8KN natural microtalc should be used in the olive oil industry to obtain extra virgin olive oils with high quality from olive varieties prone to forming difficult pastes.

Author Contributions: Conceptualization, S.S., R.P. and J.F.G.M.; methodology, S.S., M.D.L.R., R.Ó., I.O. and J.G.P.; software, validation, and formal analysis, J.F.G.M. and I.O.; investigation and data curation, R.Ó., M.D.L.R., I.O. and J.G.P.; resources, R.P., M.D.L.R., R.Ó. and I.O.; writing—original draft preparation, S.S., J.F.G.M. and I.O.; writing—review and editing, S.S., J.F.G.M. and M.D.L.R.; visualization, J.F.G.M., I.O. and J.G.P.; supervision: S.S., J.F.G.M. and I.O.; project administration, S.S. and R.P.; funding acquisition, S.S. All authors have read and agreed to the published version of the manuscript.

Funding: The authors acknowledge the industry Mondo Minerals BV (Barcelona, Spain) and the Andalusia Regional Government (Spain) for its financial support to our research group 'Bioprocesses TEP-138'.

Informed Consent Statement: Not applicable.

Acknowledgments: The authors are grateful to the industries Mondo Minerals BV (Barcelona, Spain), 'Aceites San Antonio S.L.U.' (Escañuela, Jaen, Spain), and 'Oleoalgaidas S.C.A.' (Villanueva de Algaidas, Malaga, Spain).

Conflicts of Interest: The authors declare no conflict of interest.

References

1. *IOC/T.15/NC No 3/ Rev.17/2021*; Trade Standard Applying to Olive Oils and Olive Pomace Oils. International Olive Council: Madrid, Spain, 2021; pp. 1–17.
2. Caponio, F.; Squeo, G.; Difonzo, G.; Pasqualone, A.; Summo, C.; Paradiso, V.M. Has the use of talc an effect on yield and extra virgin olive oil quality? *J. Sci. Food Agric.* **2016**, *96*, 3292–3299. [CrossRef] [PubMed]
3. Martínez, F.; Moyano, M.J.; Alba, J.; Ruiz, M.A.; Hidalgo, F.; Heredia, F.J. Método rápido de obtención de aceite de oliva virgen para determinación de acidez. *Grasas Aceites* **1999**, *50*, 122–126. [CrossRef]
4. Sánchez, S.; Rubia, M.D.; Sánchez, A.; Pereira, M.G. Evolución histórica de la utilización de coadyuvantes tecnológicos en la producción de aceite de oliva. In Proceedings of the 'I Congreso de Cultura del Olivo', Jaén, Spain, 27–29 October 2005; pp. 623–633.
5. Di Giovacchino, L. L'impiego del talco micronizzato nell' estrazione dell' olio dalle olive con il sistema continuo. *Riv. Ital. Delle Sostanze Grasse* **1988**, *65*, 623–628.
6. Alba, J.; Muñoz, E.; Martínez, J.M. Obtention of olive oil use of additives facilitating its extraction. *Alimentaria* **1982**, *138*, 25–55.
7. Alba, J. Obtención del aceite oliva. Empleo de productos que facilitan su extracción. In Proceedings of the XVIII Reunión Plenaria de la Asamblea del Instituto de la Grasa y sus Derivados, Seville, Spain, 15–17 October 1982.
8. Sadkaoui, A.; Jiménez, A.; Pacheco, R.; Beltrán, G. Micronized natural talc affects the proteins and pectic cell wall polysaccharides during "Hojiblanca" virgin olive oil extraction. *Eur. J. Lipid Sci. Technol.* **2017**, *119*, 1600039. [CrossRef]

9. Di Giovacchino, L. I sottoprodotti della lavorazione delle olive in oleificio. In *Tecnologie di Lavorazione Delle Olive in Frantoio*; Di Giovacchino, L., Ed.; Tecnica Alimentare: Milano, Italy, 2010; pp. 243–258. ISBN 978-88-481-2554-3.
10. Plasquy, E.; García Martos, J.M.; Florido Fernández, M.D.C.; Sola-Guirado, R.R.; García Martín, J.F. Adjustment of olive fruit temperature before grinding for olive oil extraction. Experimental study and pilot plant trials. *Processes* **2021**, *9*, 586. [CrossRef]
11. Beltrán, G.; Uceda, M.; Jiménez, A.; Aguilera, M.P. Olive oil extractability index as a parameter for olive cultivar characterisation. *J. Sci. Food Agric.* **2003**, *83*, 503–506. [CrossRef]
12. Commission Regulation (EEC) No 2568/91 on the characteristics of olive oil and olive-residue oil and on the relevant methods of analysis. *Off. J. Eur. Communities* **1991**, *L248*, 1–15.
13. Commission Implementing Regulation (EU) No 1348/2013 amending Regulation (EEC) No 2568/91. *Off. J. Eur. Union* **2013**, *L338*, 31–67.
14. García Martín, J.F. Potential of near-infrared spectroscopy for the determination of olive oil quality. *Sensors* **2022**, *22*, 2831. [CrossRef]
15. Gutierrez Rosales, F. Determination of oxidative stability in virgin olive oils: Comparison between active oxygen method and Rancimat method. *Grasas Aceites* **1989**, *40*, 1–5.
16. Vázquez-Roncero, A.; Janer del Valle, C.; Janer del Valle, M.L. Determinación de los polifenoles totales del aceite de oliva. *Grasas Aceites* **1973**, *24*, 350–357.
17. Pocklington, W.D.; Dieffenbacher, A. Determination of tocopherols and tocotrienols in vegetable oils and fats by high performance liquid chromatography: Results of a collaborative study and the standardised method. *Pure Appl. Chem.* **1988**, *60*, 877–892. [CrossRef]
18. Sadkaoui, A.; Jiménez, A.; Aguilera, M.P.; Pacheco, R.; Beltrán, G. Virgin olive oil yield as affected by physicochemical talc properties and dosage. *Eur. J. Lipid Sci. Technol.* **2017**, *119*, 1600112. [CrossRef]
19. Carrapiso, A.I.; García, A.; Petrón, M.J.; Martín, L. Effect of talc and water addition on olive oil quality and antioxidants. *Eur. J. Lipid Sci. Technol.* **2013**, *115*, 583–588. [CrossRef]
20. Aguilera, M.P.; Beltran, G.; Sanchez-Villasclaras, S.; Uceda, M.; Jimenez, A. Kneading olive paste from unripe "Picual" fruits: I. Effect on oil process yield. *J. Food Eng.* **2010**, *97*, 533–538. [CrossRef]
21. Caponio, F.; Monteleone, J.I.; Martellini, G.; Summo, C.; Paradiso, V.M.; Pasqualone, A. Effect of talc addition on the extraction yield and quality of extra virgin olive oils from Coratina cultivar after production and during storage. *J. Oleo Sci.* **2014**, *63*, 1125–1132. [CrossRef]
22. Cert, A.; Alba, J.; León-Camacho, M.; Moreda, W.; Pérez-Camino, M.C. Effects of talc addition and operating mode on the quality and oxidative stability of virgin olive oils obtained by centrifugation. *J. Agric. Food Chem.* **1996**, *44*, 3930–3934. [CrossRef]
23. Sánchez, S.; Pacheco, R.; La Rubia, M.D.; Sánchez, A.; Pereira, G. Aplicación de distintos talcos naturales, como coadyuvantes tecnológicos, en los procesos de extracción de aceites de oliva. In Proceedings of the XII Simposium Científico-Técnico Expoliva, Jaen, Spain, 12–13 May 2005; p. TEC-65.
24. Sadkaoui, A.; Jiménez, A.; Pacheco, R.; Beltrán, G. Micronized natural talc with a low particle size and a high carbonate rate is more effective at breaking down oil-in-water emulsion. *Eur. J. Lipid Sci. Technol.* **2016**, *118*, 545–552. [CrossRef]
25. Clodoveo, M.L.; Durante, V.; La Notte, D.; Punzi, R.; Gambacorta, G. Ultrasound-assisted extraction of virgin olive oil to improve the process efficiency. *Eur. J. Lipid Sci. Technol.* **2013**, *115*, 1062–1069. [CrossRef]
26. Martínez Moreno, J.M. *Fundamentos Físico-Químicos de la Técnica Oleícola*; C.S.I.C.—Patronato Juan de la Cierva: Madrid, Spain, 1972.
27. Nushtaeva, A.V. Emulsion stabilization with talc microparticles. *Colloid J.* **2019**, *81*, 425–430. [CrossRef]
28. Sánchez, S.; Landeta, J.M.; Olivares, I.; Relaño, P.L.; Pacheco, R.; La Rubia, M.D. Use of natural microtalcs in the elaboration of virgin olive oil of 'Hojiblanca' variety. In Proceedings of the XVI Simposium Científico-Técnico Expoliva, Jaen, Spain, 8–12 May 2013; p. IND-58.
29. Ben-David, E.; Kerem, Z.; Zipori, I.; Weissbein, S.; Basheer, L.; Bustan, A.; Dag, A. Optimization of the Abencor system to extract olive oil from irrigated orchards. *Eur. J. Lipid Sci. Technol.* **2010**, *112*, 1158–1165. [CrossRef]
30. Cruz, S.; Yousfi, K.; Pérez, A.G.; Mariscal, C.; Garcia, J.M. Salt improves physical extraction of olive oil. *Eur. Food Res. Technol.* **2007**, *225*, 359–365. [CrossRef]
31. Cayuela, J.A.; García, J.F. Sorting olive oil based on alpha-tocopherol and total tocopherol content using near-infra-red spectroscopy (NIRS) analysis. *J. Food Eng.* **2017**, *202*, 79–88. [CrossRef]
32. Lozano Sánchez, J. Aceite de Oliva Como Alimento Funcional: Nuevas Perspectivas Analíticas y Tecnológicas. Ph.D. Thesis, University of Granada, Granada, Spain, 2013.
33. Dionisi, F.; Prodolliet, J.; Tagliaferri, E. Assessment of olive oil adulteration by reversed-phase high-performance liquid chromatography/amperometric detection of tocopherols and tocotrienols. *J. Am. Oil Chem. Soc.* **1995**, *72*, 1505–1511. [CrossRef]
34. Rovellini, P.; Azzolini, M.; Cortesi, N. Tocoferoli e tocotrienoli in oli e grassi vegetali mediante HPLC. *Riv. Ital. Delle Sostanze Grasse* **1997**, *74*, 1–5.
35. Beltrán, G.; Aguilera, M.P.; Del Rio, C.; Sánchez, S.; Martínez, L. Influence of fruit ripening process on the natural antioxidant content of Hojiblanca virgin olive oils. *Food Chem.* **2005**, *89*, 207–215. [CrossRef]

processes

Review

Olive Sound: A Sustainable Radical Innovation

Maria Lisa Clodoveo [1], Pasquale Crupi [1,*] and Filomena Corbo [2]

[1] Interdisciplinary Department of Medicine, University Aldo Moro Bari, 702125 Bari, Italy; marialisa.clodoveo@uniba.it
[2] Department of Pharmacy-Drug Sciences, University Aldo Moro Bari, 702125 Bari, Italy; filomena.corbo@uniba.it
* Correspondence: pasquale.crupi@uniba.it; Tel.: +39-3471252849

Abstract: Olive Sound is the acronym of a Horizon 2020 European Project aimed at the development of a high-flow oil extraction plant, the Sono-Heat-Exchanger, which combines ultrasound and heat exchange in order to break, through a radical innovation model in the oil mill, the historical paradigm that sees as inversely correlated the oil yield and the content of bio-phenols. These compounds are biologically active molecules that transform the product, extra virgin olive oil, from a mere condiment into a functional food. The primary objective of the project, financially supported by the European Union through the "Fast Track to Innovation" program, is the development of a product "ready for the market" (TRL 9) capable of making the involved companies more competitive while increasing the competitiveness of European extra virgin olive oil in the international context.

Keywords: malaxation; sonication; competitiveness

Citation: Clodoveo, M.L.; Crupi, P.; Corbo, F. Olive Sound: A Sustainable Radical Innovation. *Processes* **2021**, *9*, 1579. https://doi.org/10.3390/pr9091579

Academic Editors: Sebastián Sánchez Villasclaras and Juan Francisco García Martín

Received: 12 July 2021
Accepted: 28 August 2021
Published: 3 September 2021

1. Introduction

Innovation is the process that allows individual companies, or entire production sectors, to create value, remain in or enter new markets, increase profitability, generate employment, and increase competitiveness [1].

Considering the trends of the last decade, it is possible to affirm, with a certain degree of awareness, that the European olive oil sector has an urgent need to innovate to generate competitive advantage; that is, the set of elements that must characterize the product with the aim to create added value and differentiate, mainly by exploiting health properties, its offer from its global competitors [2]. It is worth pointing out that, in the market, product competitiveness is a strategy pursued by companies to generate greater profit. In the world of olive oil, dominated by information asymmetry, the problem of marginal profits and below-cost sales is considered the principal reason for penalizing companies in the sector. Currently, the price strategy, which sees companies challenge each other on the lowest price and which is based on a production model that aims at maximum production yield, leads to the production of an extra virgin olive oil (EVOO) devoid of distinctive chemical and organoleptic characteristics; therefore, this is a product that does not have elements to compete on the market. Controlling and reducing overhead costs, designing an efficient production line, and differentiation are three of the company's strategies for gaining a competitive advantage.

Conway and Steward stated that "innovation can be briefly defined as the successful exploitation of new ideas". This means that to generate an innovative process it is not enough to have an idea, but it is also necessary to act so that the idea can be put into practice by transforming the effective contest where innovation finds application [3].

Technological innovation is the result of a creative process that involves a series of actors motivated by the common goal of responding to a need in a specific sector [4]. In the case of the olive oil sector, in the transformation phase, which occurred more than 40 years after the introduction of the last real radical innovation of the extraction process, the decanter [5] combined with the malaxer [6], the need for research has been expressed by the community of olive millers in all the olive-growing areas of the globe. These needs can be summarized in the following points:

- Making the extraction process effectively continuous [7];
- Contracting processing times [8];
- Increasing extraction yields by reducing fat residues in the olive pomace [9];
- Maximizing the extraction capacity of bio-phenols, molecules with a recognized health effect, with the aim to place the product in the highest range of the market certified by the application of health claims approved by the European Food Safety Authority (EFSA) [10–12];
- Using extraction temperatures (~20 °C) useful for the development of the volatile fraction without affecting the yields [13];
- Creating a sustainable process consistent with the objectives of the 2030 Agenda and in line both with the directives of the European Green Deal and with the emerging needs of consumers [14].

When a change process is stimulated by the community of future users (i.e., stakeholders), an innovation model defined as bottom-up is created, which comes from the collection of needs that really take into account the needs of the various stakeholders [15].

The mapping of these needs, collected by researchers from the Departments of Medicine and Pharmacy of the University of Bari [16–21], has been transformed into a project resource that has led to the creation of an award-winning partnership, in an extremely competitive tender, by the European Union with a commitment total income of EUR 2.5 million. The strength of the idea, which sees the collaboration of three nations (Italy, France, and Spain) and five partners (three companies and two research institutions), is in the way it is generated, which has never imposed standard solutions but has been able to aggregate people around a common and shared project born from the participation, involvement, and enhancement of the skills and knowledge of the human resources involved.

The scientific dissemination activity dedicated to the operators in the olive oil sector conducted by the University of Bari, and in particular by the professors M. L. Clodoveo and F. Corbo, in the Italian territory and also abroad, has transformed over the years into a fundamental tool for building relationships with the olive millers, sharing values, knowledge, skills, and previous experiences, to create a real community in which the various stakeholders have been constantly involved and encouraged to weave a dense network of exchanges both within the entire production system and towards academic interlocutors. In fact, in the first stages of the development of innovation (from TRL, Technology Readiness Level, 3 to TRL 7), that is, the simultaneous application of ultrasound and heat exchange in the process of extraction of extra virgin olive oil [1], the community of millers (in particular some companies located in Apulia, which is the Italian region that holds the leading production) has contributed to the co-generation of knowledge, through the development, experimentation, and optimization of Sono-Heat-Exchanger innovation within industrial olive mills [22] (Figure 1). They also favor an interactive transfer of know-how focused on the specificities of the oil sector and open an effective and constructive dialogue between all the subjects aimed at stimulating the processes of mutual learning.

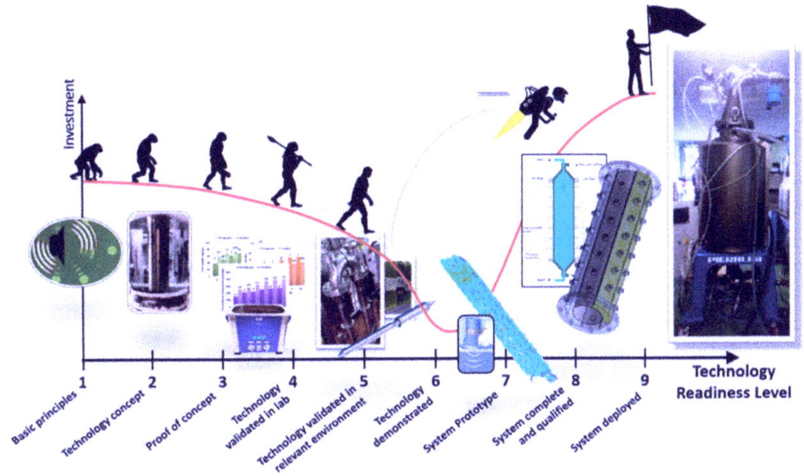

Figure 1. The stages of the TRL—technology readiness level—in the development of the Sono-Heat-Exchanger.

This approach has made it possible to undertake a preparatory path for the industrial implementation of innovation based on three principles: (1) the creation of opportunities for interaction and discussion with the various stakeholders; (2) listening to the needs of stakeholders; and (3) co-planning.

2. Fast Track to Innovation

The Fast Track to Innovation (FTI) [23], a funding instrument that supported the European Olive Sound Project, is a completely bottom-up innovation support program that promotes innovation activities close to the market.

This tool requires that the project leader be a company, and in the case of Olive Sound it is Pieralisi di Jesi, the leading company in the international market of oil machines (Figure 2).

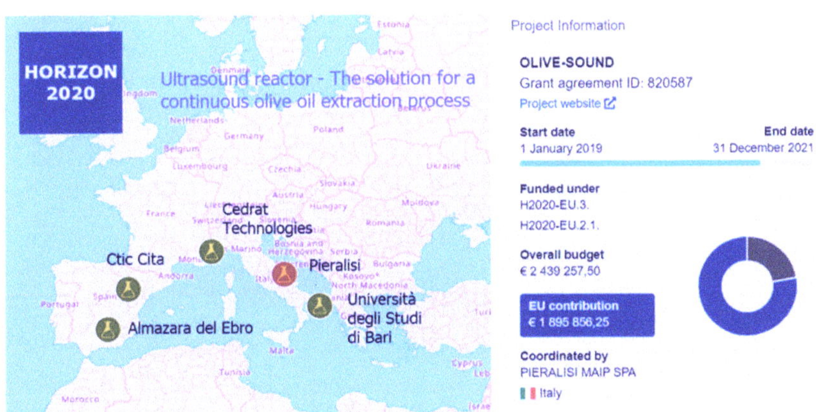

Figure 2. Composition of the partnership of the Horizon 2020 Olive Sound project.

In the European vision, companies are the engine of the European economy, essential for the creation of jobs and for economic growth and capable of ensuring social stability. Therefore, the Fast Track program aims to ensure that innovative ideas are transformed *quickly* into new products *ready for the market* that stimulate growth, create quality jobs, and contribute to addressing the challenges of European and global society. Indeed, the promoted actions, *guided by the business*, aim to give the last push to innovative ideas to ensure their rapid adoption by the market. The term *rapid* in the sense of the fast track founding program means *within a period of three years from the start of the action*. On average, the success rate for consortia that aspire to receive the funding is extremely low, less than 5%, reflecting the rigorous selection process applied by international auditors which rewards only European entrepreneurial and academic excellence.

3. Radical Innovations

According to Joseph A. Schumpeter [24], innovation is the main determinant of industrial change as a force that destroys the old competitive environment to create a completely new one. It is, therefore, "a creative response that occurs whenever an industry offers something that is outside existing practice". The Sono-Heat-Exchanger (Figure 3), which breaks the historical paradigm of the olive mill that sees as opposite the technological conditions that favor the extraction yield and those aimed at increasing quality and effectively eliminates the bottleneck represented by the now obsolete malaxer, can truly be considered a radical innovation in the field of olive oil extraction plants, and is capable of opening up opportunities for the development of new markets [1,8,25,26].

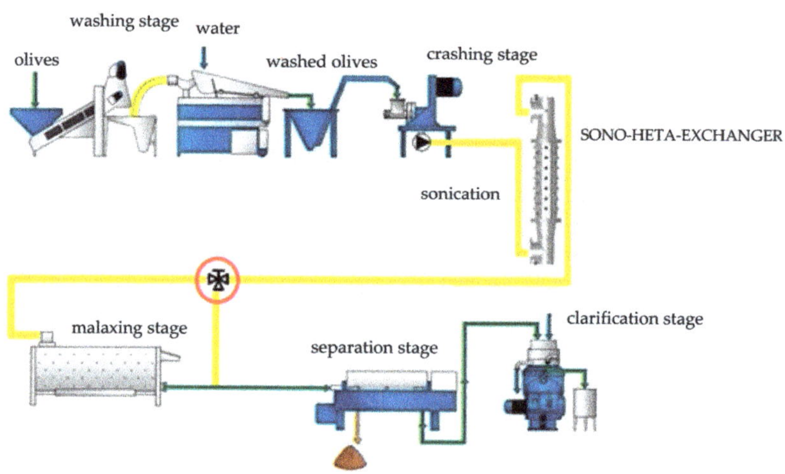

Figure 3. Olive oil processing line.

As part of the Olive Sound project, innovation management was based on a multifaceted approach divided into three development phases which involved collaboration and integration of the partners' skills:

- Planning (design);
- Physical realization (manufacturing);
- Marketing of the Sono-Heat-Exchanger.

3.1. The Design Phase

The design phase involved the definition of the geometry of the ultrasonic device, such as to ensure that the ultrasonic waves emitted by the transducers were effective in inducing the phenomenon of cavitation, responsible for the mechanical action of ultrasound, and efficient in penetrating the entire thickness of the olive paste, all without representing a threat to the wear of the components. In order for this phase to be compatible with the timing of the project, the University of Bari made use of the technical-scientific collaboration of the Polytechnic of Bari, in particular of Prof. Amirante, to benefit from the support of numerical simulation [27,28]. Numerical simulation is an indispensable tool for reducing the time to market necessary for the final product because it is an approach capable of determining a rapid orientation of the design towards the optimal design, as well as reducing design costs, including the number of prototypes and experimental tests to be performed [29].

The design instead made use of a simulation program in ANSYS Fluent environment, which, for the first time, allows the simulation of the pressure transients induced by transducers in a pipeline in which the fluid flows tangentially in order to optimize the sonication process (geometry, thicknesses, position of the transducers, operating pressures) so that it is effective and efficient. This process is also combined with the heat exchange technology for heating or cooling the oil paste (Figure 4).

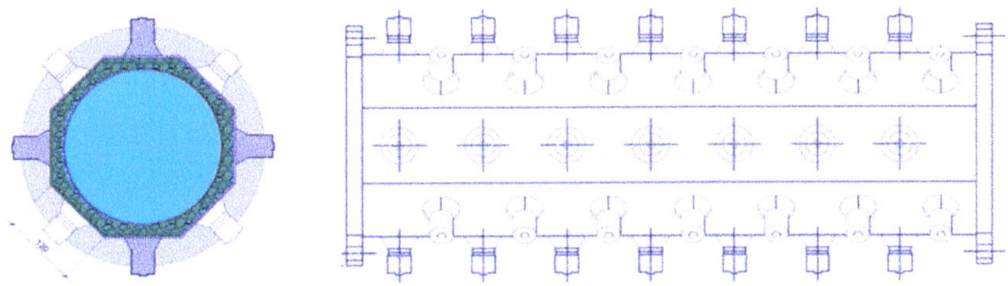

Figure 4. The device has been designed with an octagonal section equipped with a plate transducer, each of 100 W of power and 23 kHz of frequency. The inner circular section is the heat exchanger. The olive paste flows in the external annular section, while the water (cold or hot) flows in the internal annular section to modulate the temperature inside the olive paste.

The numerical simulation conducted on the Sono-Heat-Exchanger made use of an instrument designed ad hoc by the Polytechnic of Bari, which has taken into account the complex rheological characteristics of the triphasic fluid constituted by the olive paste, composed of two immiscible liquids (i.e., oil and water) and a solid (i.e., pulp and stone).

Moreover, the numerical simulation made possible the reproduction of the pressure transients induced by the action of the ultrasonic transducers (Figure 5) and the fluid dynamic profiles induced by the transport of the fluid (Figure 6), allowing for the advance knowledge of the performance or behavior of the innovative device before the first prototype had been physically built, thus speeding up the decision-making process of the design phase.

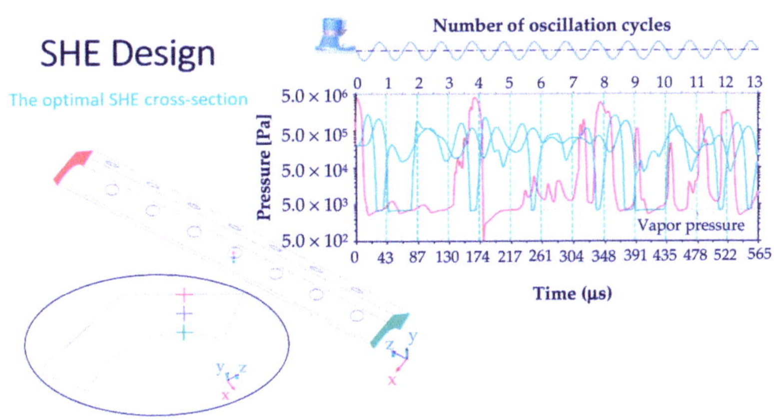

Figure 5. Determination of pressure transients inside the Sono-Heat-Exchanger—Olive Sound induced by the action of the ultrasound on the olive paste.

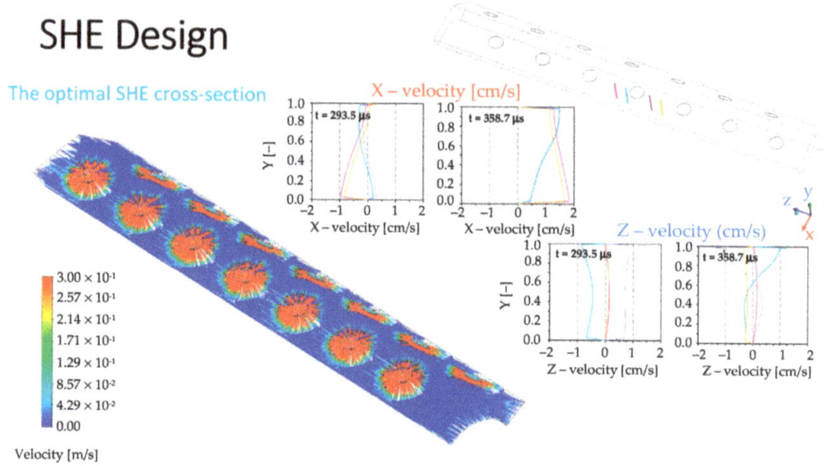

Figure 6. Fluid dynamic analysis of the olive paste inside the Sono-Heat-Exchanger—Olive Sound.

Finally, the numerical simulation made it possible to highlight two important effects of the ultrasound on the olive paste that explain the theoretical basis of the effects measured in the experimental phase. The first effect is the increase in yield and polyphenol content due to the determination of the pressure transients in the predetermined operating conditions. Indeed, the minimum pressure values reached inside the olive paste are below the vapor pressure of the water at the process temperature (~25 °C); therefore, they are compatible with effective cavitation (formation and collapse of vapor microbubbles) and useful for breaking the cells of the drupe passed intact to the crusher and for freeing the oil and minor compounds (polyphenols, tocopherols and carotenoids) trapped in them.

The second effect deals with the elimination of the malaxer because the pulsating action of the transducers causes swirling movements on the olive paste that agitate the flow of olive paste in transit in the apparatus, inducing agitation similar to that inside the malaxer and favoring coalescence phenomena among the minute drops of oil released by cavitation, which increase in diameter and are more easily separable in the centrifugal field.

3.2. *The Manufacturing of the Sono-Heat-Exchanger*

The simulation, combined with the mechanical design, has allowed the Pieralisi and Cedrat Technologies companies to create the first prototype within workshops and laboratories that specialize in piezoelectric technologies, a modular unit built for the validation phase (Figure 7) and used for the acquisition of technical and analytical data functional to the optimization preparatory phase and for the realization of the high working capacity model, which is compatible with the size of mills typical of the Spanish model.

Figure 7. Some phases of the construction of the modular unit of the Sono-Heat-Exchanger—Olive Sound built for optimization tests at the Pieralisi company.

The tests of the prototype were carried out over two oil years and in two areas (northern Spain and southern Italy) with a varietal panorama that covered 10 olive cultivars and, for each cultivar, three different ripening stages (green, partially dark, and totally uneven). The tests have made it possible to complete the modular unit with a series of sensors and actuators (Figure 8) implemented with the dual purpose of guaranteeing maximum protection of the machine and operators, and to proceed with the creation of the definitive model that will reach the international market of oil machines by January 2022 and help to create new jobs.

Figure 8. Installation of the transducers at Cedrat Technologies.

3.3. The Marketing Strategy of the Sono-Heat-Exchanger

The Fast Track to Innovation differs from other European funding programs because it provides, at the time of admission of the application, that the proposal is accompanied by a business plan that clearly describes the market potential (potential users/customers and their benefits, global markets/target Europeans, etc.), business opportunities for participants, measures that will improve the likelihood of eventual commercial adoption, and a credible marketing strategy.

The marketing strategy developed for the project focuses on two fundamental aspects: the description of the strengths of the new plant (Table 1) the demonstration that the Sono-Heat-Exchanger is the suitable solution to respond to the needs mapped out in the community of millers.

Table 1. The strengths of the new Sono-Heat-Exchanger compared to the obsolete malaxer.

Technological Effect	Malaxer	Sono-Heat Exchanger
Efficacy	The malaxer is a batch machine, which works between two continuous devices, the fruit crusher and the decanter. Its long kneading times, in addition to it constituting a threat to the quality of the oil, make this phase of mixing the oil paste at a controlled temperature the "bottleneck" of the continuous process.	The Sono-Heat-Exchanger is a continuous device. The major advantages of continuous processing are lower equipment costs, a reduced overall footprint, and the elimination of redundant plant equipment.
Efficiency	A bottleneck in the process tends to create a queue and increase the overall cycle time. The bottlenecks in production cause stalls and slowdowns in the production flow as with the same resources, production is slower and therefore smaller quantities are produced.	The elimination of the bottleneck in the oil sector translates into an effective tool if the innovation is able to increase the yields and content of antioxidants, and it is efficient if it achieves these objectives in a sustainable way, reducing energy costs with benefits regarding company economies and the impact on the environment.
Number of the devices	In the olive mill, currently, the limited working capacity of the malaxer penalizes the production efficiency of the decanter; the main plant engineering solution adopted to manage this inefficiency consists in multiplying the number of malaxers, in series or in parallel, to ensure continuity of the process, but not without an increase in investment in the crusher.	The Sono-Heat-Exchanger is a continuous machine that allows to synchronize its flow rate with the flow rate of the decanter and effectively eliminates the bottleneck of the process.
Mechanical effects	The mechanical crusher produces fruit fragments containing hundreds of cells that pass intact into the malaxer. The malaxer can be considered a finishing phase of the crushing phase by the cutting action of the stone fragments, which tears the cells passed intact to the crusher in a delicate manner but in an extremely long time (from 30 to 60 min), therefore helping to break the cells of the drupe passed intact to the crusher releasing a further amount of oil.	The ultrasounds determine a transient pressure gradient which, in some moments, reaches values lower than the saturated vapor pressure of the water contained in the olive paste, causing vaporization at low temperatures. "Cavitation bubbles" are created, which progressively increase their volume until they reach a critical value beyond which they implode, generating jets of liquid at high pressure that instantly tear the cells that have passed intact to the pressing.
Thermal effect	The malaxer is a bad heat exchanger due to an unfavorable ratio between the big volume of olive paste that should be warmed (or cooled) and the small surface for the heat exchange.	The Sono-Heat-Exchanger is equipped with an inner part that consists of a highly efficient spiral heat exchanger able to modulate the olive paste temperature (fast heating up or fast cooling of olive paste) simultaneously with the sonication treatment, adapting the results to the needs of olive millers [30].
Coalescence	The coalescence phenomena of oily drops inside the olive paste are due to hydrophobic interactions. During the malaxation, the drops of oil in the olive paste combine to form a larger drop. The role of the mixing in the hydrophobic interaction regards frequency of the collision of drops, which is one of the factors able to influence the coalescence of the oil in the olive paste. Mixing or agitation has been shown to improve coalescence by enhancing the rate of collisions.	The pulsating action of the transducers imposes swirling movements on the olive paste that agitate the flow of olive paste in transit in the apparatus, inducing agitation similar to what happens inside the malaxer, and favoring coalescence phenomena among the minute drops of oil released by cavitation, which, increasing in diameter, are more easily separable in the centrifugal field.

Table 1. *Cont.*

Technological Effect	Malaxer	Sono-Heat Exchanger
Plolyphenols	The mechanical action of the malaxer is mild and has a negligible effect on the ability to release further amounts of polyphenols. The thermal effect is limited to heating only, and the long stirring times are compatible with the activation kinetics of the oxidase enzymes [31], polyphenol oxidase, and peroxidase; therefore, the times favorable for kneading lengths to increase yields correspond to a loss of polyphenols and a lowering of oil quality.	The cavitation phenomena efficiently break the cells of the epicarp of the drupe, releasing high concentrations of polyphenols. The immediate thermal lowering of the olive paste temperature after olive crushing to lower than 20 °C permits partial inhibition of the endogenous enzymatic activity by polyphenol oxidase and peroxidase. The consequence is the reduction of degradation of phenolic compounds.
Volatile compounds	The lipoxygenase (LOX) pathway [31], responsible for the hydroperoxydation of polyunsaturated fatty acids, is activated upon crushing and grinding olive fruit tissue, which subsequently leads to the synthesis of volatile compounds. This biochemical reaction requires few seconds, if thermal condition are favorable (<24 °C). The malaxer, being a bad heat exchanger, could penalize the functionality of the pathway causing disharmonious organoleptic profiles if the temperatures of the crushed olive paste are high.	The sonicated oils have a more harmonious organoleptic profile and are rich in volatile components, as not only do the ultrasounds not damage the lipoxygenase pathway, but the rapidity of the transformation of the fruit into extra virgin olive oil contributes to a more timely stabilization of the product, thus avoiding technological threats, which could result in the loss or alteration of the profile of volatile compounds.

The strengths of the system emerged from the results of functional tests and performance optimization. Preliminary tests conducted in the two industrial mills have shown that the use of Sono-Heat-Exchanger to replace the malaxers always resulted in an increase in the extraction yield, up to 21% vs. 19%, respectively, especially at low temperature (~20 °C). However, this happened without compromising the polyphenols' content of the EVOO, which not only did not decrease in any sample analyzed, but, depending on the variety and the ripeness index of the olives, could even increase compared to the measured values in the oils obtained using the malaxers. For instance, the compounds (namely, decarboxymethyl-oleuropein aglycone in open dialdehyde form, 77 vs. 74 mg/kg, decarboxymethyl-ligstroside aglycone in open dialdehyde form, 110 vs. 105 mg/kg, lignans, 98 vs. 84 mg/kg, and oleuropein and ligstroside aglycones, 27 vs. 25 mg/kg) quantified in EVOO from Coratina and Frantoio blend green olives by the HPLC-MS/MS method as described by De Santis et al., 2021 [32], were more concentrated in the Sono-Heat-Exchanger samples than the malaxer ones.

Moreover, EVOO extracted by the Sono-Heat-Exchanger showed the identical values of the analytical parameters used for the product classification as for the oils from the same batches of olives by using the malaxers. Finally, the sensory analysis of EVOOs revealed that the new technology (i.e., Sono-Heat-Exchanger) did not just cause defects in the product, but in general allowed to obtain more intensely fruity and more harmonious oils than the samples obtained by the use of malaxers.

The functional tests have showed that the Sono-Heat-Exchanger is the suitable solution to respond to the needs mapped in the community of millers because it is a continuous system suitable for replacing the current malaxers by significantly reducing extraction times. It is also a continuous plant capable of breaking the historical paradigm about the extraction yield inversely correlated to quality, freeing millers from the dilemma of choosing whether to produce large quantities of a standard EVOO or smaller quantities of an EVOO of excellence.

Furthermore, the Sono-Heat-Exchanger is a system characterized by ease of use and maintenance, and it is sustainable, as it is built in accordance with the objectives of the 2030 Sustainability Agenda, from the selection of materials to the attention to energy saving and water consumption.

Pieralisi company has dealt with the design and construction of the Sono-Heat-Exchanger by applying integrated company management systems certified according to the international standards UNI EN ISO 9001, a management system that continuously improves the quality of products and processes, and UNI EN ISO 14001, an adequate management system that keeps the environmental impacts of its activities under control and systematically seeks improvement in a coherent, effective, and, above all, sustainable way.

The Sono-Heat-Exchanger has been designed and built according to strict quality and safety standards, ensuring compliance with the main EU directives (Machinery Directive 2006/42/EC, Low Voltage Directive 2014/35/EU, Electromagnetic Compatibility Directive 2014/30/EU, ATEX Directive 2014/34/EU) with a view to reducing the impact on the environment, protecting the health and safety of workers, and seeking the highest quality standards in products and services.

4. Conclusions

The Sono-Heat-Exchanger is the suitable solution to respond to the needs mapped in the community of millers because it is a continuous system and the best substitute for the old malaxer, it breaks the historical paradigm between yield and quality of EVOO, it is a sustainable plant solution, and it improves the health quality of the product by enhancing its polyphenols content without causing undesired sensorial defects, thus placing it in a higher production range.

All the actors involved in the innovative processes are well aware that it is possible to distinguish three fundamental moments within the research and technology transfer path: (1) the invention, or the moment in which a potentially beneficial idea arises, but is not necessarily implemented in a concrete form of product or process; (2) the innovation, which consists of transforming ideas into new or improved products and processes capable of leading to an economic and/or social benefit; and (3) the dissemination, or the phase in which the utility of an innovation is made known to the company and the sector concerned can actually receive an economic and social benefit.

The next step of the partnership will concern the dissemination actions.

Author Contributions: Conceptualization, M.L.C. and F.C.; writing—original draft preparation, M.L.C.; writing—review and editing, F.C. and P.C.; funding acquisition, M.L.C. and F.C. All authors have read and agreed to the published version of the manuscript.

Funding: This project has received funding from the European Union's Horizon 2020 research and innovation program under grant agreement No 820587.

Institutional Review Board Statement: Not applicable.

Informed Consent Statement: Not applicable.

Conflicts of Interest: The authors declare no conflict of interest.

References

1. Clodoveo, M.L. Industrial ultrasound applications in the extra-virgin olive oil extraction process: History, approaches, and key questions. *Foods* **2019**, *8*, 121. [CrossRef] [PubMed]
2. Clodoveo, M.L.; Corbo, F.; Amirante, R. Does the introduction of ultrasound in extra-virgin olive oil extraction process improve the income of the olive millers? The first technology for the simultaneous increment of yield and quality of the product. In *Technological Innovation in the Olive Oil Production Chain*; IntechOpen: Rijeka, Croatia, 2018. [CrossRef]
3. Conway, S.; Steward, F. *Managing and Shaping Innovation*; Oxford University Press: New York, NY, USA; Oxford, UK, 2009.
4. Sørensen, E.; Torfing, J. Enhancing collaborative innovation in the public sector. *Adm. Soc.* **2011**, *43*, 842–868. [CrossRef]
5. Amirante, P.; Clodoveo, M.L.; Leone, A.; Tamborrino, A.; Patel, V.B. Influence of different centrifugal extraction systems on antioxidant content and stability of virgin olive oil. In *Olives and Olive Oil in Health and Disease Prevention*; Academic Press: New York, NY, USA, 2010; pp. 85–93. [CrossRef]
6. Tamborrino, A.; Clodoveo, M.L.; Leone, A.; Amirante, P.; Paice, A.G. The malaxation process: Influence on olive oil quality and the effect of the control of oxygen concentration in virgin olive oil. In *Olives and Olive Oil in Health and Disease Prevention*; Academic Press: New York, NY, USA, 2010; pp. 77–83. [CrossRef]

7. Clodoveo, M.L. Malaxation: Influence on virgin olive oil quality. Past, present and future—An overview. *Trends Food Sci. Technol.* **2012**, *25*, 13–23. [CrossRef]
8. Clodoveo, M.L.; Durante, V.; La Notte, D. Working towards the development of innovative ultrasound equipment for the extraction of virgin olive oil. *Ultrason. Sonochem.* **2013**, *20*, 1261–1270. [CrossRef] [PubMed]
9. Clodoveo, M.L.; Dipalmo, T.; Schiano, C.; La Notte, D.; Pati, S. What's now, what's new and what's next in virgin olive oil elaboration systems? A perspective on current knowledge and future trends. *J. Agric. Eng.* **2014**, *45*, 49–59. [CrossRef]
10. Bellumori, M.; Cecchi, L.; Innocenti, M.; Clodoveo, M.L.; Corbo, F.; Mulinacci, N. The EFSA health claim on olive oil polyphenols: Acid hydrolysis validation and total hydroxytyrosol and tyrosol determination in Italian virgin olive oils. *Molecules* **2019**, *24*, 2179. [CrossRef]
11. Roselli, L.; Clodoveo, M.L.; Corbo, F.; De Gennaro, B. Are health claims a useful tool to segment the category of extra-virgin olive oil? Threats and opportunities for the Italian olive oil supply chain. *Trends Food Sci. Technol.* **2017**, *68*, 176–181. [CrossRef]
12. De Santis, S.; Cariello, M.; Piccinin, E.; Sabbà, C.; Moschetta, A. Extra virgin olive oil: Lesson from nutrigenomics. *Nutrients* **2019**, *11*, 2085. [CrossRef] [PubMed]
13. Amirante, P.; Clodoveo, M.L.; Dugo, G.; Leone, A.; Tamborrino, A. Advance technology in virgin olive oil production from traditional and de-stoned pastes: Influence of the introduction of a heat exchanger on oil quality. *Food Chem.* **2006**, *98*, 797–805. [CrossRef]
14. Colglazier, W. Sustainable development agenda: 2030. *Science* **2015**, *349*, 1048–1050. [CrossRef] [PubMed]
15. Romero, D.; Molina, A. Collaborative networked organisations and customer communities: Value co-creation and co-innovation in the networking era. *Prod. Plan. Control* **2011**, *22*, 447–472. [CrossRef]
16. Clodoveo, M.L.; Hbaieb, R.H. Beyond the traditional virgin olive oil extraction systems: Searching innovative and sustainable plant engineering solutions. *Food Res. Int.* **2013**, *54*, 1926–1933. [CrossRef]
17. Clodoveo, M.L. New advances in the development of innovative virgin olive oil extraction plants: Looking back to see the future. *Food Res. Int.* **2013**, *54*, 726–729. [CrossRef]
18. Clodoveo, M.L. An overview of emerging techniques in virgin olive oil extraction process: Strategies in the development of innovative plants. *J. Agric. Eng.* **2013**. [CrossRef]
19. Clodoveo, M.L.; Dipalmo, T.; Crupi, P.; Durante, V.; Pesce, V.; Maiellaro, I.; Lovece, A.; Mercurio, A.; Laghezza, A.; Corbo, F.; et al. Comparison between different flavored olive oil production techniques: Healthy value and process efficiency. *Plant Foods Human Nutr.* **2016**, *71*, 81–87. [CrossRef]
20. De Luca, M.; Restuccia, D.; Clodoveo, M.L.; Puoci, F.; Ragno, G. Chemometric analysis for discrimination of extra virgin olive oils from whole and stoned olive pastes. *Food Chem.* **2016**, *202*, 432–437. [CrossRef]
21. Amirante, R.; Clodoveo, M.L. Developments in the design and construction of continuous full-scale ultrasonic devices for the EVOO industry. *Eur. J. Lipid Sci. Technol.* **2017**, *119*, 1600438. [CrossRef]
22. Cecchi, L.; Bellumori, M.; Corbo, F.; Milani, G.; Clodoveo, M.L.; Mulinacci, N. Implementation of the sono-heat-exchanger in the extra virgin olive oil extraction process: End-user validation and analytical evaluation. *Molecules* **2019**, *24*, 2379. [CrossRef]
23. Rollwagen, I.; Cincera, M.; Costantino, L.; Franke, J.; Kalligatsi, K.; Mińska-Struzik, E.; Russo, E. *Fast Track to Innovation Pilot (2015–2016) Final Evaluation*; Publication Office of European Union: Brussels, Belgium, 2019. [CrossRef]
24. Malerba, F.; McKelvey, M. Knowledge-intensive innovative entrepreneurship integrating Schumpeter, evolutionary economics, and innovation systems. *Small Bus. Econ.* **2020**, *54*, 503–522. [CrossRef]
25. Jiménez, A.; Beltrán, G.; Uceda, M. High-power ultrasound in olive paste pretreatment. Effect on process yield and virgin olive oil characteristics. *Ultrason. Sonochem.* **2007**, *14*, 725–731. [CrossRef] [PubMed]
26. Bejaoui, M.A.; Beltran, G.; Aguilera, M.P.; Jimenez, A. Continuous conditioning of olive paste by high power ultrasounds: Response surface methodology to predict temperature and its effect on oil yield and virgin olive oil characteristics. *LWT-Food Sci. Technol.* **2016**, *69*, 175–184. [CrossRef]
27. Clodoveo, M.L.; Moramarco, V.; Paduano, A.; Sacchi, R.; Di Palmo, T.; Crupi, P.; Corbo, F.; Pesce, V.; Distaso, E.; Tamburrano, P.; et al. Engineering design and prototype development of a full scale ultrasound system for virgin olive oil by means of numerical and experimental analysis. *Ultrason. Sonochem.* **2017**, *37*, 169–181. [CrossRef] [PubMed]
28. Amirante, R.; Distaso, E.; Tamburrano, P.; Paduano, A.; Pettinicchio, D.; Clodoveo, M.L. Acoustic cavitation by means ultrasounds in the extra virgin olive oil extraction process. *Energy Procedia* **2017**, *126*, 82–90. [CrossRef]
29. Amirante, R.; Distaso, E.; Tamburrano, P.; Corbo, F.F.R.; Calò, G.; Clodoveo, M.L. Fluid dynamic-based Engineering design of a Full-Scale Device for the improvement of Extra Virgin Olive Oil Yield and Quality by means of Combined Acoustic Cavitation and Thermal Conditioning. *E3S Web Conf.* **2020**, *197*, 08010. [CrossRef]
30. Plasquy, E.; García Martos, J.M.; Florido Fernández, M.D.C.; Sola-Guirado, R.R.; García Martín, J.F. Adjustment of Olive Fruit Temperature before Grinding for Olive Oil Extraction. Experimental Study and Pilot Plant Trials. *Processes* **2021**, *9*, 586. [CrossRef]
31. Clodoveo, M.L.; Hbaieb, R.H.; Kotti, F.; Mugnozza, G.S.; Gargouri, M. Mechanical strategies to increase nutritional and sensory quality of virgin olive oil by modulating the endogenous enzyme activities. *Compr. Rev. Food Sci. Food Saf.* **2014**, *13*, 135–154. [CrossRef] [PubMed]
32. De Santis, S.; Liso, M.; Verna, G.; Curci, F.; Milani, G.; Faienza, M.F.; Franchini, C.; Moschetta, A.; Chieppa, M.; Clodoveo, M.L.; et al. Extra Virgin Olive Oil Extracts Modulate the Inflammatory Ability of Murine Dendritic Cells Based on Their Polyphenols Pattern: Correlation between Chemical Composition and Biological Function. *Antioxidants* **2021**, *10*, 1016. [CrossRef] [PubMed]

processes
Article

Characterization of Flavoured Olive Oils of 'Madural' Variety

María Helena Chéu-Guedes [1], M. Dolores La Rubia [2,3,*], Sebastián Sánchez [2,3], Natividad Ramos [2,3] and Rafael Pacheco [2,3,*]

1 Education and Community Intervention Research Unit—RECI, Campus of Piaget Institute of Viseu, Estrada do Alto do Gaio, 3515-776 Lordosa, Portugal
2 Department of Chemical, Environmental and Materials Engineering, University of Jaén, 23071 Jaén, Spain
3 University Institute of Research in Olive Grove ans Olive Oils, University of Jaén, GEOLIT Science and Technology Park, 23620 Mengíbar, Spain
* Correspondence: mdrubia@ujaen.es (M.D.L.R.); rpacheco@ujaen.es (R.P.)

Abstract: Flavoured oils arouse great interest among consumers in many countries due to their variety of flavours and versatility, especially in the culinary field. The aromatization of oils seeks to improve their sensory and nutritional properties, and extend their useful life due to the added substances can be beneficial as antioxidant and antimicrobial agent. In this research, olive oils of the 'Madural' variety from Trasos Montes region of Portugal have been obtained and flavoured with different aromatics herbs and condiments (flower of salt and bay leaf, garlic, rosemary and dehydrated lemon peel). The objective is to study the influence of the aromatic herbs and condiments on the physicochemical parameters of the oils: quality, purity, oxidative stability and microbiological analysis. It can be noted that the flavourings do not significantly alter the quality of the monovarietal oil, although, for some parameters, the excessive contact times can affect the category of the oil. On the other hand, the high content of antioxidants provided by flavouring agents can favour its stability and prolong its expiration. In this sense, the flavouring agent that contributes to stop the oxidation of the oil over time is salt + bay leaves, as higher oxidative stability values were detected than those obtained in monovarietal oil. However, oils flavoured with rosemary or lemon show a decrease over time for this parameter, which could indicate that this flavouring accelerates oxidation. In the case of oxidative stability referred to those flavoured with garlic, they present similar values to the monovarietal. The effect exerted by flavourings on the different parameters of the oils is complex, since it is influenced by the method followed and the operating variables established for flavouring.

Keywords: olive oils; flavoured oils; fleur de sel; bay leaf; garlic; rosemary; lemon; oxidation stability

Citation: Chéu-Guedes, M.H.; La Rubia, M.D.; Sánchez, S.; Ramos, N.; Pacheco, R. Characterization of Flavoured Olive Oils of 'Madural' Variety. *Processes* **2023**, *11*, 205. https://doi.org/10.3390/pr11010205

Academic Editor: Chi-Fai Chau

Received: 13 December 2022
Revised: 31 December 2022
Accepted: 2 January 2023
Published: 9 January 2023

1. Introduction

Nowadays, there is a great demand for innovative and specialized products, of high nutritional quality, with different sensory characteristics, such as flavoured oils.

The origin of these oils seems to come from ancient processing and conservation practices, in which the oils acquired the flavour of the aromatics herbs and condiments used, and they were later used in the preparation of dishes and salads [1]. Their versatility, ease of use, and wide variety of flavours have made flavoured oils staples for regular and unusual consumers in many countries around the world [2].

The aromatization of olive oils seeks to improve their organoleptic and nutritional properties and increase their useful life, since added flavouring agents can provide benefits through their antioxidant and antimicrobial characteristics [3].

In addition, the aromatization of olive oils deals a greater diversity of products in which virgin olive oil is the base product, allowing more choice for the consumer and improving the organoleptic characteristics of the oils [4].

Olive oils have been flavoured with their own phenolics or with phenolic compounds from other vegetable sources. Thus, different matrices have been used, namely spices and aromatic herbs (basil, pepper, garlic, and laurel), fruits (apple, banana, lemon, and orange), mushrooms, and nuts (almonds, and hazelnuts), being evaluated the impact at chemical and sensory levels [5].

The influence of flavourings on the physical-chemical and organoleptic characteristics of the oils are of vital importance to establish the amounts of flavouring to be used and the flavouring times. On the other hand, one of the most important factors in food quality control is the determination of the level of oxidation in food or food products.

The olive oil deterioration is evaluated by the peroxide index, among other quality parameters. The oxidation of fats is one of the main causes of food spoilage. It causes great concerns, mainly of an economic nature in the food industry, because it stimulates the appearance of unpleasant flavours and doors, generally identified as rancidity in foods that contain fat. These unpleasant flavours influence the organoleptic quality of the negatively [6].

The oxidation of fats leads to the formation of hydroperoxides, which are very unstable primary products that, by decomposition, form secondary products. These can include hundreds of individual compounds that negatively affect the flavour of foods. In many cases, these compounds are associated with oxidative rancidity and/or off-flavour formation [7].

Virgin olive oil is resistant to oxidation due to its low content of polyunsaturated fatty acids and the content of natural antioxidants, such as: alpha-tocopherol and phenolic compounds, hydroxytyrosol, tyrosol, and caffeic acid, among others [7].

On the other hand, oxidative stability is not considered a standard quality parameter, and it can be used as an indicator of the useful life of the oil. Normally it is evaluated by the induction time, that is, the time that elapses until the critical point of oxidation is reached, causing a sensory degradation of the oil, due to a sudden acceleration of the oxidative process. Oxidative stability is generally determined by the Rancimat method and reveals the resistance of the product to the initiation of oxidation, characterized by reactions involving free radicals [8,9].

The oxidative stability of olive oils depends on several factors, including the variety and degree of maturity of the olives used in its production. The latter plays a fundamental role in the chemical composition of the olive and, therefore, in its resistance to oxidation. The resistance to oxidation is mainly attributed to two factors: the composition of the fatty acids, which in the case of olive oils is characterized by a high value of the monounsaturated fatty acids/polyunsaturated fatty acids ratio and the presence of minor compounds with high antioxidant activity, mainly tocopherols and polyphenols, but also chlorophylls and carotenoids [10].

Thus, oleic acid (monounsaturated) confers greater resistance to oxidation. However, high percentages of polyunsaturated fatty acids (linoleic and linolenic) can contribute to shortening the useful life of the oil, due to the formation of compounds with conjugated double bonds, from which oxidation volatile products are formed with unpleasant aroma characteristic of rancidity. Being the relative oxidation values, of linolenic, linoleic and oleic acids, of approximately 20:10:1, respectively [10].

Therefore, the incorporation of flavourings into olive oil can be an advantageous procedure since it helps consumers accept olive oils by enhancing known and desirable flavours [11]. On the other hand, the flavouring agents used are usually rich in antioxidant and phenolic compounds, so they can increase shelf life and improve the nutritional, healthy and sensory composition of the oils [12–15].

However, it should be noted that, although the demand for these flavoured oils is growing regularly, there are currently no specific regulations for them that allow their correct labelling. In 2014, the International Olive Oil Council (IOOC), in its 22nd extraordinary meeting, asked all its members to "adopt the necessary measures to prevent flavoured oils from incorporating the terms "extra virgin olive oil", "virgin olive oil" and "olive oil", on their labels, given that these preparations do not comply with the IOC standard" [16], since this standard only allows the use of the aforementioned terms for products whose definition excludes the addition of any additive [17,18].

However, considering the growing interest in these products, this work has studied the variation of the physicochemical properties of olive oils flavoured with different substances. In addition, the oxidation and oxidative stability evolution of the flavoured oils, with respect to the monovarietal olive oils has been evaluated.

2. Materials and Methods

2.1. Preparation of the Samples

Olive oils of the 'Madural' (MO) variety were obtained in two successive seasons (2015/2016 and 2016/2017). In addition, natural products for flavouring were added to these oils at the time they were obtained: fleur de sel and dehydrated bay leaves (MOSB), dehydrated rosemary (MOR), dehydrated garlic (MOG) and dehydrated lemon peel (MOL). After 11 months the different flavoured olive oils were obtained. Next, the following parameters were analysed: moisture, acid index, peroxide index, ultraviolet parameter (K_{232}, K_{268} and ΔK), organoleptic analysis, fatty acid composition, waxes, sterols, erythrodiol+uvaol, phenolic compounds, tocopherols, pH, oxidative stability and microbiological analysis.

2.2. Olives Characterization

The olives of the 'Madural' variety from the Mirandela area (Portugal) were characterized by moisture (%) [19], fat content (%) [20], biometric parameters of the fruit and the pulp/stone ratio. The evaluation of these parameters allowed to determine the harvest date of the olives.

The determination of moisture and volatile matter was carried out by drying a portion of the sample (1 g) in an oven at 103 °C until constant weight [19]. The determination of the total fat content was carried out according to the Soxhlet fat extraction method [20]. The dried sample, used in the determination of moisture, is introduced into the Soxhlet extractor and the extraction of the fat with n-hexane takes place for approximately 24 h. Subsequently, in the oven at 105 ± 1 °C, the traces of solvent are eliminated. With the amount of oil recovered, the richness of dry fat is determined, and from this the richness of wet fat. The arithmetic mean of two determinations will be taken as the result.

2.3. Quality Parameters

The acidity index is determined according to the methodology described in the ISO 660:2020 standard, those being oils and fats of animal and vegetable origin, along with determination of acidity or acidity index [21]. The sample (20 g) was dissolved in a mixed solvent of ethanol and diethyl ether in a 1:1 ratio and the free fatty acids present were titrated with potassium hydroxide solution (0.1 M). The titration was carried out in the presence of phenolphthalein solution as indicator. The acidity is expressed as percentage of oleic acid. All the measurements were made in triplicate.

Peroxide value (PV) was expressed as milliequivalent of active oxygen per kilogram of oil and determined according to the methodology described in ISO 3960:2007 [22]. A sample of 1.2–2.0 g of oil was dissolved in a mixture of chloroform and acetic acid (2:3 *v/v*). Next, 1 mL of KI saturated solution was added and 75 mL of deionized water was also added after stirring and 5 min in darkness. Finally, the sample is titrated with sodium thiosulphate and starch solution as indicator. All the measurements were made in triplicate.

K_{232} and K_{268} extinction coefficients were calculated according to analytical methods described in the CEE/2568/91, CEE/1429/92 regulations and subsequent modifications of the European Commission [23,24] by spectrophotometric examination. A sample of 100 mg of oil were dissolved in 25 mL of iso-octane and the extinction of the solution is then determined at the specified wavelengths with reference to pure solvent. Specific extinctions are calculated from the spectrophotometer readings using a quartz cell with a 1 cm optical path. All the measurements were made in triplicate.

The sensorial analysis was made according to COI/T.20/Doc.N° 5 of 2007 [25]. The testing panel was constituted of 8 testers (6 women and 2 men) of age between 40 and 53 years old. They carried out the sensory characterization and linked the flavour stimuli of the oils with a numeric scale according to the standards [25]. In this analysis the positive perceptions identified were fruity, bitter and spicy and the negative perceptions were mouldy/damp earth, musty, burnt olive and vinegary.

2.4. Purity Parameters

The gas chromatography (GC) technique was used to determine the qualitative and quantitative fatty acid methyl-esters composition [24]. The sample were prepared by shaking a solution of oil in heptane (0.1 g in 2 mL) with 0.2 mL of 2 N methanolic potassium hydroxide stirred during 30 s. The methyl-esters were analysed from the supernatant solution by a gas chromatographer Shimadzu GC-2010 Plus ATF with double injection port of Split/Splitless capillary columns with digital flow control, detectors (FID, ECD-2010Plus Detector and TCD) and AOC-5000 automatic injector (PAL LHS2-SHIM) for injection of liquids (up to 10 μL) or head-space (up to 2.5 mL). The capillary column was SP-2380 de 60 m × 0.2 mm × 0.2 μm, (Supelco, Pensilvania, EEUU). The oven temperature was set at 185 °C, the injector and detector temperature were set at 220 °C. The chromatograms were read in ascending order of number of carbons and insaturations using "Shimadzu GCsolution" for the treatment of the results.

The wax content was determined according to the methodology described in Regulation No. 2568/91 [24]. The contents of the different waxes and the sum of these contents are expressed in mg/kg of oil. All the measurements were made in triplicate.

2.5. Minor Components

The tocopherols were evaluated by HPLC following the method described by Cunha [26] on a Shimadzu HPLC mod. Prominesce series 20. A Sigma-Aldrich (St. Louis, MO, USA) (250 mg/L) α-tocopherol in acetonitrile was used as internal pattern. A sample of 0.2 g of oil was dissolved in 10 mL of n-hexane and 20 μL of sample was analysed. As mobile phase hexane was used: isopropanol (97:3 *v/v*) in isocratic conditions and 1 mL/min of flow rate. The chromatograms were recorded at 275 nm. The results are expressed as mg of α-tocopherol per oil kg. All the measurements were made in triplicate.

The total phenolic compounds were determined following the method described by Vázquez-Roncero [27], using the Folin-Ciocalteau reagent. The absorption of the solution measured at 725 nm. The results were given as mg/kg of caffeic acid. All the measurements were made in triplicate.

2.6. Oxidative Stability

For the determination of the stability of olive oils a Metrohm equipment, mod. Rancimat 743 was used. A sample of 3 g of the oil sample were weighed and then placed on the heating block. The air flow is turned on and passes through the tubes containing the oil samples. The oxidation curves are recorded by ionic conductivity measurement, as a result of stability through time expressed in hours, which corresponds to the induction period. It works at 120 °C, using a volumetric air flow rate of 10 L/h, and all the samples have been performed in duplicate.

2.7. Microbiological Study

2.7.1. Content of Microorganisms at 30 °C

The methodology used in the analytical determination is based on the NP 4405:2002 standard [28].

2.7.2. Content of Coliform Bacteria at 30 °C

These bacteria are a good indicator of the state of hygiene. The determination of coliform bacteria was carried out according to NP 3788:1990 [29].

2.7.3. Coagulase+ Staphylococci Count

The procedure used for the count of coagulase-positive Staphylococci is described in ISO 6888-2:1999 [30].

3. Results

3.1. Olive Determinations

The samples of olives of the 'Madural' variety from the Tras-os-Montes region of Portugal, were collected during the 2015/2016 and 2016/2017 seasons and were characterised in terms of moisture (%), fat content (%), fat content in dry matter, biometric parameters of the fruit and the pulp/stone ratio. Figure 1 shows the results.

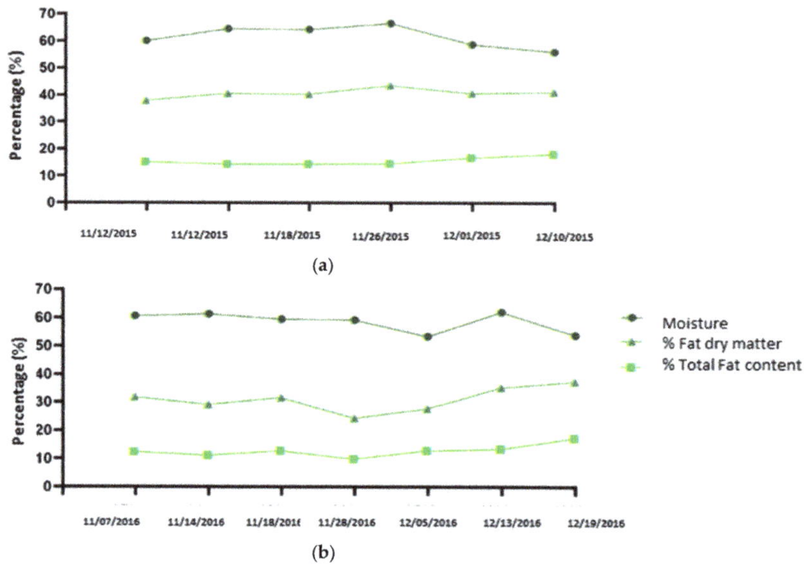

Figure 1. Evolution of fat content (%), moisture (%) and fat content in dry matter (%) in olives of the 'Madural' variety (**a**) 2015/2016 season (**b**) 2016/2017 season.

Figure 1a shows that during the 2015/2016 season, fat levels increase (14.58%) and total content moisture content decreases (66.59%) from 26 November 2015. Regarding the fat content in dry matter, it is observed that lipid synthesis is in a growth phase, with a greater increase from 18 November 2015 (40.39%) and then stabilizes. Next, the selected harvest date was on 7 December 2015.

During 2016/2017 season, the fat content increases from 28 November 2016 (9.95%), reaching a value of 17.28% on 19 December 2016. The total moisture content of the olives drops until 12/5/2016 (53.34%), with a peak on 13 December 2016 (61.95%) and then drops sharply, reaching a value of 53.83% on 19 December 2016. Regarding the fat content in the dry matter, it is observed that lipid synthesis is in the growth phase, with an increase from28 November 2016 (24.41%), then increases at the beginning of December (27.71%) reaching a value of 37.42% on 19 December 2016. Combining the evolution of the different determinations, especially the total fat and the fat content in dry matter, the olives were harvested on 16 December 2016.

Table 1 shows the morphological parameters for the characterization of the biometric parameters of the olives (weight of the pulp and endocarp) and characterization of the pulp/stone ratio.

Table 1. Mean values and standard deviation of the biometric parameters evaluated in the fruit and the pulp/stone ratio.

Season	Average Weight (g)	Average Endocarp Weight (g)	Average Pulp Weight (g)	Pulp/Stone Ratio
2015/2016	3.12 ± 0.20	0.59 ± 0.10	2.52 ± 0.16	4.23 ± 0.21
2016/2017	3.17 ± 0.26	0.61 ± 0.10	2.54 ± 0.22	4.15 ± 0.18

In both seasons, the average weight of the olives and the pulp/stone ratio were similar. The weight of the olives was considered medium (2–4 g) and the endocarp weight high (0.45–0.7 g).

Barranco et al. [31], consider that the main components of ripe olives are water and oil, and it is in the pulp where more than 95% of the total oil is found. This factor highlights the importance of the pulp/stone ratio in the total fat yield of the olive. Gouveia [32] explains that during the development and maturation of the olives there is a progressive decrease in the humidity of the pulp and, at the same time, an increase in the oil content, weight and volume of the fruit. García et al. [33] considers that the oil content of the olives increases as they mature, reaching its maximum value when the green fruits of the tree disappear. From this moment, the oil content remains practically constant.

3.2. *Quality Parameters*

The monovarietal and flavoured oils were analysed at the beginning and end of each year (2016 and 2017) to evaluate the evolution of the analytical parameters and the influence of flavouring agents on them. IOOC Regulations have been used as a reference in absence of applicable regulations for flavoured oils [17].

3.2.1. Humidity

The moisture values of the monovarietal and flavoured oils analysed in all the seasons are below 0.10. According to the IOOC standard [17], moisture and volatile matter should be less than 0.2 (% m/m).

3.2.2. Acidity

Table 2 reports the acidity values of the samples. The acidity of the oils increases over the time in all the seasons. However, the oils obtained in 2016 have acidity values higher than those obtained in 2017 and below 0.8%. On the other hand, the flavoured oils obtain similar acidity values to those obtained in monovarietal oil, though the increases over time are greater. Of the flavouring agents used, garlic and lemon produce minor increases in acidity in the two seasons. Considering the values obtained, monovarietal olive oil and flavoured oils correspond to the "Extra Virgin" Olive Oil category.

Flavourings usually have a negative effect on the acidity of the oils, as various authors reported, specifically with garlic [5,34], laurel [5], rosemary [35,36] or lemon [35].

3.2.3. Peroxide Value

The peroxide value (PV) evaluates the formation of primary oxidation products and the deterioration of natural antioxidants. The maximum limit of the peroxide value of an oil for human consumption is 20 mEq O_2/kg [17]. The PV of olive oil can be associated with the olive varieties and the edaphoclimatic conditions of the regions [7].

From Table 2, it can be seen that the monovarietal oils have low PV which increases throughout the year, reaching, in the 2015/2016 season, its highest value in the last month of the study (10.7 mEq O_2/kg). Similar behaviour it is observed for the 2016/2017 season, but with a higher value (19 mEq O_2/kg), which may be due to the fact that the starting value was also higher. In both cases, the PV is below 20 mEq O_2/kg, the maximum value allowed for oil of the "Extra Virgin" category [17].

Table 2. Quality parameters of all olive oils.

Samples	February 2016	March 2016	April 2016	December 2016	February 2017	March 2017	April 2017	December 2017
				Acidity (% oleic acid)				
MO	0.24 ± 0.00	0.25 ± 0.01	0.25 ± 0.00	0.31 ± 0.00	0.10 ± 0.00	0.11 ± 0.00	0.13 ± 0.00	0.16 ± 0.00
MOSB	0.24 ± 0.00	0.24 ± 0.00	0.26 ± 0.00	0.32 ± 0.00	0.10 ± 0.00	0.11 ± 0.00	0.12 ± 0.00	0.14 ± 0.00
MOR	0.24 ± 0.00	0.24 ± 0.00	0.25 ± 0.00	0.35 ± 0.01	0.10 ± 0.00	0.13 ± 0.00	0.13 ± 0.00	0.16 ± 0.00
MOG	0.24 ± 0.00	0.24 ± 0.00	0.23 ± 0.00	0.28 ± 0.00	0.10 ± 0.00	0.15 ± 0.00	0.15 ± 0.00	0.17 ± 0.00
MOL	0.24 ± 0.00	0.24 ± 0.00	0.24 ± 0.00	0.28 ± 0.00	0.10 ± 0.00	0.14 ± 0.00	0.14 ± 0.00	0.17 ± 0.00
				Peroxide value (mEq O_2/kg)				
MO	4.00 ± 0.00	6.40 ± 0.00	7.10 ± 0.01	10.70 ± 0.00	6.00 ± 0.00	6.20 ± 0.00	11.00 ± 0.00	19.00 ± 0.00
MOSB	4.00 ± 0.01	5.50 ± 0.00	5.90 ± 0.00	9.80 ± 0.00	6.00 ± 0.00	6.50 ± 0.01	13.00 ± 0.00	20.00 ± 0.00
MOR	4.00 ± 0.01	6.60 ± 0.01	6.80 ± 0.00	9.50 ± 0.00	6.00 ± 0.01	5.80 ± 0.00	9.00 ± 0.01	35.00 ± 0.01
MOG	4.00 ± 0.00	6.60 ± 0.01	6.90 ± 0.00	15.20 ± 0.01	6.00 ± 0.01	7.50 ± 0.01	9.00 ± 0.00	22.00 ± 0.00
MOL	4.00 ± 0.00	7.90 ± 0.00	8.40 ± 0.00	17.30 ± 0.00	6.00 ± 0.00	7.70 ± 0.00	10.00 ± 0.00	25.00 ± 0.00
				K_{232}				
MO	1.49 ± 0.01	1.51 ± 0.03	1.62 ± 0.01	1.78 ± 0.02	1.70 ± 0.01	1.81 ± 0.02	1.87 ± 0.03	2.00 ± 0.12
MOSB	1.49 ± 0.01	1.60 ± 0.02	1.64 ± 0.04	2.02 ± 0.05	1.70 ± 0.01	1.80 ± 0.02	1.85 ± 0.01	2.26 ± 0.03
MOR	1.49 ± 0.03	1.59 ± 0.01	1.61 ± 0.03	1.91 ± 0.02	1.70 ± 0.05	1.79 ± 0.03	1.81 ± 0.01	2.20 ± 0.01
MOG	1.49 ± 0.02	1.61 ± 0.02	1.62 ± 0.01	1.96 ± 0.01	1.70 ± 0.02	1.82 ± 0.01	1.84 ± 0.05	2.27 ± 0.06
MOL	1.49 ± 0.03	1.60 ± 0.02	1.69 ± 0.04	2.03 ± 0.03	1.70 ± 0.01	1.81 ± 0.04	1.88 ± 0.06	2.31 ± 0.02
				K_{270}				
MO	0.13 ± 0.00	0.13 ± 0.00	0.13 ± 0.00	0.19 ± 0.00	0.13 ± 0.00	0.15 ± 0.00	0.15 ± 0.00	0.20 ± 0.00
MOSB	0.13 ± 0.00	0.13 ± 0.00	0.13 ± 0.00	0.20 ± 0.00	0.13 ± 0.00	0.16 ± 0.00	0.16 ± 0.00	0.24 ± 0.00
MOR	0.13 ± 0.00	0.13 ± 0.00	0.13 ± 0.00	0.20 ± 0.00	0.13 ± 0.00	0.14 ± 0.00	0.14 ± 0.00	0.21 ± 0.00
MOG	0.13 ± 0.00	0.13 ± 0.00	0.13 ± 0.00	0.18 ± 0.00	0.13 ± 0.00	0.15 ± 0.00	0.15 ± 0.00	0.19 ± 0.00
MOL	0.13 ± 0.00	0.13 ± 0.00	0.13 ± 0.00	0.18 ± 0.00	0.13 ± 0.00	0.15 ± 0.00	0.15 ± 0.00	0.19 ± 0.00
				ΔK				
MO	0.00 ± 0.00	0.00 ± 0.00	0.00 ± 0.00	0.01 ± 0.00	0.00 ± 0.00	0.00 ± 0.00	0.00 ± 0.00	0.01 ± 0.00
MOSB	0.00 ± 0.00	0.01 ± 0.00	0.01 ± 0.00	0.02 ± 0.00	0.00 ± 0.00	0.01 ± 0.00	0.01 ± 0.00	0.01 ± 0.00
MOR	0.00 ± 0.00	0.01 ± 0.00	0.00 ± 0.00	0.01 ± 0.00	0.00 ± 0.00	0.01 ± 0.00	0.01 ± 0.00	0.01 ± 0.00
MOG	0.00 ± 0.00	0.01 ± 0.00	0.00 ± 0.00	0.01 ± 0.00	0.00 ± 0.00	0.00 ± 0.00	0.00 ± 0.00	0.00 ± 0.00
MOL	0.00 ± 0.00	0.01 ± 0.00	0.01 ± 0.00	0.02 ± 0.00	0.00 ± 0.00	0.01 ± 0.00	0.01 ± 0.00	0.02 ± 0.00

The results expressed as mean value ± typical deviation.

Regarding flavoured oils, they show the same trend over time as monovarietal do, and in December of the 2015/2016 season the flavoured oils with rosemary (MOR) and with salt + bay leaves (MOSB) obtained the lowest increases, reaching 9.5 and 9.8 mEq O_2/kg, respectively. Despite the high values of the rest of the flavourings, they all correspond to the Extra Virgin category according to IOOC [17]. However, at the end of the following season, significant increases were observed in all the oils, being the oil flavoured MOSB the one that obtained the lowest value, as in the previous season, maintaining the category of "Extra Virgin". On the other hand, those flavoured with rosemary reached the highest value in this season (35.0 mEq O_2/kg) and, such as those flavoured with garlic and lemon, which also exceeded 20 mEq O_2/kg, all of them can be included. in the "Virgin" category.

The influence of flavouring agents on PV is complex and could be related to chemical effects. The influence of rosemary on PV may be due to its chemical composition, rich in polyphenolic compounds and derivates of caffeic acid (rosmarinic acid). Laurel leaves have antioxidant properties and, fleur de sel is a natural preservative and antibacterial agent, and when it comes to lemon-flavoured oil, it always obtains the highest PV values, regardless of the season under study, with the exception of the oil flavoured with rosemary from December of 2017.

On the other hand, the PV obtained can be related not only to the flavouring agent used, but also with the time of aromatization.

In the literature it can be seen that the influence of flavouring agent on PV is not clear. While some authors reported similar effects for garlic [1] and rosemary [34,35] decreasing PV and increasing the oxidative stability, other authors reported the opposite [36]. However, other authors consider that PV increase with the flavouring agents [1] and the oxidative stability decreases [36]. Regarding dry laurel, some authors [5], state that it decreases PV and increases oxidative stability, and as for dry lemon rind, Ayadi et al. [37] considers that it decreases the PV and increases the oxidative stability.

3.2.4. K_{232}, K_{270} and ΔK

During the two seasons, K_{232} values (Table 2), both in monovarietal and flavoured oils, increase over time, but remain below 2.5%, which is the maximum value allowed for the olive oil in the "Extra Virgin" category. Regarding the flavoured oils, it can be seen that, in the two seasons, all the oils correspond to the "Extra Virgin" category. The olive oil flavoured with rosemary obtained the lowest value in the two seasons, while the oil flavoured with lemon reached the highest value.

Similarly, it can be seen that the values of K_{270}, in monovarietal oils, increase over time, but remain below 0.22% (maximum value allowed for olive oil in the "Extra Virgin" category).

For the flavoured oils, in the studied seasons, the lowest values correspond to the oils with garlic and lemon, while the highest are those with salt, laurel leaf and rosemary. Based on the obtained results, it seems that all the oils correspond to the "Extra Virgin" category, with the exception of flavoured oil with salt and laurel leaf, from the 2016/2017 season, which are included in the "Virgin" category.

There is agreement between the results obtained with the bibliographic data consulted, according to which dried rosemary causes increases in the extinction coefficients [38,39] although the amounts of flavouring and contact have been different. Finally, regarding ΔK, in all cases, the values are equal or less than 0.01, so they correspond to "Extra Virgin" category. In the case of flavoured oils, in all cases they are considered "extra virgin" category except the oils with salt and bay leaf from December 2016 and the oils with lemon from the two seasons.

From all of the above, we conclude that monovarietal oils have good protection due to their natural antioxidants, with no significant alterations in their quality. Regarding the flavoured oils, the agents that cause the greatest alteration are salt, bay leaf and lemon, which cause losses of their initial category.

3.2.5. Sensory Evaluation

Table 3 shows the results of sensory evaluation of the monovarietal olive oil.

Table 3. The organoleptic characteristics of the monovarietal olive oil.

Samples	February 2016	July 2016	December 2016	February 2017	July 2017	December 2017
MO	7.5	7.4	6.7	7.0	6.7	6.6

The oils have a score higher than 6.5, which correspond to "Extra Virgin" category. Figure 2 shows the sensory profile of the monovarietal oils at different season.

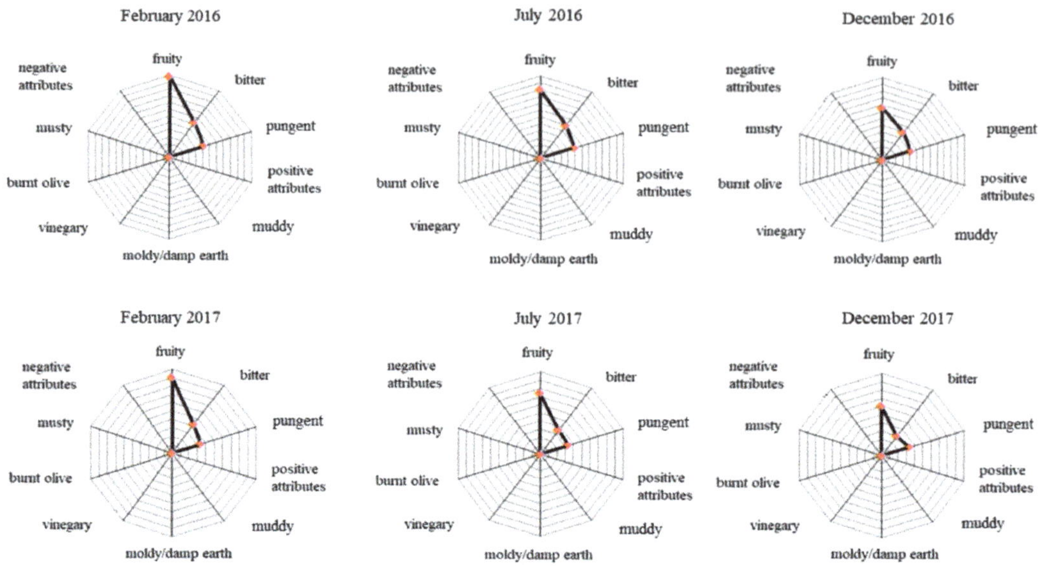

Figure 2. The sensory profiles of the olive oils at different seasons.

The oil from the 2015/2016 season has, in February, fruitiness of 5.9, bitterness of 3.1 and pungent of 2.6. In July, there was a slight decrease in fruit intensity (5.0) and, practically, the remaining positive attributes constant at 3.0 and 2.5, respectively. In December, it shows show a small decrease in the intensity of the positive attributes, fruity (3.8), bitter (2.5) and pungent (2.1). Considering that the median fruitiness is greater than zero (Mf > 0) and the median of defects is equal to zero (Md = 0), this oil corresponds to the "Extra Virgin" category.

The oils from the 2016/2017 season have, in February, fruit intensity score of 4.5, bitterness of 2.1 and spiciness of 1.7. In July there was a slight decrease in the positive attributes of fruitiness (3.7) and bitterness (1.8), while the intensity of pungent (1.7) was maintained. In December there was a small decrease in the intensity of the positive attributes, fruity (3.0) and bitter (1.5) and pungent (1.7) remained constant. The median fruitiness is greater than zero (Mf > 0) and the median defects is equal to zero (Md = 0), therefore, this oil falls into the "Extra Virgin" category.

Regarding the organoleptic analysis of the flavoured oils, Table 4 presents the results of the organoleptic characteristics. In all cases, the tasters identified the flavouring agents both in the taste and in the smell of the oils, being this perception lower in the case of rosemary (taste) and lemon (taste and smell). This perception is constant over time in the two seasons. It can be seen that in all cases the bitter and pungent characteristics decrease over time, the highest decrease occurs in the flavoured oil with lemon. Probably the intensity and character of the flavourings condition the perceptions of the superior bitter and pungent attributes. Similarly, the intensity of the flavour decreases over time, with the exception of the garlic oil which is maintained. It should be noted that seems to be a coincidence with Rodrigues et al. [35], for the garlic and rosemary flavourings, when they detect an increase in fruitiness and bitterness in the oils.

Table 4. Sensorial profile of the flavoured oils.

Samples	Description	2015–2016		2016–2017	
		July	December	July	December
MOSB	Aroma (garlic)	VN	VN	VN	VN
	Taste (garlic)	VN	VN	VN	VN
	Bitter and pungent	3-5	3-4	3-3	2-3
	Flavour intensity	5	3	5	4
MOR	Aroma (rosemary)	VN	VN	VN	VN
	Taste (rosemary)	N	N	N	N
	Bitter and pungent	2-3	2-2	2-3	2-1
	Flavour intensity	4	4	4	4
MOG	Aroma (garlic)	VN	VN	VN	VN
	Taste (garlic)	VN	VN	VN	VN
	Bitter and pungent	3-4	2-3	3-3	2-3
	Flavour intensity	4	4	5	4
MOL	Aroma (lemon)	N	N	N	N
	Taste (lemon)	N	N	N	N
	Bitter and pungent	3-2	2-2	2-2	2-2
	Flavour intensity	4	3	4	2

Notation: LN: Slightly noticeable; N: noticeable; VN: Very noticeable.

3.3. Purity Parameters

3.3.1. Fatty Acid Profile

Table 5 shows the most representative fatty acid composition values, according to the season.

Table 5. Most representative fatty acids (%) of all olive oils.

Samples		February 2016	December 2016	February 2017	December 2017
MO	C16:0	11.70± 0.12	11.3 ± 0.10	10.80 ± 0.09	10.0 ± 0.05
	C16:1	0.50± 0.01	0.50 ± 0.01	0.40 ± 0.02	0.40 ± 0.01
	C18:0	2.20 ± 0.03	2.20 ± 0.02	2.70 ± 0.01	2.50 ± 0.03
	C18:1	71.20 ± 0.02	70.60 ± 0.02	71.10 ± 0.03	73.40 ± 0.08
	C18:2	12.30 ± 0.03	11.90 ± 0.04	12.10 ± 0.02	11.70 ± 0.03
	C18:3	1.10 ± 0.02	1.00 ± 0.01	1.00 ± 0.01	1.00 ± 0.01
MOSB	C16:0	11.70 ± 0.03	11.20 ± 0.02	10.8 ± 0.01	10.0 ± 0.01
	C16:1	0.50 ± 0.01	0.50 ± 0.06	0.40 ± 0.03	0.50 ± 0.01
	C18:0	2.20 ± 0.02	2.20 ± 0.01	2.70 ± 0.02	2.50 ± 0.01
	C18:1	71.20 ± 0.12	70.70 ± 0.09	71.10 ± 0.01	73.60 ± 0.03
	C18:2	12.30 ± 0.03	11.90 ± 0.02	12.10 ± 0.02	11.80 ± 0.02
	C18:3	1.10 ± 0.01	1.00 ± 0.01	1.00 ± 0.02	1.10 ± 0.03
MOR	C16:0	11.70 ± 0.01	11.50 ± 0.02	10.80 ± 0.10	10.10 ± 0.03
	C16:1	0.50 ± 0.01	0.50 ± 0.01	0.40 ± 0.02	0.50 ± 0.01
	C18:0	2.20 ± 0.03	2.20 ± 0.02	2.70 ± 0.01	2.50 ± 0.01
	C18:1	71.20 ± 0.08	71.30 ± 0.03	71.10 ± 0.02	73.80 ± 0.01
	C18:2	12.30 ± 0.03	11.7 ± 0.01	12.10 ± 0.02	11.60 ± 0.03
	C18:3	1.10 ± 0.01	1.00 ± 0.02	1.00 ± 0.03	1.00 ± 0.02
MOG	C16:0	11.70 ± 0.01	11.30 ± 0.03	10.80 ± 0.05	10.10 ± 0.01
	C16:1	0.50 ± 0.02	0.50 ± 0.05	0.40 ± 0.02	0.50 ± 0.01
	C18:0	2.20 ± 0.02	2.20 ± 0.04	2.70 ± 0.07	2.50 ± 0.01
	C18:1	71.20 ± 0.04	71.10 ± 0.02	71.10 ± 0.01	73.60 ± 0.02
	C18:2	12.30 ± 0.01	11.90 ± 0.02	12.10 ± 0.04	11.70 ± 0.03
	C18:3	1.10 ± 0.02	1.00 ± 0.03	1.00 ± 0.01	1.10 ± 0.02
MOL	C16:0	11.70 ± 0.02	11.40 ± 0.02	10.80 ± 0.01	10.10 ± 0.05
	C16:1	0.50 ± 0.01	0.50 ± 0.01	0.40 ± 0.02	0.50 ± 0.02
	C18:0	2.20 ± 0.03	2.20 ± 0.05	2.70 ± 0.04	2.50 ± 0.01
	C18:1	71.20 ± 0.02	70.70 ± 0.02	71.10 ± 0.05	73.70 ± 0.04
	C18:2	12.30 ± 0.01	11.80 ± 0.01	12.10 ± 0.02	11.80 ± 0.02
	C18:3	1.10 ± 0.04	1.00 ± 0.01	1.00 ± 0.01	1.10 ± 0.01

Regarding of monovarietal oil, very similar percentages of palmitic acid (C16:0) were observed in all cases, with a slight decrease, in the different seasons. Regarding palmitoleic acid (C16:1), it remained constant (0.5 and 0.4%) in all seasons. The percentage of stearic acid (2.2%) also remained constant in the 2015/2016 season, with a decrease of 0.2% during the 2016/2017 season. There is a slight decrease in the percentage of oleic acid (C18:1), with the exception of the 2016/2017 season, in which there was an increase in this parameter (2.3%). Linoleic acid (C18:2) has values within the preference limit (3.5–21.0%) stipulated by the IOOC. They suffered a slight decrease, in the seasons under study, over time. Being in the 2015/2016 season when there was a higher value (12.3%). The percentage of linolenic acid (C18:3) is 1%, except for the 2015/2016 season in which the result obtained in February was 1.1%, with a slight decrease throughout of 2016.

All the flavoured oils have similar percentages of palmitic, palmitoleic, stearic, oleic, linoleic and linolenic acid during the two seasons.

Regarding the oleic acid, during the 2016/2017 season, an increase of the value was observed from 2.5 to 2.7, in all the flavoured oils. The rosemary is the most contributing herb to this parameter in 2015/2016 and 2016/2017 seasons (0.1 and 2.7%, respectively).

Sousa et al. [5] consider that rosemary and lemon have a strong influence on the fatty acids profile of the flavoured olive oils. These ingredients increase the oleic acid content and slight decrease of the saturated fatty acid concentration. These results agree with the obtained in this study (Table 5).

In conclusion, the flavoured agents influence the lipidic composition of the flavoured olive oils.

Table 6 shows the fatty acid composition of the olive oils.

All the fatty acid percentages obtained are within the ranges established by the EU Regulations [17], with the exception of the linolenic acid which reach the maximum limit in the monovarietal oil (1%) and in some of the flavoured oils.

Regarding the trans fatty acid values in Table 6, the monovarietal olive oil and the flavoured olive oils have values below of 0.05% which is the limit for the extra virgin category.

For C18:3 trans + C18:2 trans fatty acids, according to Delegated Regulation (EU) 2016/2095 [24] is the sum of the trans forms of linoleic and linolenic acids must be less than 0.05% for olive oil of the "Extra Virgin" category. In this case, the levels are within the legally accepted values, observing the same behaviour and identical values, as in the case of trans-oleic acids. There is now consistent scientific evidence on the harmful effects of consuming industrially produced trans fats. Given this situation, measures have been developed to reduce trans fats in processed foods. Another way is to replace these fats with olive oil, since it has been proven that the results obtained in the levels of fatty acids in the trans forms are within the legally accepted values.

3.3.2. Wax

Table 7 shows the composition of waxes determined in the oil samples of the 'Madural' variety and the respective flavoured oils during the campaigns under study.

Table 7 shows a maximum value can be observed for the monovarietal oil in December 2016 (49 mg/kg oil) and the minimum value observed corresponds to the oil analysed in February 2017 (34 mg/kg oil).

The maximum value allowed for the extra virgin category is 150 mg/kg oil, so these oils fall into this category. Finally, in terms of the wax content of the flavoured oils have similar values to those of the monovarietal oil, so they can be included in the same category. The oil flavoured with salt and bay leaf obtained lower values, especially in the 2016/2017 season. It can be concluded that the flavourings have an influence on the presence of waxes in the oils, since there is a decrease in values over time.

Table 6. The percentage of fatty acids (%) of all olive oils.

Samples		February 2016	December 2016	February 2017	December 2017
MO	C14:0	<0.03	<0.03	<0.03	<0.03
	C18:3	1.10 ± 0.01	1.10 ± 0.02	1.10 ± 0.01	1.10 ± 0.01
	C20:0	0.30 ± 0.00	0.30 ± 0.00	0.40 ± 0.00	0.30 ± 0.00
	C20:1	0.30 ± 0.00	0.30 ± 0.00	0.30 ± 0.00	0.30 ± 0.00
	C22:0	<0.10	<0.10	0.10 ± 0.00	0.10 ± 0.01
	C24:0	<0.1	<0.10	0.10 ± 0.00	<0.10
	C18:1 trans	<0.02	<0.02	<0.02	<0.02
	C18:3 trans + C18:2 trans	<0.02	<0.02	<0.02	<0.02
MOSB	C14:0	<0.03	<0.03	<0.03	<0.03
	C18:3	1.10 ± 0.01	1.00 ± 0.01	1.00 ± 0.02	1.1 ± 0.03
	C20:0	0.30 ± 0.00	0.40 ± 0.00	0.40 ± 0.00	0.40 ± 0.00
	C20:1	0.30 ± 0.00	0.30 ± 0.00	0.30 ± 0.00	0.30 ± 0.00
	C22:0	<0.10	<0.10	0.10 ± 0.00	0.10 ± 0.00
	C24:0	<0.10	<0.10	0.10 ± 0.00	<0.10
	C18:1 trans	<0.02	<0.02	<0.02	<0.02
	C18:3 trans + C18:2 trans	<0.02	<0.02	<0.02	<0.02
MOR	C14:0	<0.03	<0.03	<0.03	<0.03
	C18:3	1.10 ± 0.02	1.00 ± 0.01	1.00 ± 0.01	1.10 ± 0.01
	C20:0	0.30 ± 0.01	0.30 ± 0.00	0.40 ± 0.00	0.40 ± 0.01
	C20:1	0.30 ± 0.00	0.30 ± 0.00	0.30 ± 0.01	0.30 ± 0.00
	C22:0	<0.10	0.10 ± 0.00	0.10 ± 0.00	0.10 ± 0.00
	C24:0	<0.10	<0.10	0.10 ± 0.00	<0.10
	C18:1 trans	<0.02	<0.02	<0.02	<0.02
	C18:3 trans + C18:2 trans	<0.02	<0.02	<0.02	<0.02
MOG	C14:0	<0.03	<0.03	<0.03	<0.03
	C18:3	1.10 ± 0.02	1.00 ± 0.03	1.00 ± 0.12	1.10 ± 0.10
	C20:0	0.30 ± 0.01	0.40 ± 0.01	0.40 ± 0.01	0.40 ± 0.01
	C20:1	0.30 ± 0.00	0.30 ± 0.00	0.30 ± 0.00	0.30 ± 0.01
	C22:0	<0.10	0.10 ± 0.01	0.10 ± 0.00	0.10 ± 0.01
	C24:0	<0.10	<0.10	0.1 ± 0.010	<0.10
	C18:1 trans	<0.02	<0.02	<0.02	<0.02
	C18:3 trans + C18:2 trans	<0.02	<0.02	<0.02	<0.02
MOL	C14:0	<0.03	<0.03	<0.03	<0.03
	C18:3	1.10 ± 0.04	1.00 ± 0.02	1.00 ± 0.01	1.10 ± 0.03
	C20:0	0.30 ± 0.00	0.40 ± 0.00	0.40 ± 0.00	0.40 ± 0.00
	C20:1	0.30 ± 0.01	0.30 ± 0.00	0.30 ± 0.00	0.20 ± 0.00
	C22:0	<0.10	0.10 ± 0.00	0.10 ± 0.00	0.10 ± 0.00
	C24:0	<0.10	<0.10	0.10 ± 0.00	<0.10
	C18:1 trans	<0.02	<0.02	<0.02	<0.02
	C18:3 trans + C18:2 trans	<0.02	<0.02	<0.02	<0.02

Table 7. Wax content (mg/kg oil) of all olive oils.

Samples	February 2016	December 2016	February 2017	December 2017
MO	45.00 ± 0.20	49.00 ± 0.21	34.00 ± 0.11	37.00 ± 0.12
MOSB	45.00 ± 0.12	40.00 ± 0.15	34.00 ± 0.20	31.00 ± 0.14
MOR	45.00 ± 0.13	42.00 ± 0.13	34.00 ± 0.12	34.00 ± 0.23
MOG	45.00 ± 0.10	40.00 ± 0.14	34.00 ± 0.10	32.00 ± 0.11
MOL	45.00 ± 0.12	41.00 ± 0.21	34.00 ± 0.22	32.00 ± 0.12

3.3.3. Sterols

The qualitative and quantitative analysis of sterols is part of the criteria of purity and authenticity of the quality of olive oil.

Due to its importance, Tables 8 and 9 show the values of the individual sterols and total sterols, respectively, during the two seasons.

Table 8. Individual sterols (%) of all olive oils.

		February 2016	December 2016	February 2017	December 2017
MO	Cholesterol	0.10 ± 0.00	0.10 ± 0.00	0.10 ± 0.00	0.00 ± 0.00
	Brassicasterol	<0.10	<0.10	<0.10	0.10 ± 0.01
	Campesterol	2.20 ± 0.02	2.30 ± 0.01	2.20 ± 0.01	2.40 ± 0.01
	Estigmasterol	0.50 ± 0.01	0.50 ± 0.02	0.50 ± 0.01	0.40 ± 0.00
	Clerosterol	1.00 ± 0.01	1.00 ± 0.02	1.00 ± 0.03	0.90 ± 0.01
	β-sitosterol	84.30 ± 0.20	83.40 ± 0.18	86.30 ± 0.15	84.50 ± 0.21
	5Δ-avenasterol	10.40 ± 0.20	10.60 ± 0.10	9.80 ± 0.01	10.30 ± 0.11
	5,24Δ-estigmastadienol	0.40 ± 0.00	0.60 ± 0.00	<0.10	0.20 ± 0.00
	7Δ-avenasterol	0.10 ± 0.01	0.20 ± 0.01	0.10 ± 0.01	0.50 ± 0.01
MOSB	Cholesterol	0.10 ± 0.00	0.10 ± 0.00	0.10 ± 0.00	0.10 ± 0.00
	Brassicasterol	<0.10	<0.10	<0.10	<0.10
	Campesterol	2.20 ± 0.02	2.30 ± 0.03	2.20 ± 0.01	2.30 ± 0.02
	Estigmasterol	0.50 ± 0.01	0.40 ± 0.01	0.50 ± 0.01	0.40 ± 0.01
	Clerosterol	1.00 ± 0.01	0.90 ± 0.01	1.00 ± 0.01	0.90 ± 0.02
	β-sitosterol	84.30 ± 0.10	83.60 ± 0.17	86.30 ± 0.32	84.60 ± 0.22
	5Δ-avenasterol	10.40 ± 0.15	10.60 ± 0.21	9.80 ± 0.17	10.30 ± 0.10
	5,24Δ-estigmastadienol	0.40 ± 0.01	0.60 ± 0.00	<0.10	0.80 ± 0.00
	7Δ-avenasterol	0.10 ± 0.00	0.20 ± 0.00	0.10 ± 0.00	0.50 ± 0.00
MOR	Cholesterol	0.10 ± 0.00	0.10 ± 0.00	0.10 ± 0.00	0.10 ± 0.00
	Brassicasterol	<0.10	<0.10	<0.10	<0.10
	Campesterol	2.20 ± 0.02	2.20 ± 0.02	2.20 ± 0.02	2.50 ± 0.02
	Stigmasterol	0.50 ± 0.00	0.40 ± 0.00	0.50 ± 0.00	0.40 ± 0.00
	Clerosterol	1.00 ± 0.01	1.00 ± 0.03	1.00 ± 0.01	0.90 ± 0.01
	β-sitosterol	84.30 ± 0.11	83.10 ± 0.13	86.30 ± 0.12	83.60 ± 0.14
	5Δ-avenasterol	10.40 ± 0.12	11.00 ± 0.06	9.80 ± 0.03	10.90 ± 0.12
	5,24Δ-stigmastadienol	0.40 ± 0.01	0.60 ± 0.010	<0.10	0.70 ± 0.02
	7Δ-avenasterol	0.10 ± 0.00	0.20 ± 0.00	0.10 ± 0.00	0.40 ± 0.00
MOG	Cholesterol	0.10 ± 0.00	0.10 ± 0.00	0.10 ± 0.00	0.10 ± 0.00
	Brassicasterol	<0.10	<0.10	<0.10	<0.10
	Campesterol	2.20 ± 0.01	2.30 ± 0.02	2.20 ± 0.04	2.50 ± 0.02
	Stigmasterol	0.50 ± 0.00	0.30 ± 0.00	0.50 ± 0.01	0.40 ± 0.00
	Clerosterol	1.00 ± 0.02	0.80 ± 0.04	1.00 ± 0.01	0.80 ± 0.04
	β-sitosterol	84.30 ± 0.10	83.80 ± 0.20	86.30 ± 0.06	84.00 ± 0.11
	5Δ-avenasterol	10.40 ± 0.12	10.60 ± 0.07	9.80 ± 0.02	10.70 ± 0.08
	5,24Δ-stigmastadienol	0.40 ± 0.01	0.50 ± 0.01	<0.10	0.60 ± 0.01
	7Δ-avenasterol	0.10 ± 0.00	0.30 ± 0.00	0.10 ± 0.00	0.40 ± 0.00
MOL	Cholesterol	0.10 ± 0.00	0.1 ± 0.001	0.10 ± 0.00	0.1 ± 0.001
	Brassicasterol	<0.10	<0.10	<0.10	<0.10
	Campesterol	2.20 ± 0.02	2.10 ± 0.01	2.20 ± 0.01	2.30 ± 0.01
	Stigmasterol	0.50 ± 0.00	0.50 ± 0.00	0.50 ± 0.00	0.40 ± 0.00
	Clerosterol	1.00 ± 0.01	0.90 ± 0.02	1.00 ± 0.02	0.90 ± 0.01
	β-sitosterol	84.30 ± 0.12	83.60 ± 0.11	86.30 ± 0.13	84.50 ± 0.12
	5Δ-avenasterol	10.4±00.02	10.70 ± 0.03	9.80 ± 0.02	10.40 ± 0.01
	5,24Δ-stigmastadienol	0.40 ± 0.00	0.50 ± 0.00	<0.10	0.80 ± 0.00
	7Δ-avenasterol	0.10 ± 0.00	0.30 ± 0.00	0.10 ± 0.00	0.50 ± 0.00

Table 9. Total sterols (mg/kg) of all olive oils.

Samples	February 2016	December 2016	February 2017	December 2017
MO	1915 ± 23	1978 ± 15	1933 ± 11	1956 ± 05
MOSB	1915 ± 11	1896 ± 10	1933 ± 25	1916 ± 19
MOR	1915 ± 11	1954 ± 25	1933 ± 17	1966 ± 21
MOG	1915 ± 01	1922 ± 05	1933 ± 15	1947 ± 13
MOL	1915 ± 03	1893 ± 12	1933 ± 10	1916 ± 05

The values of individual sterols are within the permitted legal limits, for olive oil and flavoured oils of the "Extra Virgin" category, in the two seasons. Note that the apparent β-Sitosterol corresponds to: Δ-5,23-stigmastadienol + clerosterol + -sitosterol + sitostanol + Δ-5-avenasterol + Δ-5,24-stigmastadienol and must be ≥93.0% [17].

On the other hand, the percentage of stigmasterol has to be lower than campesterol, which occurs in both seasons. The percentage of β-sitosterol decreases over time, in the samples of monovarietal MO and flavoured oils MOSB and MOR, in the two seasons.

Regarding the total sterols, the values are within the legal limits (≥1000), for olive oil of the "Extra Virgin" category (Table 9) [17]. MOR and MOG show higher values, while MOSB and MOL show minor differences, which can be attributed to the composition of these flavouring agents.

3.3.4. Erythrodiol and Uvaol

Table 10 shows the resulting values from the determination of erythrodiol and uvaol in the monovarietal variety oil and flavoured oils, in the 2015/2016 and 2016/2017 seasons.

Table 10. Erythrodiol and uvaol (%) and total phenolic compounds (mg/kg oil) of all olive oils.

	Erythrodiol and Uvaol (%)			
Samples	**February 2016**	**December 2016**	**February 2017**	**December 2017**
MO	1.20 ± 0.01	0.90 ± 0.01	1.50 ± 0.01	0.80 ± 0.01
MOSB	1.20 ± 0.01	0.90 ± 0.02	1.50 ± 0.01	1.00 ± 0.02
MOR	1.20 ± 0.03	1.10 ± 0.04	1.50 ± 0.05	1.10 ± 0.03
MOG	1.20 ± 0.02	1.20 ± 0.01	1.50 ± 0.02	1.10 ± 0.01
MOL	1.20 ± 0.01	1.00 ± 0.03	1.50 ± 0.01	1.00 ± 0.04
	Total Phenolic Compounds (mg/kg oil)			
	February 2016	**December 2016**	**February2017**	**December 2017**
MO	222.00 ± 1.01	206.00 ± 1.10	249.00 ± 1.80	234.00 ± 1.23
MOSB	222.00 ± 1.02	237.00 ± 2.33	249.00 ± 2.04	269.00 ± 1.60
MOR	222.00 ± 1.12	241.00 ± 1.21	249.00 ± 1.22	200.00 ± 2.02
MOG	222.00 ± 1.23	233.00 ± 2.30	249.00 ± 2.23	256.00 ± 1.23
MOL	222.00 ± 1.22	208.00 ± 1.20	249.00 ± 1.23	236.00 ± 1.27

The results show that both monovarietal and flavoured olive oils have erythrodiol and uvaol values lower than the maximum limit (4.5%) [17]. These values are within the interval of 1.10 and 3.96% reported by Seabra et al. [38] and they are similar to the values reported by Chéu-Guedes Vaz [9] of 1.26% for 'Madural' variety olive oil from Mirandela region.

The flavoured oils have a sharper decrease in values over time, which became more pronounced in the last season. The olive oil flavoured with garlic acquires the highest values and the flavoured with salt and bay leaf obtains the lowest results, in the two seasons.

3.3.5. Phenolic Compounds

The content of total polyphenols in monovarietal oil decreases over time (Table 10). In the 2015/2016 and 2016/2017 seasons, the decreases in total polyphenols are 16 and 15 mg/kg, respectively (7% and 6% reduction, from February to December).

With respect to flavoured oils, in general, the content of total phenolic compounds, in December, increases in relation to the MO, in that month of each season. Although in the case of rosemary flavoured there is a decrease in the 2016/2017 season, and the MOL oil is the one that has the least increase, compared to the MO, in those months.

However, if we consider the periods from February to December, the total polyphenols content increases over the time in MOSB, MOG and MOR in the first season, decreasing in 2016/2017, while polyphenol content decreases in MOL in both seasons.

The fact that MOL shows these decreases compared to monovarietal oil may be due to the contribution of the polyphenols present in the composition of bay, rosemary and garlic. This would be in accordance with what different research groups indicated for garlic [36,38–40] and lemon [39] but contrary to what was stated by Sousa et al. [5] for the laurel, when they indicate that it decreases, and according to the results obtained it increases, although the operating variables variety, quantity, time and temperature differ from those of this investigation. As for those flavoured with rosemary, the results for both campaigns are contradictory, since it increases in the first campaign, according to same studies [36,41,42] and decreases in the second, similar to other studies [34,35].

There are several authors and research groups that pay attention to the anti-oxidant effects of polyphenols present in certain aromatic plants. Thus, Mestre et al. [43], mentions that the compounds extracted from aromatic plants are often attributed interesting antioxidant properties. In addition, therefore, the addition of flavourings to olive oil improves its nutritional properties and its beneficial effects on health, especially in terms of preventing oxidation. Furthermore, Middleton et al. [44], indicate that, of all the antioxidants, polyphenols have a wide range of biological effects generally attributed to antioxidant activity. Seabra et al. [38] mention that the antioxidant properties of rosemary extract have received a lot of attention in recent years, because of its antioxidant composition, among other constituents. Finally, Gambacorta et al. [1], consider that the addition of different concentrations (10–40 g/L) of garlic, pepper, oregano and rosemary increased the long-term stability of the oils.

3.3.6. Tocopherols

Table 11 shows the results of the total composition of tocopherols, for the 'Madural' monovarietal oil and its flavoured oils, for the different seasons under study. In all cases, the levels are within the established values, although in the 2015/2016 campaign the results obtained were higher.

Table 11. Tocopherols (mg/kg oil) of all olive oils.

Samples		February 2016	December 2016	February 2017	December 2017
MO	Total	313.00 ± 2.30	355.40 ± 2.80	203.00 ± 1.70	299.20 ± 1.90
	α- tocopherol	307.00 ± 1.30	348.20 ± 2.30	203.00 ± 2.20	291.20 ± 1.30
	β- tocopherol	3.00 ± 0.03	2.30 ± 0.01	<1.00	2.60 ± 0.02
	γ- tocopherol	3.00 ± 0.02	4.90 ± 0.04	<0.10	5.40 ± 0.04
MOSB	Total	313.00 ± 3.10	344.10 ± 2.40	203.00 ± 1.30	273.80 ± 1.90
	α- tocopherol	307.00 ± 1.30	338.20 ± 1.30	203.00 ± 1.30	266.10 ± 1.30
	β- tocopherol	3.00 ± 0.02	1.90 ± 0.01	<1.00	2.50 ± 0.01
	γ- tocopherol	3.00 ± 0.04	3.90 ± 0.06	<1.00	5.20 ± 0.02
MOR	Total	313.00 ± 1.70	366.20 ± 2.30	203.00 ± 1.80	288.0 ± 1.70
	α- tocopherol	307.00 ± 0.30	359.30 ± 1.60	203.00 ± 2.30	280.10 ± 0.30
	β- tocopherol	3.00 ± 0.08	2.20 ± 0.23	<1.00	2.60 ± 0.02
	γ- tocopherol	3.00 ± 0.14	4.60 ± 0.03	<1.00	5.30 ± 0.01
MOG	Total	313.00 ± 1.20	376.50 ± 2.30	203.00 ± 1.90	317.60 ± 1.30
	α- tocopherol	307.00 ± 1.50	369.30 ± 3.30	203.00 ± 2.30	309.30 ± 1.00
	β- tocopherol	3.00 ± 0.03	2.40 ± 0.06	<1.00	2.70 ± 0.02
	γ- tocopherol	3.00 ± 0.09	4.70 ± 0.05	<1.00	5.60 ± 0.03
MOL	Total	313.00 ± 1.30	361.60 ± 1.30	203.00 ± 1.30	304.60 ± 1.30
	α- tocopherol	307.00 ± 1.90	354.40 ± 0.70	203.00 ± 1.50	296.70 ± 2.10
	β- tocopherol	3.00 ± 0.02	2.40 ± 0.02	<1.00	2.60 ± 0.02
	γ- tocopherol	3.00 ± 0.01	4.80 ± 0.03	<1.00	5.30 ± 0.04

It should be noted that the results show an increase in the total tocopherol content, going from 313.0 to 355.4 and from 203.0 to 299.2 mg/kg oil, respectively, for the 2015/16 and 2016/17 seasons, mainly due to the contribution of α and γ tocopherols, since the β form has minor influence.

Thus, it is observed that, in the 2015/2016 campaign, the α-tocopherols content [26,27] in MO ranged between 307.0 and 348.2 mg/kg oil, while in the 2016/2017, the values were lower, ranging between 203.0 and 291.2 mg/kg oil.

In this sense, and taking previous studies as a reference, the α-tocopherol content of the oils analysed from the 'Madural' variety, in the Mirandela region, varied between

150.9 and 456.2 mg/kg, obtaining mean levels of tocopherols, in the α form, of 285.8 mg/kg oil. It can be deduced that tocopherol content depends on the variety, the location of the olive groves and, most likely, technological factors [9].

On the other hand, in the first season, the β and γ forms evolved with an increase in tocopherols in the γ form, but a decrease in the β form. However, in the second season, tocopherols increased in both forms, obtaining results higher than those of the previous season.

Regarding the flavoured oils, the total tocopherol content increases with respect to MO with the different flavourings, except MOSB, which coincides with the bibliography, which indicates an increase in all the isoforms of vitamin E when rosemary is used [34,35], as it occurs with fresh garlic [5], which differs from the dried garlic used in this research. On the other hand, together with the increase in tocopherols, they also state that there is an increase in the oleic acid content. In this sense, when the fatty acid profile was analysed, a slight increase already appeared, caused by some herbs and spices used in flavouring [5]. Finally, it should be noted that no bibliographical evidence has been found to compare the behaviour obtained with salt and bay leaves or lemon flavourings.

In 2015/2016 season, flavoured oils with garlic, rosemary and lemon obtained increases of 20.30%, 17.03% and 15.43% in α-tocopherol, respectively, compared the MO. In the campaign 2016/2017, oils flavoured with lemon and garlic had an increase in α-tocopherol of 52.36% and 46.15%, respectively. In general, tocopherols, in the α form, increased with time, in all seasons.

Regarding β and γ tocopherols, in the 2015/2016 season, the values are similar to those in MO, with slight decreases with all flavourings. However, in the 2016/20/17 season, these forms of tocopherols obtained higher values than MO, highlighting the olive oil flavoured with garlic.

3.4. Oxidative Stability

Table 12 shows the oxidative stability resistance values for the monovarietal oil and the flavoured oils, determined by Rancimat at 120 °C, in the two seasons.

Table 12. Oxidative stability (h) of all olive oils.

Samples	February 2016	December 2016	February 2017	December 2017
MO	8.8 ± 0.1	6.0 ± 0.1	8.9 ± 0.1	6.0 ± 0.2
MOSB	8.8 ± 0.1	7.0 ± 0.2	8.9 ± 0.4	6.5 ± 0.4
MOR	8.8 ± 0.2	4.3 ± 0.1	8.9 ± 0.3	4.4 ± 0.1
MOG	8.8 ± 0.3	6.0 ± 0.5	8.9 ± 0.1	5.9 ± 0.2
MOL	8.8 ± 0.1	5.7 ± 0.1	8.9 ± 0.1	5.6 ± 0.1

From the analysis of the Table, it can be deduced that in the two campaigns, the oil of the 'Madural' variety has low resistance to oxidation and it decreases during the year.

As in monovarietal oil, oxidative stability of flavoured oils also decreases over time. The oils flavoured with garlic, are similar to MO and it could be said that garlic does not influence this parameter. However, there are some bibliographical references that indicate that garlic increases oxidative stability [36,38,39] perhaps the discrepancies in the results may be due to the operating conditions.

In the case of oils flavoured with lemon, the values are somewhat lower than the monovarietal, indicating that these flavouring agents act by slightly decreasing oxidative stability. However, Ayadi et al. [31] detected an increase in stability for different operating conditions, with contact times of 15 days, which suggests that the contact time used in this investigation was too long.

Regarding the salt + bay leaf flavoured oil, in the two seasons, it can be seen that this flavouring agent slows down oxidation when stability values were observed higher than those obtained in monovarietal oil, which agrees with what was indicated by Sousa et al. [5]

for contact times of 90 days, and Taleb et al. [38] who used fresh laurel and contact times very different from those of this research.

Considering the oil flavoured with rosemary, the results indicate that there is a significant decrease in oxidative stability over time, with very similar values in each season, in view of which, it could be said that this flavouring accelerates oxidation. Coinciding with the results indicated by Soares et al. [36], for contact times of 7 and 15 days, but not with what is indicated by other research groups [34–36,41], which operate with amounts of rosemary and contact conditions different from those of our research.

3.5. Microbiological Analysis

Considering that in this research dehydrated flavourings were added to the oils we intend to understand if there was any risk of microbial spoilage.

Olive oil does not contain water in its composition [9] which is a vital source for the existence of microorganisms. Nevertheless, determinations were made of microorganism counts at 30 °C (evaluation of hygienic quality) of coliform bacteria at 30 °C (evaluation of the existence of pathogenic microorganisms) and of coagulase + Staphylococci (evaluation of the quality of the products food), in the samples of the first and last campaign of the study to verify the veracity of the results. Table 13 shows the results obtained in the two seasons.

Table 13. Microbiological analysis (CFU/g) of all olive oils.

Samples				
MO	**February 2015**	**December 2015**	**February 2017**	**December 2017**
Microorganisms	<1	<1	<1	<1
Coliform bacteria	<1	<1	<1	<1
coagulase + Staphylococci	<1	<1	1	<1
MOSB	**February 2015**	**December 2015**	**February 2017**	**December 2017**
Microorganisms	<1	<1	<1	<1
Coliform bacteria	<1	<1	<1	<1
coagulase + Staphylococci	<1	<1	<1	<1
MOR	**February 2015**	**December 2015**	**February 2017**	**December 2017**
Microorganisms	<1	<1	<1	<1
Coliform bacteria	<1	<1	<1	<1
coagulase + Staphylococci	<1	<1	<1	<1
MOG	**February 2015**	**December 2015**	**February 2017**	**December 2017**
Microorganisms	<1	<1	<1	<1
Coliform bacteria	<1	<1	<1	<1
coagulase + Staphylococci	<1	<1	<1	<1
MOL	**February 2015**	**December 2015**	**February 2017**	**December 2017**
Microorganisms	<1	<1	<1	<1
Coliform bacteria	<1	<1	<1	<1
Coagulase + Staphylococci	<1	<1	<1	<1

The results indicate that there is no risk of microbial presence in monovarietal and flavoured oils during both seasons because the existence of microorganisms is practically null. This situation also occurs because the flavouring agents used were all dehydrated.

4. Conclusions

The olive oil obtained from olives of the 'Madural' variety, in the studied seasons, is a balanced oil in its composition, which complies with current regulations and can be classified as "Extra Virgin olive oil".

The incorporation of flavouring agents does not significantly affect to the quality of the monovarietal oils. Although flavourings such as garlic, bay leaf, rosemary or lemon usually have negative effects on the acidity of the oils, in this case, the effects are not significant. Regarding to the peroxide value, it can be said that the obtained values may be associated with the flavouring agent used and the aromatization time, since, in general, it can be seen that when the contact times are short (one to three months), the results are close to those of monovarietal oil. However, at longer times, the differences are so high that the category changes for some flavoured oils. In this sense, the ultraviolet absorption parameters allow classifying flavoured oils with the "Extra Virgin" category, except for those flavoured with salt and laurel, which would be "Virgin" category. The sensory profile shows that the aromas, and flavours are correctly detected in all the oils without affecting to others sensory attributes.

The fatty acid profile is not altered by the presence of flavourings and only it is observed an increase in the oleic acid content in all the flavoured oils, the one using rosemary standing out slightly. On the other hand, the total sterol contents show higher values for the oils flavoured with rosemary and garlic, while those flavoured with salt + bay leaf and lemon present less differences, which can be attributed to the composition of said flavourings.

In relation to the phenolic compounds, in general, an increase can be seen with respect to the monovarietal oil, perhaps due to the contribution of the phenolic compounds present in the composition of bay leaves, rosemary and garlic, although in the case of lemon its content decreases in both seasons. On the other hand, the total tocopherol content increases with respect to monovarietal oil with the different flavourings, except with salt + bay leaf, the increases that are mainly due to the α isoform.

Regarding oxidative stability, the flavouring that presents the greatest protection against oxidation is salt and bay leaf, and the least is rosemary, which seems to accelerate oxidation. Garlic and lemon show similar stability values, with small differences between them with respect to the monovarietal oil.

On the other hand, the total sterol contents show higher values for the oils flavoured with rosemary and garlic, while those flavoured with salt and bay leaf and lemon have smaller differences, which can be attributed to the composition of flavouring agents.

With regard to microbial deterioration in monovarietal and flavoured oils, the presence of detected microorganisms is practically null. These results are due to dehydration of the flavouring agents and the hygienic conditions during the obtention and storage of the oils.

Finally, the effect exerted by the flavourings, on the different parameters of the oils, depends on the method and the operating variables established for the flavouring operation: olive variety, quantity, physical characteristics and composition, flavouring, time and temperature of operation.

Author Contributions: Conceptualization, R.P. and N.R.; methodology, S.S., M.D.L.R., R.P. and N.R.; software, validation, and formal analysis, M.H.C.-G. and M.D.L.R.; investigation and data curation, M.H.C.-G., R.P. and N.R.; resources, R.P., N.R. and M.D.L.R.; writing—original draft preparation, M.H.C.-G., R.P., N.R. and M.D.L.R.; writing—review and editing, M.H.C.-G., R.P., N.R. and M.D.L.R.; supervision: S.S., R.P. and N.R. All authors have read and agreed to the published version of the manuscript.

Funding: Research group 'Bioprocesses TEP-138' of the Andalusia Regional Government (Spain) and Piaget Institute (Portugal).

Data Availability Statement: Data are contained within the article.

Acknowledgments: The authors acknowledge the Andalusia Regional Government (Spain) for its financial support to our research group 'Bioprocesses TEP-138' and Piaget Institute (Portugal).

Conflicts of Interest: The authors declare no conflict of interest.

References

1. Gambacorta, G.; Faccia, M.; Pati, S.; Lamacchia, C.; Baiano, A.; La Notte, E. Changes in the chemical and sensorial profile of extra virgin olive oils flavoured with herbs and spices during storage. *J. Food Lipids* **2007**, *14*, 202–215. [CrossRef]
2. Baiano, A.; Gambacorta, G.; La Notte, E. Aromatization of olive oil. *Transw. Res. Netw.* **2010**, *661*, 1–29.
3. Abo, B.; Bevan, J.; Greenway, S.; Healy, B.; McCurdy, S.M.; Peutz, J.; Wittman, G. Acidification of Garlic and Herbs for Consumer Preparation of Infused Oils. *Food Prot. Trends* **2013**, *34*, 247–257.
4. Moustakime, Y.; Hazzoumi, Z.; Joutei, K.A. Aromatization of virgin olive oil by seeds of Pimpinella anisum using three different methods: Physico-chemical change and thermal stability of flavoured oils. *Grain Oil Sci. Technol.* **2021**, *4*, 108–124. [CrossRef]
5. Sousa, A.; Casal, S.; Malheiro, R.; Lamas, R.; Bento, A.; Pereira, J.A. Aromatized olive oils: Influence of flavouring in quality, composition, stability, antioxidants, and antiradical potential. *LWT Food Sci. Technol.* **2015**, *60*, 22–28. [CrossRef]
6. Fennema, O.R. *Química de Los Alimentos*, 2nd ed.; ACRIBIA: Zaragoza, Spain, 2000.
7. Morales, M.A.; Przbylski, R. Olive Oil Oxidation. In *Chemistry and Technology*, 2nd ed.; Boskou, D., Ed.; Department of Chemistry, Aristotle University of Thessaloniki: Thessaloniki, Greece, 2003.
8. Matos, L.C.; Pereira, J.A.; Andrade, P.; Oliveira, M. Evaluation of a numerical method to predict the polyphenols content in monovarietal olive oils. *Food Chem.* **2007**, *102*, 976–983. [CrossRef]
9. Guedes-Vaz, C. *Azeite de Trás-os-Montes. Influência da Localização do Olival e das Cultivares nas Características do Azeite. Instituto Piaget*; Colecção Estudos e Documentos: Lisboa, Portugal, 2011.
10. Custódio, T. Azeites Virgem Extra Comerciais: Composição em Compostos Voláteis e Relação com Parâmetros Químicos de Qualidade. Master's Thesis, Faculdade de Farmácia da Universidade do Porto, Porto, Portugal, 2009.
11. Lamas, S.; Rodrigues, N.; Peres, A.M.; Pereira, J.A. Flavoured and fortified olive oils—Pros and cons. *Trends Food Sci. Technol.* **2022**, *124*, 108–127. [CrossRef]
12. Baiano, A.; Terracone, C.; Gambacorta, G.; Notte, E.L. Changes in quality indices, phenolic content and antioxidant activity of flavoured olive oils during storage. *JAOCS J. Am. Oil Chem. Soc.* **2009**, *86*, 1083–1092. [CrossRef]
13. Bittencourt Fagundes, M.; Ballus, C.A.; Perceval Soares, V.; de Freitas Ferreira, D.; Sena Vaz Leães, Y.; Sasso Robalo, S. Characterization of olive oil flavoured with Brazilian pink pepper (Schinus terebinthifolius Raddi) in different maceration processes. *Food Res. Int.* **2020**, *137*, 109593. [CrossRef]
14. Reboredo-Rodríguez, P.; Figueiredo-González, M.; González-Barreiro, C.; Simal-Gándara, J.; Salvador, M.D.; Cancho-Grande, B. State of the art on functional virgin olive oils enriched with bioactive compounds and their properties. *Int. J. Mol. Sci.* **2017**, *18*, 668. [CrossRef]
15. Sena-Moreno, E.; Alvarez-Ortí, M.; Serrano-Díaz, J.; Pardo, J.E.; Carmona, M.; Alonso, G.L. Olive oil aromatization with saffron by liquid–liquid extraction. *J. Food Sci. Technol.* **2018**, *55*, 1093–1103. [CrossRef] [PubMed]
16. International Olive Council Reunión Extraordinaria. Madrid. 2014. Available online: https://www.olimerca.com/noticiadet/el-coi-se-pronuncia-sobre-los-aceites-de-oliva-con-aditivos/62dbad318a564a84781316b96554b9b6 (accessed on 8 July 2014).
17. *COI/T.15/NC No 3/Rev*; Trade Standard Applying to Olive Oils and Olive Pomace Oils. International Olive Council: Madrid, Spain, 18 June 2022.
18. Regulation N° 136/66/EEC Establishing a Common Organization of the Market in Oils and Fats. *Off. J.* 3025–3066. Available online: https://eur-lex.europa.eu/legal-content/EN/TXT/?uri=celex%3A31966R0136 (accessed on 1 January 2023).
19. *UNE 55031:1973 Fats*; Determination of Humidity and Volatile Matters in Olive Grease. AENOR: Madrid, Spain.
20. *UNE 55030:1961*; Determination of the Content in Total Fat of Olives. AENOR: Madrid, Spain.
21. *ISO 660:2020*; Animal and Vegetable Fats and Oils—Determination of Acid Value and Acidity. International Organization for Standardization: Geneva, Switzerland.
22. *ISO 3960:2017*; Animal and Vegetable Fats and Oils—Determination of Peroxide Value—Iodometric (Visual) End Point Determination. International Organization for Standardization: Geneva, Switzerland.
23. Commission Regulation (EEC) No 2568/91 on the Characteristics of Olive Oil and Olive-Residue Oil and on the Relevant Methods of Analysis. *Off. J. Eur. Communities* **1991**, 1–15. Available online: http://extwprlegs1.fao.org/docs/pdf/eur40621.pdf (accessed on 1 January 2023).
24. Commission Delegated Regulation (EU) 2016/2095 Amending Regulation (EEC) No. 2568/91 on the Characteristics of Olive Oil and Olive-Residue oil and on the Relevant Methods of Analysis. Available online: https://eur-lex.europa.eu/legal-content/EN/TXT/PDF/?uri=CELEX:32016R2095&rid=5 (accessed on 1 January 2023).
25. COI/T.20/Doc.N° 5 of 2007. Sensory Analysis of Oils. International Olive Council: Madrid, Spain.
26. Cunha, S.C.; Amaral, J.S.; Fernandes, J.O.; Oliveira, M.B.P.P. Quantification of tocopherols and tocotrienols in Portuguese olive oils using HPLC with three different detection systems. *J. Agric. Food Chem.* **2006**, *54*, 3351–3356. [CrossRef] [PubMed]
27. Vázquez-Roncero, A.; Janer del Valle, C.; Janer del Valle, M.L. Determinación de polifenoles totales del aceite de oliva. *Grasas Y Aceites* **1973**, *22*, 350–355.

28. *NP 4405;* Microbiologia alimentar: Regras Gerais para a Contagem de Microrganismos. Contagem de Colónias a 30 °C. Instituto Português da Qualidade, Ministério da Economia, da Inovação e do Desenvolvimento: Caparica, Portugal, 2002.
29. *NP 3788;* Regras Gerais para Contagem de Coliformes a 30°C. Instituto Português da Qualidade, Ministério da Economia, da Inovação e do Desenvolvimento: Caparica, Portugal, 1990.
30. *ISO 6888–1;* Microbiologia dos Alimentos. Método para a Contagem de Estafilococos Coagulase. International Organization for Standardization: Geneva, Switzerland, 1999.
31. Barranco, D.; Fernández-Escobar, R.; Rallo, L.; Rallo, L.E.c.d.o. *Coedición. Junta de Andalucia—Consejeria de Agricultura y Pesca;* Ediciones Mundi-Prensa: Madrid, Spain, 1999; pp. 511–596.
32. Gouveia, J. *Azeites Virgens do Alto Alentejo: Comportamentos Químico, Tecnológico e Sensorial. Dissertação para Obtenção do grau de Doutor;* Instituto Superior de Agronomia de Lisboa: Lisboa, Portugal, 1995.
33. García, J.M.; Yousfi, K. Non-destructive and objective methods for the evaluation of the maturation level of olive fruit. *Eur. Food Res Technol.* **2005**, *221*, 538–541. [CrossRef]
34. Kasimoglu, Z.; Tontul, I.; Soylu, A.; Gulen, K.; Topuz, A. The oxidative stability of flavoured virgin olive oil: The effect of the water activity of rosemary. *J. Food Meas. Charact.* **2018**, *12*, 2080–2086. [CrossRef]
35. Rodrigues, N.; Silva, K.; Veloso, A.C.A.; Pereira, J.A.; Peres, A.M. The use of electronic nose as alternative non-destructive technique to discriminate flavoured and unflavoured olive oils. *Foods* **2021**, *10*, 2886. [CrossRef]
36. Soares, V.P.; Fagundes, M.B.; Guerra, D.R.; Leães YS, V.; Speroni, C.S.; Robalo, S.S. Ultrasound assisted maceration for improving the aromatization of extra-virgin olive oil with rosemary and basil. *Food Res. Int.* **2020**, *135*, 109305. [CrossRef]
37. Ayadi, M.A.; Grati-Kamoun, N.; Attia, H. Physico-chemical change and heat stability of extra virgin olive oils flavoured by selected Tunisian aromatic plants. *Food Chem. Toxicol.* **2009**, *47*, 2613–2619. [CrossRef]
38. Seabra, L.M.J.; Damasceno, K.S.F.S.C.; Andrade, S.A.C.; Dantas, M.M.G.; Soares, N.K.M.; Pedrosa, L.F.C. Effect of Rosemary on the quality characteristics of white shrimp (Litopenaeus Vannemei). *J. Food Qual.* **2011**, *34*, 363–369. [CrossRef]
39. Baiano, A.; Previtali, M.A.; Viggiani, I.; Varva, G.; Squeo, G.; Paradiso, V.M. As oil blending affects physical, chemical, and sensory characteristics of flavoured olive oils. *Eur. Food Res. Technol.* **2016**, *242*, 1693–1708. [CrossRef]
40. Benmoussa, H.; Farhat, A.; Elfalleh, W.; di Maio, I.; Servili, M.; Romdhane, M. A rapid application to flavour the olive oil with dried rosmarinus officinalis L. leaves: Microwave-assisted maceration. *J. Food Process. Preserv.* **2016**, *41*, e12885. [CrossRef]
41. Damechki, M.; Sotiropoulou, S.; Tsimidou, M. Antioxidant and pro-oxidant factors in oregano and rosemary gourmet olive oils. *Grasas Y Aceites* **2001**, *52*, 207–213.
42. Taleb, S.A.; Boutoial, K.; Kzaiber, F.; Oussama, A. Effect of aromatization by aromatic plants on the physicochemical, sensorial and oxidative stability of Moroccan virgin olive oil. *Int. J. Chem. Mater. Environ. Res.* **2016**, *3*, 73–77.
43. Mestre, J.J.M. *Extra Virgin Oil Macerated with Tuber Melanosporum, Boletus Edulis and Addition of Edible Gold Particles. Dissertation to obtain the Master's Degree in Food Engineering presented at the Escola Superior Agrária of the Polytechnic;* Institute of Beja: Beja, Portugal, 2017.
44. Middleton, E.; Kandaswami, C.; Theoharides, T.C. The effects of plant flavonoids on mammalian cells: Implications for inflammation, heart disease, and cancer. *Pharmacol. Rev.* **2000**, *52*, 673–751. [PubMed]

processes

MDPI

Article

Optimizing the Extraction Process of Value-Added Products from Olive Cake Using Neuro-Fuzzy Models

Emilio J. Lozano, Gabriel Blázquez *, Mónica Calero, María Ángeles Martín-Lara, Salvador Pérez-Huertas * and Antonio Pérez

Department of Chemical Engineering, University of Granada, Avenida Fuente Nueva s/n, 18071 Granada, Spain; emiliolp@ugr.es (E.J.L.); mcaleroh@ugr.es (M.C.); marianml@ugr.es (M.Á.M.-L.); aperezm@ugr.es (A.P.)
* Correspondence: gblazque@ugr.es (G.B.); shuertas@ugr.es (S.P.-H.)

Abstract: The use of olive cake, an abundant residue in the olive oil industry, has been studied by developing a biorefinery scheme. The aim was to develop a novel, efficient, and environmentally friendly strategy for the valorization of olive cake, contributing to sustainable agriculture. A special extraction procedure based on a combination of hydrothermal treatments with liquid/liquid extractions was designed to produce value-added products, along with solids that can be used for energy or adsorbent production. The optimal extraction conditions were determined by exploring the influence of the operating variables (temperature, extraction time, solvent type, solvent/extract ratio, extraction stages, and pH) on the extraction yield. The decision about the optimal conditions was made by adjusting the experimental results to a neuro-fuzzy model. Glucose and inositol showed similar response surfaces, allowing simultaneous concentration in a single process. Under optimal extraction conditions, the concentration of inositol increased by up to 70%, while glucose and fructose increased by 70 and 30 times, respectively, compared to the initial feed. The proposed methodology successfully extracted significant amounts of bioactive polyols (mainly inositol) (1126 mg/L), saccharides (15,960 mg/L glucose, 385 mg/L xylose, 5550 mg/L fructose, 165 mg/L lactose, and 248 mg/L sucrose), and polyphenols (4792 mg/L) under mild conditions, i.e., 30 °C and 30 min. Thus, olive cake extracts have a great unexploited potential for application in several industrial sectors, including, but not limited to, food and pharmaceuticals.

Keywords: olive cake; extraction; sugars; inositol; polyphenols; neuro-fuzzy models

Citation: Lozano, E.J.; Blázquez, G.; Calero, M.; Martín-Lara, M.Á.; Pérez-Huertas, S.; Pérez, A. Optimizing the Extraction Process of Value-Added Products from Olive Cake Using Neuro-Fuzzy Models. *Processes* **2024**, *12*, 317. https://doi.org/10.3390/pr12020317

Academic Editor: Agnieszka Zgoła-Grześkowiak

Received: 20 December 2023
Revised: 24 January 2024
Accepted: 27 January 2024
Published: 2 February 2024

1. Introduction

Virgin olive oil is a high-quality vegetable oil extracted from the fruits of olive trees (*Olea europaea* L.). For centuries, olive oil has played an essential role not only in Mediterranean cuisine but also in its culture, due to its pleasant organoleptic properties and proven health advantages [1]. Olive oil production is indeed one of the largest agri-food industries throughout the Mediterranean basin, expanding to other countries, such as the USA and Argentina. However, this industry generates a large quantity of by-products and residues, for which integral use has not yet been considered, and its management represents a considerable challenge [2]. One of these by-products is alperujo, which is the aqueous phase resulting from olive oil extraction (≈60%). Alperujo is characterized by its high moisture content and is primarily used to obtain pomace oil. The latter process also generates a solid residue known as olive cake (≈20–25%), which is considered the major waste product of the olive oil extraction process [3]. Different approaches have been proposed to minimize the harmful environmental impacts of these by-products. For instance, olive cake can be used as an agricultural biofertilizer [4], an additive in animal feed [5], or as a source for bio-oil production [6]; however, these applications have certain limitations that have hindered their widespread success [7]. Thus, further research is required to explore novel technological approaches for profitable utilization.

On the other hand, there is an emerging trend in the use of bioactive compounds in the food, nutraceutical, and cosmetic industries, driven by their recognized health benefits [8]. These bioactive compounds are currently obtained via synthesis reactions (generally from glucose) and laboratory-scale methods. Olive cake is primarily composed of polyalcohols, sugars, and polyphenols, making it a valuable natural source of antioxidants. Therefore, the recovery and isolation of bioactive compounds from olive cake could be an interesting ecological and economical alternative to provide this by-product with commercial value. Additionally, olive cake is considered an exploitable biofuel source due to its high energy content [9]. In light of this, developing a biorefinery scheme could offer significant advantages in obtaining commercially valuable products and enhancing the efficiency and sustainability of energy production from olive cake. Industrial implementation of this process would yield both socioeconomic and technical benefits. In short, this scheme must begin with a simple, cost-effective, and environmentally friendly extraction stage, such as hydrothermal treatment, where only water is used as the extracting agent, and the temperature serves to enhance the extraction process. The resulting solid can then be subjected to more complex processes, e.g., chemical or high-pressure treatments, to obtain other products and solids with a high calorific value [10]. Furthermore, certain fermentable compounds can be obtained for the subsequent production of bioalcohols [11]. Consequently, the integral exploitation of this waste will require the design of an optimal extraction stage that allows a high-yield recovery of value-added products from olive cake and improves the energy characteristics of the resulting solid for its subsequent energy utilization.

It is worth noting that the present research group has conducted several studies exploring different hydrothermal treatments to obtain sugars and antioxidants from olive cake, reporting interesting results [10,12]. In those works, a novel and ecofriendly approach was proposed to extract glucose, xylose, polyphenols, and oligomers from olive cake and enhance the calorific value of the resulting solid residue. This approach involves a specific hydrothermal pretreatment of olive cake, followed by an autohydrolysis treatment of the resulting solid. In these works, the influence of the extraction process parameters, i.e., temperature, time, particle diameter and solid/liquid ratio in the glucose, xylose, polyphenols, and oligomers on the extraction yield was explored. To our knowledge, the extraction procedure performed in these studies has not been previously studied by other authors. Hence, this study aims to develop a special extraction procedure that combines hydrothermal treatments with liquid/liquid extractions to achieve maximum yield recovery of value-added products from olive cake within a biorefinery scheme. The hydrothermal treatment was conducted at temperatures and times lower than those used in previous studies [12], where temperatures reached up to 90 °C, and treatment times extended up to 120 min. In this way, it is intended to enhance the energetic properties of olive cake and obtain highly value-added products under mild conditions. The target products were saccharides, (i.e., glucose, fructose, xylose, lactose, and sucrose), polyphenols, and inositol, with a special focus on the latter. Inositol is an organic compound belonging to the polyol family. It is relatively scarce, but of great functional importance. For instance, inositol supplements are frequently used to treat anxiety and stress while also exhibiting beneficial effects in other disorders. Despite some authors having reported the extraction of inositol from natural sources, such as mung beans and lettuce [13,14], its potential extraction from olive cake has not been previously studied by other authors. The extraction process was optimized by identifying the most suitable solvent and exploring the influence of the operating variables (i.e., temperature, time, solvent/extract ratio, number of extraction stages, and pH) on the extraction yield. Finally, the optimal extraction conditions were determined by adjusting the experimental results to a neuro-fuzzy model, which is one of the most powerful and precise prediction tools for process modelling [15].

2. Materials and Methods

2.1. Raw Material and Work Scheme

The olive waste, i.e., olive cake, was provided by a company located in Granada (Spain). The olive cake was ground and sieved to eliminate coarse particles and impurities. The selected particle size ranged from 2 to 0.25 mm, which, according to previous findings, is the optimal diameter for effective hydrothermal extraction [12]. Prior to the hydrothermal extraction, the olive cake was dried in an oven at 80 °C for 24 h (final moisture content ≈ 0%).

To address the aim of this study, the work consisted of four phases: (i) Determination of the optimal hydrothermal conditions; a series of washings was performed on olive cake under different working conditions to maximize the extraction of the compounds of interest. (ii) Determination of the most efficient solvent; the liquid phase resulting from the optimal hydrothermal treatment was subjected to several liquid/liquid extractions to extract and enrich the compounds of interest using different solvents. The extraction efficiencies of the target compounds were assessed to determine the most efficient solvent. (iii) Extraction process variables analysis; the influence of the process variables, i.e., extraction times (t), solvent/extract ratio (r), number of extraction steps (e), and pH (p), on the extraction of polyphenols, sugars, and polyols was explored. (iv) Data curation and process optimization; the concentrations of the extracted compounds and their variations after solvent treatment were determined. A neuro-fuzzy model was developed to fit the calculated data and predict non-experienced data. Finally, the response surfaces were plotted to determine the optimal extraction conditions.

2.2. Hydrothermal Treatment (Washing)

A series of isothermal hydrothermal treatments were carried out to assess the impact of the washing conditions on the extraction of the target compounds. The extraction process parameters were defined based on prior studies [12], i.e., temperatures of 20, 25, and 30 °C; processing times ranging from 30 min to 1 h; particle diameter from 2–0.25 mm, and solid residue/water ratio of 1:3. Thus, a total of six experiments were carried out. The hydrothermal treatment was conducted in a 1 L jacketed glass reactor, equipped with double two-bladed turbine impellers, and connected to a temperature-controlled thermostatic bath. At the end of each extraction, the solid residue was recovered through filtration and subsequently washed with distilled water to determine the gravimetric yield. Aliquots of the liquid fractions were taken to determine the contents of saccharides, polyols (mainly inositol), and polyphenols extracted from the olive cake using HPLC. Figure 1 depicts a schematic of the hydrothermal extraction process for olive cake.

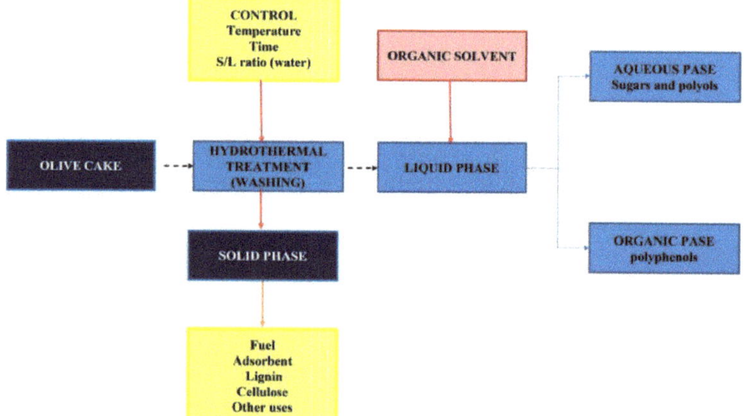

Figure 1. Schematic of the extraction process.

2.3. Liquid/Liquid Extraction

A special procedure was performed to enrich and extract the compounds of interest present in the liquid phase. This procedure was based on a simple and optimized liquid/liquid extraction separation technique. Firstly, polyphenols were separated using different organic solvents, i.e., ethyl acetate, hexane, and dichloromethane. These solvents have low polarity and are thus poorly miscible with water, which facilitates the extraction of polyphenols into an organic phase [16,17]. It is important to note that solvent selection was carried out in accordance with the Spanish Ministry of Health, Consumption and Social Welfare, the Spanish Agency for Food Safety and Nutrition (AECOSAN), and Royal Decree 1101/2011, of 22 July [18], which regulates the use of some solvents for food production. In order to determine the most suitable solvent, a series of liquid/liquid extractions was carried out with different solvent/extract ratios, i.e., 1:1, 1:2, and 1:3. Thus, each solvent was subjected to three experiments, each involving three steps. In stage I, 100 mL of solvent and 33.3 mL of liquid phase (extract) were added to the first beaker, 100 mL of solvent and 50 mL of extract were added to the second beaker, and 100 mL of solvent and 100 mL of extract were added to the third beaker. Then, the beakers were shaken for 24 h. Subsequently, the organic phase (rich in polyphenols) was separated from the aqueous phase (rich in polyols and sugars) by means of decanting process. A sample from each organic phase was collected and stored in a refrigerator for later analysis, while the aqueous phase proceeded to stage II. In stage II, the same procedure was followed using the aqueous phase from stage I. Aqueous phase I was mixed with the solvent, and after stirring, a sample from the organic phase was taken and stored in a refrigerator. The aqueous phase from stage II was subjected to subsequent experiments. In the last stage, the same procedure was repeated and samples from both the organic and aqueous phases were collected and stored in a refrigerator for later analysis. Therefore, four samples of each solvent (i.e., organic phases I, II, and III; aqueous phase) were collected to determine the most efficient solvent. This was achieved by determining the total amount of phenols present in each aqueous sample using the Folin–Ciocalteu method [19]. A low or zero phenol content in the aqueous phase (resulting from stage III) would indicate that the polyphenols have mostly moved to the organic phase, thereby confirming the effectiveness of the solvent.

Liquid/Liquid Extraction: Optimization Extraction Process

A factorial experimental design was performed based on the approach described in [20]. Briefly, the model was designed with four input variables (number of extraction steps (e), extraction time (t), solvent/extract ratio (r), and pH (p)), one response variable (compound concentration), and a central point. Each variable was assigned three levels, namely low (−1), intermediate (0), or high (1). The number of required experiments was determined using the following equation:

$$n = 2^{k-p} + 2k + n_c \tag{1}$$

where n is the number of experiments, n_c the number of central points, p is the constant for values of k, and k is the number of independent variables (if $k < 5$; $p = 0$, if $k > 5$, $p = 1$), so:

$$n = 2^{5-1} + 2 \times 5 + 1 = 27$$

Therefore, a total of 27 experiments were required to elucidate the operation and to predict the outputs. These experiments were conducted following a similar procedure to the previous extraction tests, but in this case, varying the extraction times, number of extraction stages, solvent/extract ratio, and pH (Figure 2). The latter variable is particularly important because the concentrations of polyphenols and sugars vary depending on pH [21].

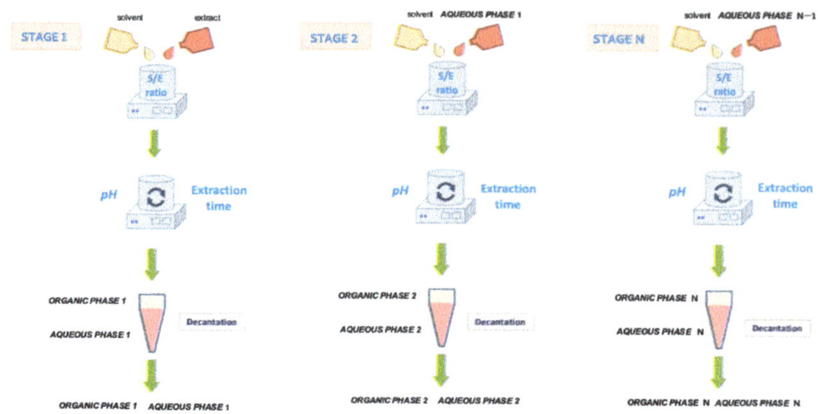

Figure 2. Schematic representation of the procedure followed for the optimization of the extraction process.

First, the organic solvent was mixed with the liquid phase (extract) according to the defined solvent/extract ratio (stage 1). The solution was then stirred for the specified extraction time, and the pH was adjusted to the defined level. Subsequently, the organic and the aqueous phases were separated by means of decanting process. In stage 2 (and subsequent stages), the aqueous phase obtained from the previous stage was mixed with the solvent and stirred under the same conditions as the previous stage. Then, the organic and liquid phases were separated by decanting. This sequence was repeated based on the predetermined number of extractions. Finally, the resulting aqueous phase was stored in the refrigerator until subsequent chromatographic analysis. Table 1 lists the operating variables assigned for the identification of the most effective extraction conditions.

Table 1. Extraction operating conditions with factors.

Time (t)	S/E Ratio (r)	No. Stages (e)	pH (p)	Code
30 min	100/100	2	3.5	−1
75 min	75/100	3	4	0
120 min	50/100	4	4.5	+1

It is important to note that these data are derived from a thorough analysis of results from several dozen experiments, as well as data curation from prior studies [10,12]. For instance, extractions conducted with a solvent/extract ratio lower than 1:2 were not suitable for the present purpose. Similarly, performing more than four extraction stages or extractions for longer than 120 min did not improve the concentration of the target compounds. Consequently, ratios ranging from 1:1 to 1:2 or extractions shorter than 120 min were further considered. The factor parameters were coded by the values −1 and +1, representing the maximum and minimum values within the defined domain, respectively. Parameter 0 represents the central value, with experiments 10, 17, and 27 being the central points. Table 2 presents the 27 experiments conducted for the designed model.

Table 2. Experimental extraction conditions designed for the mathematical model.

Exp.	Time min	S/E Ratio	Number Stages	pH	Exp.	Time min	S/E Ratio	Number Stages	pH
1	30	100/100	2	4.5	15	75	75/100	4	4
2	30	100/100	2	3.5	16	75	75/100	2	4
3	30	75/100	3	4	17 *	75	75/100	3	4
4	30	50/100	4	4.5	18	120	50/100	4	4.5
5	30	50/100	4	3.5	19	120	50/100	4	3.5
6	30	100/100	4	3.5	20	120	75/100	3	4
7	30	100/100	4	4.5	21	120	50/100	2	4.5
8	30	50/100	2	4	22	120	50/100	2	3.5
9	30	50/100	2	3.5	23	120	100/100	4	3.5
10 *	75	75/100	3	4	24	120	100/100	4	4.5
11	75	75/100	3	4.5	25	120	100/100	2	4.5
12	75	75/100	3	3.5	26	120	100/100	2	3.5
13	75	50/100	3	4	27 *	75	75/100	3	4
14	75	100/100	3	4					

* Central points.

2.4. Neuro-Fuzzy Model

A neuro-fuzzy model was used with the aim of fitting the data. This mathematical model combines the advantages of fuzzy logic systems and neural networks, offering a powerful prediction tool [22]. It is based on the following equation, with two independent variables, the use of rules (μ), a constant, and a Gaussian dependence function:

$$y_e = \frac{\sum_{l=1}^{m} y^l \cdot \left[\prod_{i=1}^{n} \mu_{F_i}^{l}\left(x_i, \theta_i^l\right)\right]}{\sum_{l=1}^{m}\left[\prod_{i=1}^{n} \mu_{F_i}^{l}\left(x_i, \theta_i^l\right)\right]} \quad (2)$$

where y_e is the estimated value of the property to be modelled; μ represents a fuzzy rule; x_i, θ_i indicate the values of time (t); solvent/extract ratio (r); number of stages (e); pH (p). A Gaussian dependence function with three levels (low, medium, and high) was used for one of the variables and a Gaussian dependence function with two levels (low and high) for the other three. Thus, with 4 variables, n was 4, and m the number of fuzzy rules. Taking this into account, the numerator and denominator would contain 24 terms, respectively. The Gaussian dependence function would be as follows:

$$\mu(\text{low}) = \exp\left(-0.5 \times \left(\frac{c - c_{low}}{L}\right)^2\right) \quad (3)$$

$$\mu(\text{medium}) = \exp\left(-0.5 \times \left(\frac{c - c_{medium}}{L}\right)^2\right) \quad (4)$$

$$\mu(\text{high}) = \exp\left(-0.5 \times \left(\frac{c - c_{high}}{L}\right)^2\right) \quad (5)$$

where c is the absolute value of the variable and L is the width of its Gaussian distribution. The parameters and constants of the above equations were estimated using the ANFIS (Adaptive Neural Fuzzy Inference System) Edit tool. Finally, the rates of increase/in of compounds concentrations were fitted to a neuro-fuzzy model and the response surfaces were prepared to better understand the influence of the variables in the extraction.

Relative Value and Relative Increase

Based on the poor correlation between the response variable (concentrations of the compounds of interest) and the operating ones (time (t), S/E ratio (r), number of stages (e) and pH (p)) found in previous works [10,12], the data obtained were adjusted before performing the fit. For this purpose, relative values and relative increments were calculated using the concentration of each component in the extract. Relative values (Equation (6)) were calculated for compounds that were present in the extract and relative increments (Equation (7)) for the compounds not extracted in the hydrothermal treatment.

$$\text{Relative value} = \frac{(C_{ext} - C_h)}{C_h} \tag{6}$$

$$\text{Relative increase} = \frac{(C_{ext} - Cmin_{ext})}{(Cmax_{ext} - Cmin_{ext})} \tag{7}$$

where C_{ext} is the concentration of the compound present in the aqueous phase after L/L extraction, C_h is the concentration of the compound present in the extract after hydrothermal extraction, and $Cmín_{ext}$ and $Cmax_{ext}$ are the minimum and maximum concentrations of the compound after L/L extraction, respectively. The parameters and constants of Equation (2) were estimated using a Gaussian dependence function with three levels (low, medium, and high) for one independent variable and Gaussian dependence functions with two levels (low and high) for three operational variables. We tested which combination of levels provides the most similar values to the experimental ones with higher R^2. The estimation was determined using ANFIS Edit tool. It was found that the variable with three levels makes the model more effective. Outliers were discarded.

2.5. Determination of Sugars and Oligomers

Chromatographic determination was performed to quantify the sugars, i.e., glucose, xylose, fructose, lactose, sucrose, and inositol, content using an HPLC 940 professional IC Vario (Metrohm, Herisau, Switzerland) equipped with a conductivity meter as detector, and a column Metrosep Carb 2—250/4.0., under the following conditions: mobile phase composed of 100 mM NaOH and 10 mM NaAc; a flow rate of 0.500 mL/s; and an operating temperature of 30 °C. The retention times were determined using the reference chemical standard of each compound. The retention times for each target compound are given in Table 3.

Table 3. Retention times of each measured compounds.

Compound	Retention Times (min)
Inositol	5.242
Glucose	16.284
Xylose	17.183
Fructose	19.381
Lactose	27.923
Sucrose	33.357

The chromatography analyses were repeated four times for each sample, and the results were reported as arithmetic means. Any results that deviated by 5% or more from the mean were excluded.

2.6. Determination of Polyphenols: Folin–Ciocalteu Method

The Folin–Ciocalteu method measures the total phenolic compounds present in vegetable products by assessing the ability of phenols to react with oxidizing agents. It is based on the capability of phenolic compounds to react with Folin–Ciocalteu reagent at basic pH, producing a blue color that can be measured spectrophotometrically at 765 nm. The reagent contains molybdate and sodium tungstate which react with phenols to form phosphomolybdic–phosphotungstic complexes. At a basic pH, the phosphomolybdic–phosphotungstic complexes are reduced to deep blue chromogenic oxides of tungsten (W_8O_{23}) and molybdenum (Mo_8O_{23}) through electron transfer. The intensity of the resulting color is directly proportional to the number of hydroxyl groups present in the molecule. The concentration of total phenols is expressed as milligrams of Gallic Acid Equivalents

(GAE) per unit weight. The following reagents were used: 98% w/v, Na_2CO_3, (Panreac, Castellar del Vallès, Spain), Folin–Ciocalteu reagent 2N (Sigma-Aldrich, St. Louis, MO, USA), ethanol, and gallic acid.

The gallic acid standard curve (see Figure S1 in the Supplementary Materials) was established by dissolving 0.5 g of gallic acid in 10 mL of ethanol and then diluting it to 100 mL with distilled water. Solutions with concentrations of 50, 100, 150, 250, 500, 1000, and 2000 mg/L of gallic acid were prepared from this solution. The total phenols content was determined using the following steps. Initially, 100 µL of the sample was mixed with 8 mL of distilled water. Subsequently, 500 µL of Folin's reagent was added, and the resulting solution stirred using a vortex. Then, 1.5 mL of carbonate solution was added, and the solution was stirred again. Finally, the solution was placed in an oven at 40 °C for 30 min. After cooling, the absorbance at 765 nm was measured.

3. Results and Discussion

3.1. Hydrothermal Treatment

Table 4 presents the concentrations of inositol and saccharides obtained from the liquid fractions of various hydrothermal treatments of the olive cake.

Table 4. Results of hydrothermal treatment of olive cake at different conditions.

Temperature °C	Time min	Inositol mg/L	Glucose mg/L	Xylose mg/L	Fructose mg/L	Sucrose mg/L
20	30	645.6	257.9	51.7	182.5	162.2
25	30	657.1	218.3	43.1	188.1	186.1
30	30	661.8	228.3	-	185.6	168.7
20	60	627.7	208.7	-	298.3	220.1
25	60	661.1	218.3	-	310.7	215.9
30	60	655.2	213.1	-	297.9	203.5

Optimal operating conditions were found to be 30 °C and 30 min, as they produced the highest inositol concentration. Furthermore, analysis of the liquid fractions revealed the presence of saccharides, such as glucose, sucrose, fructose, and to a lesser extent xylose. In any case, the extraction yield of the main component (inositol) was larger than that of all saccharides (Table 4). Therefore, the hydrothermal treatment proposed in this study proved to be more efficient and cost-effective than that in previous studies [12], where higher temperatures (50, 70, and 90 °C) and longer treatment durations (60–120 min) were needed to extract the compounds of interest (e.g., sugars and antioxidants). The liquid phase resulting from optimal hydrothermal treatment was subjected to a special procedure to assess the potential for further enrichment of these compounds (see Figure 2).

3.2. Liquid/Liquid Extraction

Table 5 shows the phenol concentrations in several aqueous phases resulting from liquid/liquid extractions with different solvents.

Table 5. Total phenol concentration in different aqueous phases. Study of the most suitable solvent.

Solvent	S/E Ratio	Absorbance	Phenol Concentration mg/L
None		0.681	4792
Ethyl acetate	100/100	0.25	1713
Ethyl acetate	100/50	0.444	3099
Ethyl acetate	100/33.3	0.468	3270
Dichloromethane	100/100	0.332	2299
Dichloromethane	100/50	0.636	4470
Dichloromethane	100/33.3	0.66	4642
Hexane	100/100	0.351	2435
Hexane	100/50	0.662	4656
Hexane	100/33.3	0.695	4892

It was found that the lowest phenol concentration, i.e., 1713 mg/L, was obtained using ethyl acetate as solvent. Consequently, ethyl acetate was the most efficient solvent for the present purpose. As mentioned above, these solvents are poorly miscible with water, which facilitates the extraction of polyphenols into the organic phase. After mixing the liquid fraction resulting from hydrothermal treatment with these solvents, two phases appeared: an organic phase (enriched in polyphenols) and an aqueous phase (enriched in inositol and saccharides). Since the phenol concentration of the native aqueous phase was 4792 mg/L (Table 5), a decrease in phenol concentration in the aqueous phase indicates the migration of polyphenols to the organic phase. Thus, the lower the phenol content, the more efficient the solvent. Regarding the S/E ratio, the phenol content tended to increase as the S/E ratio decreased. The highest phenol concentrations were found at an S/E ratio of 1:3, regardless of the solvent used. Consequently, the optimal conditions involve extraction using ethyl acetate with an S/E ratio of 1:1.

A series of liquid/liquid extractions were carried out to explore the influence of the number of extraction stages, pH, S/E ratio, and extraction time on phenol concentration. The extractions were conducted using ethyl acetate as the solvent. Table 6 shows the extraction conditions and phenol concentrations in the resulting aqueous phase.

Table 6. Additional experiments, total phenol concentrations.

Exp.	S/E Ratio	Stage	Time h	pH	Absorbance	[Phenol] mg/L
1	100/100	1	1	5.18	0.417	2906
		2	1	5.19	0.404	2813
		3	1	5.35	0.381	2649
2	100/33.3	1	1	5.37	0.42	2927
		2	1	5.54	0.422	2942
		3	1	5.73	0.473	3306
3	100/100	1	3	5.07	0.501	3506
		2	3	5.15	0.419	2920
		3	3	5.28	0.44	3070
4	100/33.3	1	3	5.3	0.425	2963
		2	3	5.48	0.443	3092

Increasing the extraction time from 1 to 3 h did not decrease the phenol concentration, regardless of the number of stages performed or the S/E ratio. At S/E ratios of 1:1 (exp. 1 and 3), the aqueous phase exhibited the highest concentration of phenols (indicating lower phenol extraction) in the first extraction stage. In contrast, at the S/E ratio of 1:3 (exp. 2 and 4), the highest phenol concentration was observed in the last extraction stage. Therefore, extractions using S/E ratios of 1:1 improve when performing two or three stages, as opposed to extractions using S/E ratios of 1:3, which worsened after the first extraction stage. The lowest concentration of phenols (2649 mg/L) was found after performing three stages of 1 h using an S/E ratio of 1. Finally, the pH increased after each extraction, regardless of the S/E ratio or extraction time. Subsequent extractions were conducted based on these findings.

3.3. Study of the Variables of the Extraction Process

This section presents the proposed models designed to optimize the extraction process and enrich target compounds. Neuro-fuzzy models were validated using the coefficient of correlation (R^2). The response surfaces were used to determine the optimal output values (compound concentrations) and related input values (extraction time, number of extraction stages, pH, and solvent/extract ratio).

3.3.1. Inositol

Table 7 shows the results of the experiments conducted to optimize of the inositol extraction process. A total of 27 experiments were designed based on previous findings (see Table 2). The experimental data were introduced in the ANFIS Edit Tool to estimate the parameters and constants for Equation (2), which are also given in Table 7.

Table 7. Experimental results of inositol extractions, estimated ANFIS values and model constants.

Exp.	Experimental mg/L	Relative Value	Estimation ANFIS	% Error	Exp.	Experimental mg/L	Relative Value	Estimation ANFIS	% Error
1	641	−0.03	−0.03	1.14	15	719	0.09	0.09	0.92
2	561	−0.15	−0.15	0.54	16	524	−0.21	−0.21	0.29
3	823	0.24	0.25	1.81	18	698	0.05	0.06	15.01
4	916	0.38	0.38	0.54	19	671	0.01	0.01	5.64
5	618	−0.07	−0.07	0.52	20	703	0.06	0.04	31.82
6	962	0.45	0.45	0.19	21	685	0.03	0.04	16.32
7	690	0.04	0.04	1.80	22	630	−0.05	−0.05	3.31
8	794	0.20	0.20	1.08	23	754	0.14	0.14	0.02
9	823	0.24	0.24	0.02	24	644	−0.03	−0.03	2.78
11	652	−0.02	−0.02	0.24	25	700	0.06	0.06	0.37
12	588	−0.11	−0.11	0.43	26	694	0.05	0.05	0.49
13	608	−0.08	−0.08	0.23	10 * 17 * 27 *	637.49	−0.04	−0.03	8.40

Constants				Variables		Value L	Molecule
a1	0.314	a13	0.051		30	19.1097	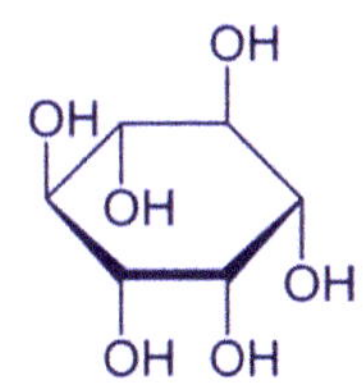
a2	0.205	a14	0.512	t (min)	75	19.1097	
a3	−0.132	a15	0.167		120	19.1097	
a4	0.407	a16	0.848	r	0.5	0.2298	
a5	−0.326	a17	−0.059		1	0.1261	
a6	−0.104	a18	0.063	e	2	0.7648	
a7	0.641	a19	0.022		4	0.8928	
a8	−0.055	a20	0.063	p	3.5	0.2778	
a9	−0.150	a21	0.049		4.5	0.4594	
a10	−0.366	a22	0.042	Coefficient of determination R^2			
a11	−0.172	a23	0.184	0.999			
a12	0.061	a24	−0.096				

* Central point.

The data from the 27 experiments were fitted to a Gaussian dependence of 3 × 2 × 2 × 2. As shown, there was a minimal difference between the experimental and ANFIS estimated values, showing a low error rate with an R^2 coefficient of 0.999. An error rate equal to or lower than 5% is considered acceptable, indicating that ANFIS can be effectively employed to provide a reliable and precise prediction model. The response surface was plotted as a function of pH, number of stages, extraction times, and solvent/extract ratios. The inositol concentration was expressed in relation to that obtained in the hydrothermal treatment, i.e., 661 mg/L (Table 4). The surface plot from the neuro-fuzzy fitting of the optimal conditions to obtain inositol is presented below (Figure 3).

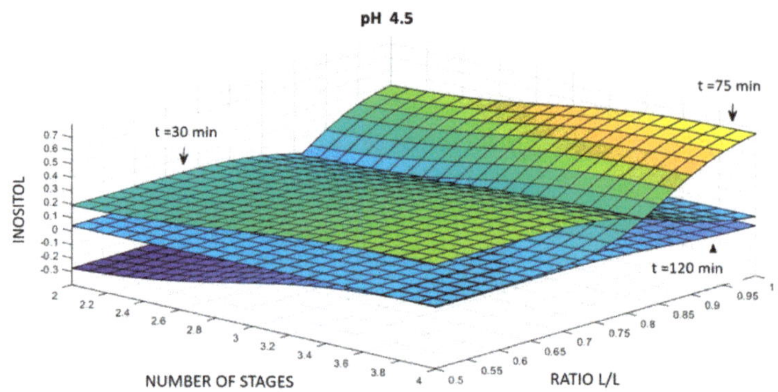

Figure 3. Surface plot of the most effective inositol extractions as a function of the studied variables.

Figure 3 displays the three-dimensional illustration of the effect of each input (number of stages, solvent/extract ratio, and the extraction time) on the output (inositol concentration) at pH 4.5 found by the neuro-fuzzy model. This shows that a higher amount of inositol was obtained (50–70%) after conducting more than three stages of 75 min with ratios exceeding 0.75. The inositol concentration also increased (20%) when using an extraction time shorter than 75 min. Under these conditions, neither the number of stages nor the ratio influenced the extraction yield, as the inositol concentration remained virtually constant (Figure 3). Additionally, extraction times longer than 75 min were not effective, as they resulted in lower inositol yields compared to those obtained in the hydrothermal treatment, i.e., 562 and 661 mg/L, respectively.

The response surface obtained at pH 4 was similar to that obtained at pH 4.5 (see Figure S2). A higher inositol concentration (1057 mg/L) was achieved by conducting four extraction stages of 75 min at a ratio of 1. Moreover, extraction for 30 min increased the inositol concentration by up to 20%, while extraction for 120 min reduced it. At pH 3.5, the most enriched inositol extract was achieved by performing extractions for 30 min, either with more than three stages and ratios higher than 0.75 (up to 962 mg/L), or with less than three stages and ratios lower than 0.5 (up to 823 mg/L) (Table 7). Performing longer extractions was not effective, as the inositol concentration remained virtually constant with respect to that obtained in the hydrothermal treatment (Figure S2).

Consequently, the optimal conditions for the extraction of inositol involved a solvent/extract ratio of 1, pH 4.5, extraction time of 75 min, and four extraction stages. Extractions under these conditions led to a significant increase in inositol concentration of up to 70%, i.e., 1126 mg/L (Figure 4). Zuluaga et al. [14] obtained inositol from different lettuce types using a microwave-assisted extraction process. The highest inositol concentration (5.42 mg/g dry sample) was obtained using a liquid–solid ratio of 100:1, performing one extraction for 30 min at 40 °C, with an ethanol–water mixture as the solvent. Ruiz-Aceituno et al. [23] developed a pressurized liquid extraction method to obtain inositol from pine nuts and reported an inositol concentration of 5.7 mg/g. Optimal conditions were as follows: 50 °C, 18 min, three cycles of 1.5 mL water each, at 10 MPa. Therefore, the extraction process developed in this study has proven to be an efficient and cost-effective strategy to obtain enriched extracts of inositol from olive cake waste under mild conditions.

3.3.2. Glucose

Table 8 shows the experimental results for the optimization of the glucose extraction process, parameters, constants, and the ANFIS model fit.

Table 8. Experimental results of glucose extractions, estimated ANFIS values, and model constants.

Exp.	Glucose mg/L	Relative Value	Estimation ANFIS	% Error	Exp.	Glucose mg/L	Relative Value	Estimation ANFIS	% Error
1	755	2.31	3.04	31.59	15	10,029	42.93	42.93	0.01
2	672	1.94	1.88	3.20	16	6679	28.26	28.24	0.06
3	1031	3.52	3.48	1.14	18	9376	40.07	40.03	0.09
4	12,798	55.06	55.42	0.67	19	7720	32.81	32.50	0.94
5	8582	36.59	36.58	0.04	20	7901	33.61	34.43	2.45
6	13,078	56.28	56.34	0.09	21	8983	38.35	38.29	0.13
7	9351	39.96	39.95	0.01	22	8284	35.28	34.86	1.21
9	11,368	48.79	48.79	0.01	23	9692	41.45	41.47	0.03
11	8900	37.98	37.98	0.01	24	8144	34.67	34.66	0.04
12	8040	34.21	33.97	0.71	25	9228	39.42	39.42	0.01
13	8316	35.43	34.41	2.88	26	8353	35.59	35.58	0.00
14	8954	38.22	38.22	0.00	10 * 17 * 27 *	7779	33.08	34.41	4.03
Constants				Variables		Value L		Molecule	
a1	50.548	a13	46.5	t (min)	30	19.1097		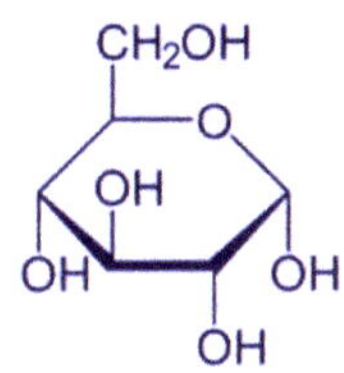	
a2	−735.60	a14	14.997		75	19.1094			
a3	35.061	a15	36.887		120	19.1097			
a4	127.85	a16	7.227	r	0.5	0.2221			
a5	−10.921	a17	35.357		1	0.0737			
a6	119.23	a18	37.515	e	2	0.888			
a7	67.376	a19	31.718		4	0.7697			
a8	29.409	a20	38.647	p	3.5	0.4412			
a9	27.995	a21	34.806		4.5	0.2602			
a10	56.012	a22	41.797	Coefficient of determination R^2					
a11	41.325	a23	43.996	0.999					
a12	79.067	a24	33.736						

* Central point.

The data from the 27 experiments were fitted to a Gaussian dependence of 3 × 2 × 2 × 2. Similar to the previous case, a good fit with a low error rate (<5%) was obtained. The coefficient of determination was R^2 = 0.999. The response surface was plotted as a function of pH, number of stages, extraction times, and S/E ratios. The glucose concentration was expressed in relation to that achieved during the optimal hydrothermal treatment, i.e., 228 mg/L (Table 4). The surface plot from the neuro-fuzzy fitting of the optimal conditions to obtain glucose is shown in Figure 4.

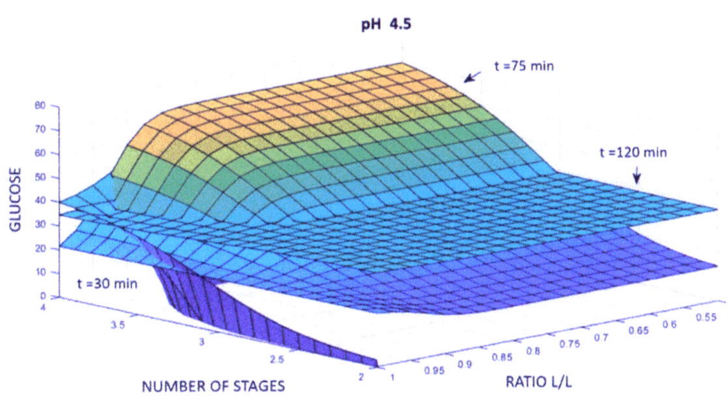

Figure 4. Surface plot of the most effective glucose extractions as a function of the studied variables.

A higher concentration of glucose was achieved (45–70-fold increase) by performing more than three stages with extraction times of 75 min, a ratio lower than 0.9, and a pH of 4.5 (Figure 4). Additionally, extractions longer than 75 min resulted in an extract with a concentration 35 times greater than that obtained in the hydrothermal treatment, regardless of the number of stages or solvent/extract ratio used.

The response surfaces obtained at pH 3.5 and 4 were similar (Figure S3). The most enriched glucose extracts (13,078 and 12,210 mg/L) were achieved after four stages of 30 min and a ratio of 1 at pH levels of 3.5 and 4, respectively. Extractions longer than 30 min resulted in a significant increase in the glucose concentration (40-fold), which remained almost constant with respect to the number of stages and solvent/extract ratio.

Therefore, the most efficient extraction conditions included a solvent/extract ratio ranging from 0.9 to 0.5, a pH of 4.5, and four extraction stages of 75 min. These specific conditions resulted in an extract containing 15,960 mg/L of glucose, indicating that glucose was concentrated by approximately 70 times compared to the initial feed. The glucose concentration obtained from olive cake was larger than that obtained from other materials using similar extraction techniques. For instance, López et al. [24] conducted sugar extraction from sunflowers stalks using autohydrolysis, and the resulting liquor contained 960 mg/L of glucose, whereas Casas-Godoy et al. [25] reported lower concentrations of reduced sugars extracted from blue agave bagasse using water as the solvent (680 mg/L). Furthermore, Zoubiri et al. [26] obtained 8.14 mg/mL of fructose from apricot waste through hydrolysis at 80 °C for 30 min. These differences can be attributed to the distinct nature of the precursors. It is worth noting that glucose exhibited a similar response to inositol, with the optimal extraction conditions being quite similar for both compounds. This provides the opportunity for their simultaneous concentration within a single process.

3.3.3. Xylose, Fructose, Lactose, and Sucrose

The surface plots from the neuro-fuzzy fitting of the optimal conditions to obtain xylose, fructose, lactose, and sucrose are given in Figure 5.

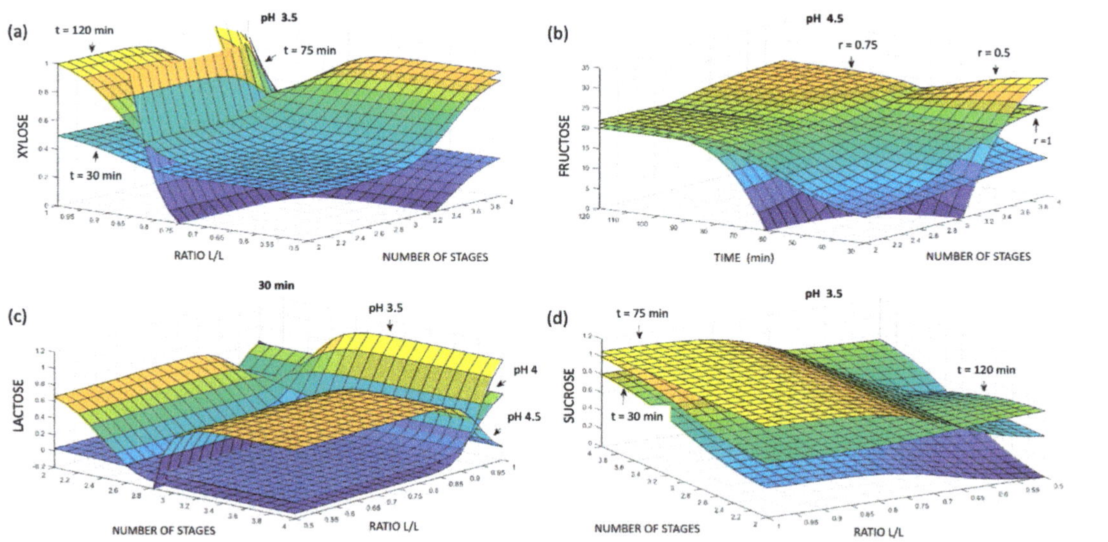

Figure 5. Surface plot of the most effective extractions of (**a**) xylose, (**b**) fructose, (**c**) lactose, and (**d**) sucrose as a function of the studied variables.

Generally, a strong fit with high coefficient of determination, i.e., R^2 = 0.999–0.983, was obtained for these saccharides (see Tables S1–S4). The surface plot from the neuro-fuzzy fitting of xylose extractions at a pH of 3.5 is shown in Figure 5a. The xylose concentration was expressed in relation to the relative increments calculated from Equation (7). It was found that a greater xylose content was achieved by conducting less than four stages, with extraction times longer than 75 min, and employing a ratio larger than 0.75 at a pH of 3.5 (up to 385 mg/L). At pH 3.5 and an extraction time of 30 min, the solvent/extract ratio did not exhibit a significant influence on the response variable, as it remained nearly constant. This was also observed at pH values of 4 and 4.5, with extractions of 120 min (Figure S4). It is worth noting that no xylose content was found in the liquid phase resulting from the hydrothermal treatment (see Table 4), but it was detected after liquid/liquid extraction using ethyl acetate as the solvent. This could be attributed to several factors, e.g., a potentially low extracted concentration that might fall below the detection limit of the chromatograph or the possible degradation of xylose during the extraction process. Additionally, hydrothermal extraction conditions might not have been suitable for the effective extraction of this compound (monomer), necessitating the use of an organic solvent for its extraction. Similar findings have been reported in other studies, where no xylose content was found after hydrothermal treatment of several woods but was subsequently identified in the liquid phase after post-treatment [24]. In conclusion, the optimum conditions to obtain extracts enriched in xylose were to perform two stages of 75 min with a ratio of 1 at pH 3.5.

Figure 5b shows the surface plot from the neuro-fuzzy fitting of the optimal conditions to obtain fructose. The fructose concentration was expressed in relation to that achieved during the optimal hydrothermal treatment, i.e., 185 mg/L (Table 4). A higher amount of fructose was achieved when performing four stages with an extraction time either shorter than 75 min and a ratio of 0.5 at pH 4.5 (30-fold increase) (Figure 5b), or longer than 75 min, and a ratio of 0.75 at pH 3.5 (25-fold increase) (Figure S5). The minimum values of fructose were obtained when fewer than three steps were performed with times shorter than 75 min, regardless of pH. Therefore, the optimum conditions included a solvent/extract ratio of 0.5, pH 4.5, extraction times of 30 min, and four stages, resulting in an extract containing 5550 mg/L fructose. These findings are interesting because a low quantity of solvent, low acidification rate (4.5 pH of the extract), and short extraction times can result in a substantial increase in the fructose concentration (approximately 30 times greater).

Similar to xylose, lactose was not detected in the extract resulting from the hydrothermal treatment (Table 4). Lactose is a dimer resulting from the union of glucose and galactose molecules, and hydrothermal treatment may not be suitable for its formation [27]. However, the L/L extraction process probably facilitated their bonding, leading to the formation of lactose molecules during this process. This may explain why lactose was not detected in the extract of some experiments (Table S3). Figure 5c shows the effect of the number of stages, solvent/extract ratio, and pH on the lactose concentration for 30 min extractions determined with the neuro-fuzzy model. In this case, the resulting response surfaces were very similar to each other (Figure S6). This indicates that the lactose concentration remained almost constant for extraction times of 30, 75, and 120 min. At pH 4.5, a higher lactose content was achieved by performing three to four stages with a ratio ranging from 0.5 to 0.75, regardless of the extraction time. At pH 3.5, a higher lactose concentration was obtained by conducting more than three stages using ratios lower than 0.8. Thus, the optimal conditions included performing four extraction stages for 30 min with a solvent/extract ratio of 1.

Figure 5d displays the surface plot from the neuro-fuzzy fitting of the optimal sucrose extraction conditions. At pH 3.5, higher extraction of sucrose was achieved by performing extractions for 75 min with ratios larger than 0.75, regardless of the number of stages. Extractions of 120 min were also effective when conducting four stages with a ratio of 1. The response surfaces for extractions at pH levels of 4 and 4.5 were nearly identical; thus, the sucrose content remained constant at these pH levels. In both cases, higher sucrose extraction was obtained after four stages of 30 min at a ratio of 0.5 (Figure S7). For longer extraction times, a higher amount of sucrose was extracted by either performing two stages with a ratio of 0.5 (75 min extraction times) or four stages with a ratio larger than 0.8

(120 min extraction time). The optimum conditions included a ratio of l, pH 3.5, extraction times of 75 min, and two extraction stages.

3.3.4. Polyphenols

The data were fitted to a Gaussian dependence of 3 × 2 × 2 × 2. The experimental data and the ANFIS estimates were practically identical; thus, a strong fit with a low error rate (<5%), and a high coefficient of determination (R^2 = 0.999) was achieved (Table S5). In this case, the optimal extraction conditions involved the lowest phenol concentration. It means that the polyphenols migrated to the organic phase, resulting in an aqueous phase enriched in inositol and saccharides. The response surface was plotted as a function of pH, number and duration of extraction stages, and solvent/extract ratios. The polyphenols concentration was expressed in relation to that obtained in the liquid phase resulting from the hydrothermal treatment, i.e., 4792 mg/L (Table 4). Figure 6 shows the surface plot resulting from the neuro-fuzzy fitting of the least effective polyphenol extractions.

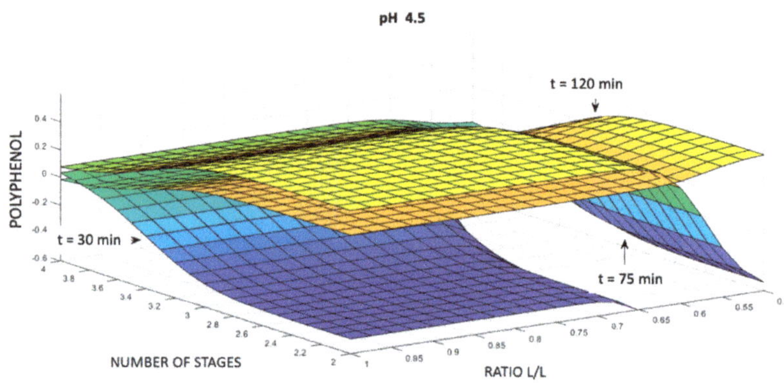

Figure 6. Surface plot of the polyphenol extractions as a function of the studied variables.

Generally, the extraction time was the most influential variable, whereas the pH had almost no influence. The minimum values were obtained by performing fewer than three stages of 30 min with ratios greater than 0.65. Additionally, a low concentration of phenols was achieved by performing two to three stages of 75 min with a ratio of 0.5. Conducting four stages with ratios larger than 0.65 was the most efficient conditions for longer extractions. In any case, the optimum polyphenol extraction conditions (lowest concentration of polyphenols, 2865 mg/L) involved performing two stages of 30 min with a ratio of 0.65, regardless of the pH.

4. Conclusions

The use of olive cake was studied by developing a biorefinery scheme. The aim was to produce value-added products of potential food and pharmaceutical interest, such as saccharides, polyphenols, and polyols, as well as solids that can be used for energy or adsorbent production. All experimental results were perfectly fitted to neuro-fuzzy models. The optimal inositol extraction conditions involved a solvent/extract ratio of 1, pH 4.5, extraction time of 75 min, and four extraction stages. Glucose and inositol showed similar response surfaces, allowing simultaneous concentration in a single process. The designed extraction conditions allowed the recovery of a significant amount of the target compounds, i.e., inositol—1126 mg/L, glucose—15,960 mg/L, xylose—385 mg/L, fructose—5550 mg/L, lactose—165 mg/L, sucrose—248 mg/L, and polyphenols—4792 mg/L, under mild conditions. The results indicate that olive cake waste is a promising source of bioactive compounds with potential applications as ingredients in functional foods or nutraceuticals. However, research covering extract purification would be desirable since the potential applications of the final products are determined by their purity.

Supplementary Materials: The following supporting information can be downloaded at: https://www.mdpi.com/article/10.3390/pr12020317/s1, Figure S1: Gallic acid standard curve Folin-Ciocalteu method (λ = 765 nm); Figure S2: Surface plot of inositol extractions at pH of 3.5 (left) and 4 (right) as a function of the studied variables; Figure S3: Surface plot of glucose extractions at pH of 4 (left) and 3.5 (right) as a function of the studied variables; Figure S4: Surface plot of xylose extractions at pH of 4.5 (left) and 4 (right) as a function of the studied variables; Figure S5: Surface plot of fructose extractions at pH of 3.5 as a function of the studied variables; Figure S6: Surface plot of lactose extractions for 75 min (left) and 120 min (right) as a function of the studied variables; Figure S7: Surface plot of sucrose extractions at pH of 4.5 (left) and 4 (right) as a function of the studied variables; Figure S8: Surface plot of polyphenols extractions at pH of 4 (left) and 3.5 (right) as a function of the studied variables; Table S1: Experimental results for the optimization of the xylose extraction process, parameters, constants, and the ANFIS model fit; Table S2: Experimental results for the optimization of the fructose extraction process, parameters, constants, and the ANFIS model fit; Table S3: Experimental results for the optimization of the lactose extraction process, parameters, constants, and the ANFIS model fit; Table S4: Experimental results for the optimization of the sucrose extraction process, parameters, constants, and the ANFIS model fit; Table S5: Experimental results for the optimization of the polyphenols extraction process, parameters, constants, and the ANFIS model fit.

Author Contributions: Conceptualization, A.P. and G.B.; methodology, E.J.L.; software, A.P.; validation, A.P., S.P.-H. and G.B.; formal analysis, A.P.; investigation, E.J.L.; resources, M.C.; writing—original draft preparation, E.J.L. and S.P.-H.; writing—review and editing, S.P.-H., M.Á.M.-L. and M.C.; supervision, G.B.; project administration, M.Á.M.-L. and M.C.; funding acquisition, M.Á.M.-L. and M.C. All authors have read and agreed to the published version of the manuscript.

Funding: This work was supported by Spanish Ministry of Economy, Industry and Competitiveness (Project CTM2016-75977-R). S.P.-H. is funded by a Juan de la Cierva Fellowship (FJC2021-048044-I, funded by MCIN/AEI/10.13039/501100011033 and the EU "NextGenerationEU/PRTR").

Data Availability Statement: The data presented in this study are available on request from the corresponding author (S.P.-H.) upon reasonable request.

Conflicts of Interest: The authors declare no conflicts of interest.

References

1. Oliveras-López, M.J.; Berná, G.; Jurado-Ruiz, E.; de la Serrana, H.L.G.; Martín, F. Consumption of extra-virgin olive oil rich in phenolic compounds has beneficial antioxidant effects in healthy human adults. *J. Funct. Foods* **2014**, *10*, 475–484. [CrossRef]
2. Souilem, S.; El-Abbassi, A.; Kiai, H.; Hafidi, A.; Sayadi, S.; Galanakis, M. *Olive Oil Production Sector: Environmental Effects and Sustainability Challenges*; Galanakis, C., Ed.; Olive Mill Waste; Academic Press: Cambridge, MA, USA, 2017; pp. 1–28.
3. Melgar, R.; Benitez, E.; Nogales, R. Bioconversion of wastes from olive oil industries by vermicomposting process using the epigeic earthworm Eisenia andrei. *J. Environ. Sci. Health Part B* **2009**, *44*, 488–495. [CrossRef] [PubMed]
4. Jeguirim, M.; Goddard, M.L.; Tamosiunas, A.; Berrich-Betouche, E.; Azzaz, A.A.; Praspaliauskas, M.; Jellali, S. Olive mill wastewater: From a pollutant to green fuels, agricultural water source and bio-fertilizer. *Renew. Energy* **2020**, *149*, 716–724. [CrossRef]
5. Al-Harthi, M.A. The efficacy of using olive cake as a by-product in broiler feeding with or without yeast. *Ital. J. Anim. Sci.* **2016**, *15*, 512–520. [CrossRef]
6. Demirbas, A. Producing bio-oil from olive cake by fast pyrolysis. *Energy Sources Part A Recovery Util. Environ. Eff.* **2007**, *30*, 38–44. [CrossRef]
7. Rajpoot, L.; Tagade, A.; Deshpande, G.; Verma, K.; Geed, S.R.; Patle, D.S.; Sawarkar, A.N. An overview of pyrolysis of de-oiled cakes for the production of biochar, bio-oil, and pyro-gas: Current status, challenges, and future perspective. *Bioresour. Technol. Rep.* **2022**, *19*, 101205. [CrossRef]
8. Ko, K.; Dadmohammadi, Y.; Abbaspourrad, A. Nutritional and Bioactive Components of Pomegranate Waste Used in Food and Cosmetic Applications: A Review. *Foods* **2021**, *10*, 657. [CrossRef]
9. Caballero, B.M.; López-Urionabarrenechea, A.; Pérez, B.; Solar, J.; Acha, E.; de Marco, I. Potentiality of "orujillo" (olive oil solid waste) to produce hydrogen by means of pyrolysis. *Int. J. Hydrogen Energy* **2020**, *45*, 20549–20557. [CrossRef]
10. Gálvez-Pérez, A.; Pérez, A.; Calero, M.; Martín-Lara, M.A.; Blázquez, G. Integral exploitation from olive cake for energy production in a biorefinery scheme. *Process Saf. Environ. Prot.* **2019**, *131*, 135–143. [CrossRef]
11. Lemes, A.C.; Álvares, G.T.; Egea, M.B.; Brandelli, A.; Kalil, S.J. Simultaneous production of proteases and antioxidant compounds from agro-industrial by-products. *Bioresour. Technol.* **2016**, *222*, 210–216. [CrossRef]

12. Pérez, A.; Blázquez, G.; Iáñez Rodríguez, I.; Osegueda, O.; Calero, M. Optimization of the sugar hydrothermal extraction process from olive cake using neuro-fuzzy models. *Bioresour. Technol.* **2018**, *268*, 81–90. [CrossRef]
13. Yoshimitsu, U.; Akira, H.; TSuneki, T. Isolation of 0-methyl-styllo-inositol from mung bean seeds. *Carbohydr. Res.* **1973**, *29*, 520–521.
14. Zuluaga, A.M.; Mena-García, A.; Chito-Trujillo, D.; Rada-Mendoza, M.; Sanz, M.L.; Ruiz-Matute, A.I. Development of a microwave-assisted extraction method for the recovery of bioactive inositols from lettuce (Lactuca sativa) byproducts. *Electrophoresis* **2020**, *41*, 1804–1811. [CrossRef]
15. Nauck, D.; Rudolf, K. Neuro–Fuzzy Systems. In *Handbook of Fuzzy Computation;* CRC Press: Boca Raton, FL, USA, 2020.
16. Rubio-Senent, F.; Fernández Bolaños, J.; García-Borrego, A.; Lama Muñoz, A.; Rodríguez Gutiérrez, G. New phenolic compounds hydrothermally extracted from the olive oil by product alperujo and their antioxidative activities. *J. Agric. Food Chem.* **2012**, *60*, 1175–1186. [CrossRef] [PubMed]
17. Rubio-Senent, F.; Fernández Bolaños, J.; García-Borrego, A.; Lama Muñoz, A.; Rodríguez Gutiérrez, G. Chemical characterization and properties of a polymeric phenolic fraction obtained from olive oil waste. *Food Res. Int.* **2013**, *54*, 2122–2129. [CrossRef]
18. Real Decreto 1101/2011, de 22 de Julio, Por el Que se Aprueba la Lista Positiva de los Disolventes de Extracción Que se Pueden Utilizar en la Fabricación de Productos Alimenticios y de Sus Ingredientes. Available online: https://www.boe.es/buscar/pdf/2011/BOE-A-2011-14223-consolidado.pdf (accessed on 1 October 2023).
19. Folin, O.; Ciocalteu, V. On tyrosine and tryptophane determinations in proteins. *J. Biol. Chem.* **1927**, *73*, 627–650. [CrossRef]
20. Montgomery, D.C. *Diseño y Análisis de Experimentos*, 2nd ed.; Wiley: México, 2004.
21. Rubio-Senent, F.; Fernández Bolaños, J.; García Borrego, A.; Lama Muñoz, A.; Rodríguez Gutiérrez, G. Influence of pH on the antioxidant phenols solubilised from hydrothermally treated olive oil by-product (alperujo). *Food Chem.* **2017**, *219*, 339–345. [CrossRef] [PubMed]
22. de Campos Souza, P.V. Fuzzy neural networks and neuro-fuzzy networks: A review the main techniques and applications used in the literature. *Appl. Soft Comput.* **2020**, *92*, 106275. [CrossRef]
23. Ruiz-Aceituno, L.; Rodríguez-Sánchez, S.; Sanz, J.; Sanz, M.L.; Ramos, L. Optimization of pressurized liquid extraction of inositols from pine nuts (*Pinus pinea* L.). *Food Chem.* **2014**, *153*, 450–456. [CrossRef] [PubMed]
24. López, F.; García, J.C.; Pérez, A.; Feria, M.J.; Zamudio, M.A.M.; Gil, G. Chemical and energetic characterization of species with a high-biomass production: Fractionation of their components. *Environ. Prog. Sustain.* **2010**, *29*, 499–509.
25. Casas-Godoy, L.; Campos-Valdez, A.R.; Alcázar-Valle, M.; Barrera-Martínez, I. Comparison of Extraction Techniques for the Recovery of Sugars, Antioxidant and Antimicrobial Compounds from Agro-Industrial Wastes. *Sustainability* **2022**, *14*, 5956. [CrossRef]
26. Zoubiri, F.Z.; Rihani, R.; Bentahar, F. Golden section algorithm to optimise the chemical pretreatment of agro-industrial waste for sugars extraction. *Fuel* **2020**, *266*, 117028. [CrossRef]
27. Gagić, T.; Perva-Uzunalić, A.; Knez, Ž.; Škerget, M. Hydrothermal treatment of sugars to obtain high-value products. *J. Serbian Chem. Soc.* **2020**, *85*, 97–109. [CrossRef]

processes MDPI

Article

Emulsion Formation and Stabilizing Properties of Olive Oil Cake Crude Extracts

Firdaous Fainassi [1], Noamane Taarji [1,2,*], Fatiha Benkhalti [3], Abdellatif Hafidi [1], Marcos A. Neves [4], Hiroko Isoda [4] and Mitsutoshi Nakajima [4]

[1] Department of Biology, Faculty of Sciences-Semlalia, Cadi Ayyad University, Marrakech 40001, Morocco; fainassi.f@gmail.com (F.F.); a.hafidi@uca.ac.ma (A.H.)
[2] Food and Medicinal Resource Engineering Open Innovation Laboratory, National Institute of Advanced Industrial Science and Technology (AIST), Tsukuba 305-0821, Japan
[3] Sustainable Development and Health Research Laboratory, Cadi Ayyad University, Marrakech 40000, Morocco; benkhalti.f@gmail.com
[4] Faculty of Life and Environmental Sciences, University of Tsukuba, Tsukuba 305-8572, Japan; marcos.neves.ga@u.tsukuba.ac.jp (M.A.N.); isoda.hiroko.ga@u.tsukuba.ac.jp (H.I.); nakajima.m.fu@u.tsukuba.ac.jp (M.N.)
* Correspondence: taarji.noamane@aist.go.jp

Abstract: The surface-active and emulsifying properties of crude aqueous ethanolic extracts from untreated olive oil cake (OOC) were investigated. OOC extracts contained important concentrations of surface-active components including proteins, saponins and polyphenols (1.2–2.8%, 7.8–9.5% and 0.7–4.5% (*w/w*), respectively) and reduced the interfacial tension by up to 46% (14.0 ± 0.2 mN m^{-1}) at the oil–water interface. The emulsifying ability of OOC extracts was not correlated, however, with their interfacial activity or surface-active composition. Eighty percent aqueous ethanol extract produced the most stable oil-in-water (O/W) emulsions by high-pressure homogenization. The emulsions had average volume mean droplet diameters of approximately 0.4 μm and negative ζ-potentials of about −45 mV, and were stable for up to 1 month of storage at 5, 25 and 50 °C. They were sensitive, however, to acidic pH conditions (<5) and NaCl addition (≥25 mM), indicating that the main stabilization mechanism is electrostatic due to the presence of surface-active compounds with ionizable groups, such as saponins.

Keywords: olive oil extraction by-product; interfacial tension; oil-in-water emulsion; protein; saponin; polyphenol

Citation: Fainassi, F.; Taarji, N.; Benkhalti, F.; Hafidi, A.; Neves, M.A.; Isoda, H.; Nakajima, M. Emulsion Formation and Stabilizing Properties of Olive Oil Cake Crude Extracts. *Processes* **2021**, *9*, 633. https://doi.org/10.3390/pr9040633

Academic Editors: Sebastián Sánchez Villasclaras and Juan Francisco García Martín

Received: 2 March 2021
Accepted: 30 March 2021
Published: 4 April 2021

1. Introduction

Emulsifiers are one of the most important ingredients in food industry. They are utilized as stabilizers and/or encapsulants in many food products and beverages in which they provide multiple properties such as good dispersibility, prolonged stability and improved bioavailability of other ingredients [1]. Many emulsifiers used nowadays derive from the same chemical and/or enzymatic reactions that were primarily designed in the past for mass and economic production of these substances [1,2]. The current trend towards natural and sustainable production of food emulsions, however, has led manufacturers to find new natural alternatives to synthetic emulsifiers [2]. Plant-derived compounds such as proteins, saponins, polysaccharides and phospholipids have shown promising emulsifying properties in various food applications [2,3]. They can provide similar stabilizing properties as synthetic emulsifiers, but they are also criticized because of their extensive preparation/manufacturing procedures [4–6].

A new trend in the utilization of food emulsifiers suggest the use of crude plant surface-active extracts, obtained via simple extraction/fractionation steps, to produce stable emulsions. Agro-industrial by-products contain various surface-active substances with strong emulsifying properties and can be utilized, therefore, as a source of these ingredients.

Sugar beet, Panax ginseng, oat bran, argan and bagasse by-products extracts have been successfully used, for example, to prepare stable oil-in-water (O/W) emulsions [7–12]. The active ingredients in these "crude" emulsifiers differ from one extract to another but they are, generally, attributed to a mixture of various surface-active components synergizing for an enhanced effect.

Olive oil processing by-products come in different forms and chemical profiles, depending on the olive ripening stage and the oil extraction method. In the traditional three-phase decanter systems, the solid waste is dry and contains significant contents of residual oil (2–4%, *w/w*), while in the case of two-phase decanter systems, a high-humidity (68–71%, *w/w*) semi-solid waste "Alperujo", here referred to as olive oil cake "OOC", is generated [13]. The major components found in olive solid by-products are polysaccharides (lignin 26–30%, hemicellulose 7–9% and cellulose 7–9%, *w/w*), proteins 5–7% (*w/w*), fatty acids 5–8% (*w/w*) and polyphenols (oleuropein, hydroxytyrosol and tyrosol) [14]. Such chemical composition makes solid olive processing wastes stand out as a promissory source of bioactive and nutritional compounds, especially in response to the rising trend of finding natural alternatives to synthetic ingredients.

Despite their abundance and valuable composition, studies investigating the emulsifying capacities of olive solid wastes are scarce. Filotheou et al. [15] used alcohol-insoluble olive processing wastes as a source of surface-active components to prepare acidic and non-acidic model food emulsions. The emulsifying efficiency was mainly attributed to the synergy between the surface-active macromolecular components and smaller surfactants. Koliastasi et al. [16,17] investigated the impact of partial and total composting on the surface-active composition and concluded that such treatments can produce emulsifiers with enhanced performances as compared to the extracts of non-composted materials. However, none of the previous studies have considered the use of crude extracts involving basic solid–liquid extraction from raw untreated material as emulsifiers. In this study, we evaluated the surface-active and emulsifying properties of various aqueous ethanolic extracts from olive oil extraction by-products. Our aim is to produce stable oil-in-water (O/W) emulsions using these extracts as sole emulsifiers.

2. Materials and Methods

2.1. Materials

Olive oil cakes (Picholine marocaine cultivar) obtained from a two-phase decanter system were provided by an industrial olive oil extraction unit in the region of Marrakech, Morocco. All other products were purchased from Wako Pure Chemical Industries, Ltd. (Osaka, Japan) unless stated elsewhere.

2.2. Sample Preparation and Extraction Yields

The samples (100 g) were added to fresh mixtures (300 mL) of distilled water and ethanol and stirred for 3 h at room temperature. The suspensions were then centrifuged at 2000× *g* for 30 min, filtered using Whatman filter paper type 111A and vacuum evaporated at 40 °C and 16 hPa (Eyela EVP-1100, Shanghai Co., Ltd., Shanghai, China) to remove solvent. The dried extracts were finally dispersed in distilled water, re-filtered using 0.45-µm hydrophobic PTFE membrane filters (RephiLe Bioscience, Co., Ltd., Shuangbai, China) and freeze-dried at −80 °C and 4 Pa (Eyela FDH-2110, Tokyo Rikakikai Co., Ltd., Tokyo, Japan) to remove water. The yields of extraction (EY) were calculated as follows:

$$\text{EY (\%, wet basis)} = W_1/W_0 \times 100 \tag{1}$$

where W_1 represents the weight of the extract after the freeze-drying and W_0 represents the weight of the fresh olive oil cake.

2.3. Physicochemical Characterization

Protein content was determined using a total nitrogen analyzer (UNICUBE, Elementer Ltd., Yokohama, Japan) with a nitrogen-to-protein conversion factor of 6.25. Saponins content was determined following the method of Zhaobao et al. [18], using oleanolic acid as standard (0–192 μg). Phenolic compounds content was determined following the Folin–Ciocalteu method [19] using gallic acid as standard (0.1–0.5 g L^{-1}). Interfacial tension was measured using the pendant drop method (PD-W, Kyowa Interface Science Co., Ltd., Saitama, Japan) at the soybean oil/water interface [10]. Particle size was determined using a dynamic light scattering particle size analyzer (Nano ZS, Malvern Instruments Ltd., Worcestershire, UK).

2.4. Emulsion Preparation

Coarse emulsions were prepared by homogenizing 0.1–5% (*w/w*) extract and 2.5–10% (*w/w*) soybean oil at 10,000 rpm for 5 min (Polytron PT-3000, Kinematica Inc., Luzern, Switzerland). The coarse emulsions were then passed through a high-pressure homogenizer (NanoVater NV200, Yoshida Kikai Co., Ltd., Nagoya, Japan) at 100 MPa for four passes and stored at 25 °C until analysis.

2.5. Emulsion Stability Evaluation

2.5.1. Effect of pH

Emulsions were diluted with phosphate buffer solutions (10 mM) to obtain a final oil content of 2.5% (*w/w*). The pH was then readjusted to appropriate level (2–9), using 1 M HCl or NaOH, and the samples were stored for 24 h at 5 °C before analysis [9].

2.5.2. Effect of Ionic Strength

Emulsions were diluted with phosphate buffer solutions (10 mM) of appropriate NaCl concentration (50–800 mM) to obtain a final oil content of 2.5% (*w/w*). The pH was then readjusted to 7 and the emulsions were stored at 5 °C until analysis [10].

2.5.3. Effect of Long-Term Storage

Sodium azide was added to the freshly prepared emulsions at a final concentration of 0.02% (*w/w*). The emulsions were then incubated for 30 days at 5, 25 or 50 °C prior to analysis.

2.6. Droplet Characterization

Volume mean droplet diameter ($d_{4,3}$) was measured using a static laser diffraction particle size analyzer (LS 13 320, Beckman Coulter, Brea, USA). Droplet charge was measured using a ς-potential analyzer (Zetasizer, Nano ZS, Malvern Instruments Ltd., Worcestershire, UK). In the case of ς-potential measurements, the emulsions were diluted (1/100) with deionized water or phosphate buffer solutions (10 mM) of appropriate pH and salt concentration, prior to analysis, to prevent multiple scattering effects. When using the Beckman Coulter, the emulsions were directly injected into the measurement module without dilution.

2.7. Data Analysis

All measurements were performed in duplicate using two independently prepared samples. The reported values represent means and standard deviations. The measurements were subjected to analysis of variance (ANOVA) using the "Tukey" test to assess significant differences among variables at 95% confidence level using Excel (Microsoft, Redmond, WA, USA).

3. Results and Discussion

3.1. Physicochemical Properties of OOC Extracts

3.1.1. Surface-Active Composition

Proteins and saponins are known to have surface-active properties due to the presence of hydrophilic and hydrophobic moieties in their chemical structures [20,21]. Polyphenols are also known to exhibit surface-active properties, contributing to the formation and stabilization of O/W emulsions [22]. Moreover, they can act as antioxidants in many food products and beverages, providing additional properties such as chemical stability and antioxidant activity to the prepared emulsions [23]. In this section, we evaluated the effect of extraction solvent on the concentration of proteins, saponins and phenolic compounds in OOC extracts. As shown in Table 1, increasing the concentration of ethanol did not have a strong effect on protein and saponin contents; even in the absolute ethanol extract where it was expected that the concentration of proteins would be reduced appreciably due to protein precipitation, the concentration was close to that obtained by the water extract. Phenolic compound concentration, on the other hand, was appreciably reduced in the absolute ethanol extract (Table 1).

Table 1. Extraction yields (EY) and content in surface-active components (%, *w/w*) of olive oil cake (OOC) extracts prepared using various aqueous-ethanolic solvent mixtures. Means in the same line followed by the same letter are not significantly different ($\alpha = 0.05$).

	OOC 0%	OOC 20%	OOC 40%	OOC 60%	OOC 80%	OOC 100%
EY % (*w/w*)	10.2 ± 0.9 [de]	11.6 ± 1.1 [cd]	9.6 ± 0.9 [e]	12.0 ± 1.2 [c]	16.6 ± 1.9 [b]	22.4 ± 2.3 [a]
Proteins	1.79 ± 0.14 [c]	2.86 ± 0.12 [a]	2.27 ± 0.19 [b]	2.23 ± 0.14 [b]	2.19 ± 0.14 [b]	1.20 ± 0.17 [d]
Saponins	7.80 ± 0.38 [c]	8.07 ± 0.58 [bc]	8.71 ± 0.67 [ab]	9.01 ± 1.07 [ab]	8.38 ± 0.86 [bc]	9.29 ± 0.68 [a]
Polyphenols	4.58 ± 0.26 [a]	4.20 ± 0.22 [a]	3.23 ± 0.14 [b]	4.27 ± 0.23 [a]	3.49 ± 0.11 [b]	0.72 ± 0.04 [c]

Note: The shared letter(s) in the respective column indicate(s) no significant difference ($p < 0.05$).

3.1.2. Particle Size Distribution

Proteins, saponins and phenolic compounds can interact by the effect of low molecular forces such as hydrogen bonds, electrostatic and hydrophobic interactions, in the bulk water phase, prior to adsorption or directly at the oil/water interface following adsorption of the slowly diffusing component on the pre-adsorbed layer of the other component [24–27]. To evaluate the structural organization of surface-active components in our extracts, and to gain better insight about their adsorption mechanisms, we evaluated the particle size and particle size distribution of aqueous phases containing OOC extracts. As shown in Figure 1, OOC extracts aqueous solutions (1%, *w/w*) had broad particle size distributions with average particle sizes of more than 100 nm. Moreover, except for 40% (*w/w*) aqueous ethanol extract, all samples had no particles smaller than 50 nm, indicating the presence of relatively large structures. Small molecule surfactants, such as saponins, form micelles of approximately 10 nm [28]. We suggest, therefore, that OOC extracts contain relatively large aggregates that result from the intermolecular interactions between various surface-active components, including proteins, saponins and phenolic compounds.

3.1.3. Interfacial Activity

Interfacial tension of emulsifiers plays an important role in determining their ability to form and stabilize emulsions. Usually, the lower the interfacial tension is, the greater the emulsifying properties are [29]. We measured, therefore, the interfacial tension of OOC extracts at the oil/water interface and compared their surface-active properties to commonly used emulsifiers. As shown in Figure 2, all extracts reduced their interfacial tension at the soybean oil/water interface, independently of extract composition. Extract 40% (*w/w*) aqueous-ethanol, for example, reduced its interfacial tension to approximately 13 mN m^{-1}, while 80% (*w/w*) aqueous ethanolic extract was less efficient at about 16 mN m^{-1}. These values are lower than the ones recorded from either non-composted or composted olive waste derived extracts [15,17]. They are in the same range, however, of previously evaluated extracts from sugar beet, bagasse, liquorice and argan press-cake extracts as well as purified natural emulsifiers, including proteins, saponins and phospholipids [7,10,12,30–33].

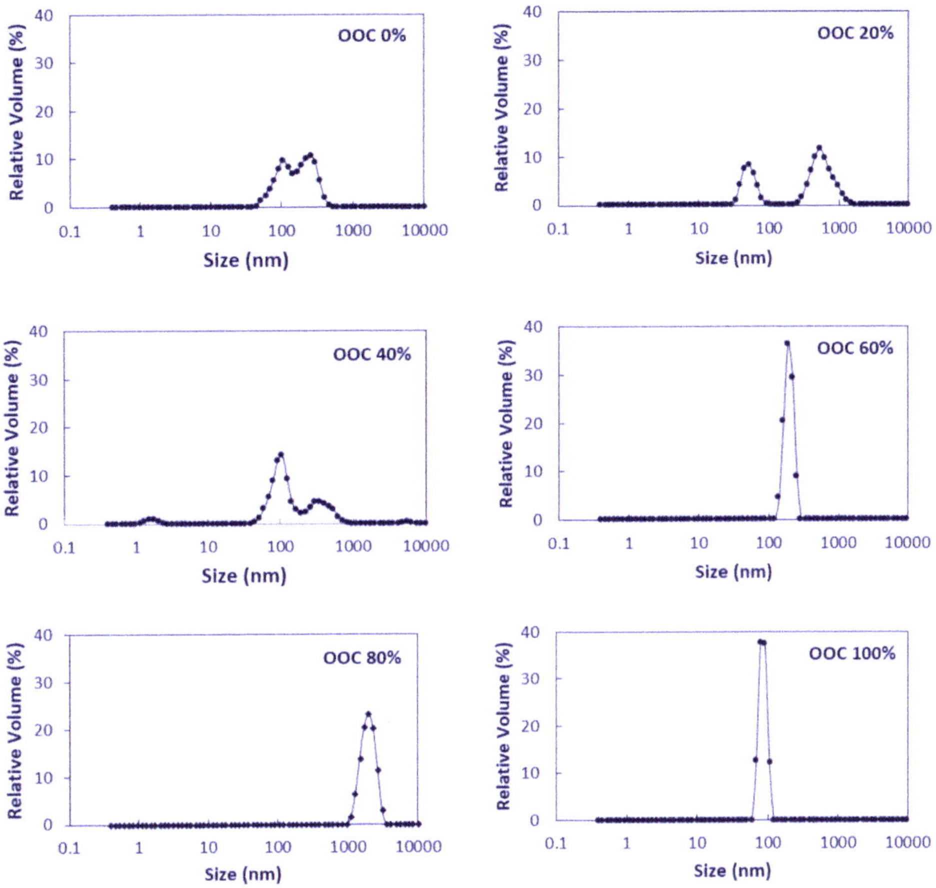

Figure 1. Particle size distribution of aqueous phases containing 1% (*w/w*) OOC extracts.

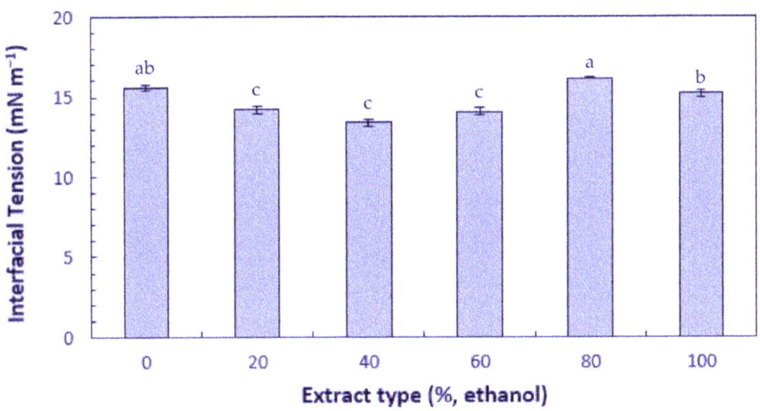

Figure 2. Interfacial tension of aqueous phases containing crude OOC extracts (1%, *w/w*). The interfacial tension between soybean oil and Milli-Q water was approximately 26 mN m^{-1}. The shared letter(s) indicate(s) no significant difference ($p < 0.05$).

3.2. Effect of Extract Type on the Formation Characteristics of O/W Emulsion

3.2.1. Droplet Size

In order to evaluate the emulsifying performance of OOC extracts, O/W emulsions were prepared using aqueous solutions (1%, *w/w*) of each extract and 5% (*w/w*) soybean oil at standard homogenization conditions (100 MPa, four passes). Except for emulsions prepared using absolute ethanol extract, which showed immediate oiling-off after homogenization, all emulsions prepared using OOC extracts were successfully produced with average droplet sizes ($d_{4,3}$) of 0.3–0.4 μm (Figure 3a). When stored at 5 °C, the emulsions showed no evidence of droplet growth or phase separation for up to 15 days with a main peak around 0.4–0.6 μm (Figure 4). At 25 °C, the peak around 0.4 μm was maintained but for the emulsions prepared using 0, 20, 40 and 60% (*w/w*) ethanol-to-water-ratios, droplet growth occurred with the emergence of lesser peaks around 4.5–15 μm. Interestingly, emulsions prepared using 80% (*w/w*) aqueous ethanolic extract were stable at all studied temperatures despite their close surface-active composition and interfacial activity to other extracts. With a main peak around 0.4 μm and a lesser peak around 0.08 μm (Figure 4), OOC 80% appears to be a better emulsifier and was used, therefore, for further experiments.

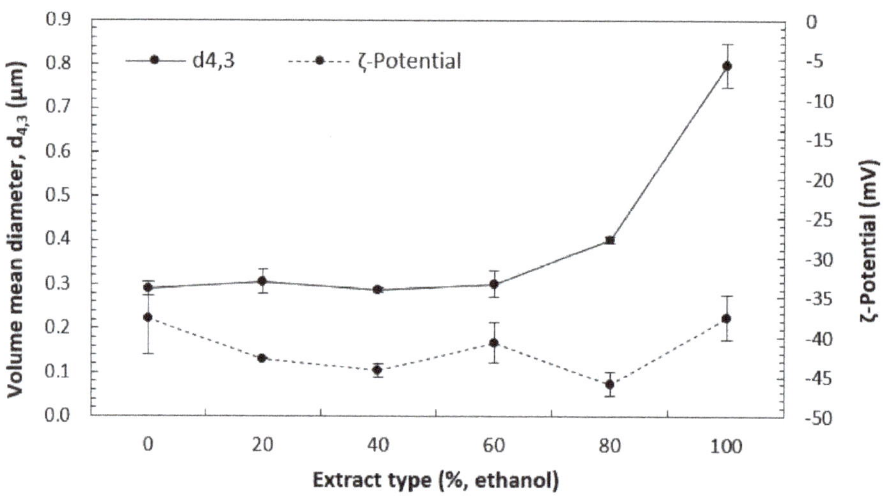

(a)

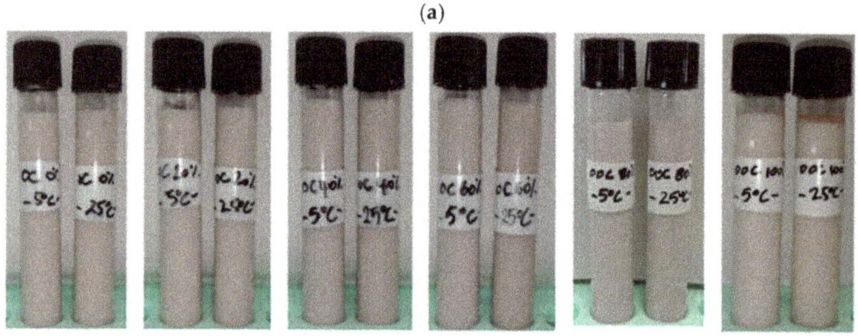

OOC 0% OOC 20% OOC 40% OOC 60% OOC 80% OOC 100%

(b)

Figure 3. (**a**) Volume mean droplet diameter ($d_{4,3}$) and ζ-potential of soybean oil-in-water (O/W) emulsions prepared using OOC aqueous ethanolic extracts, by high-pressure homogenization (100 MPa, 4 passes). (**b**) Visual appearance of OOC extract-stabilized emulsions after 15 days of storage at 5 °C or 25 °C.

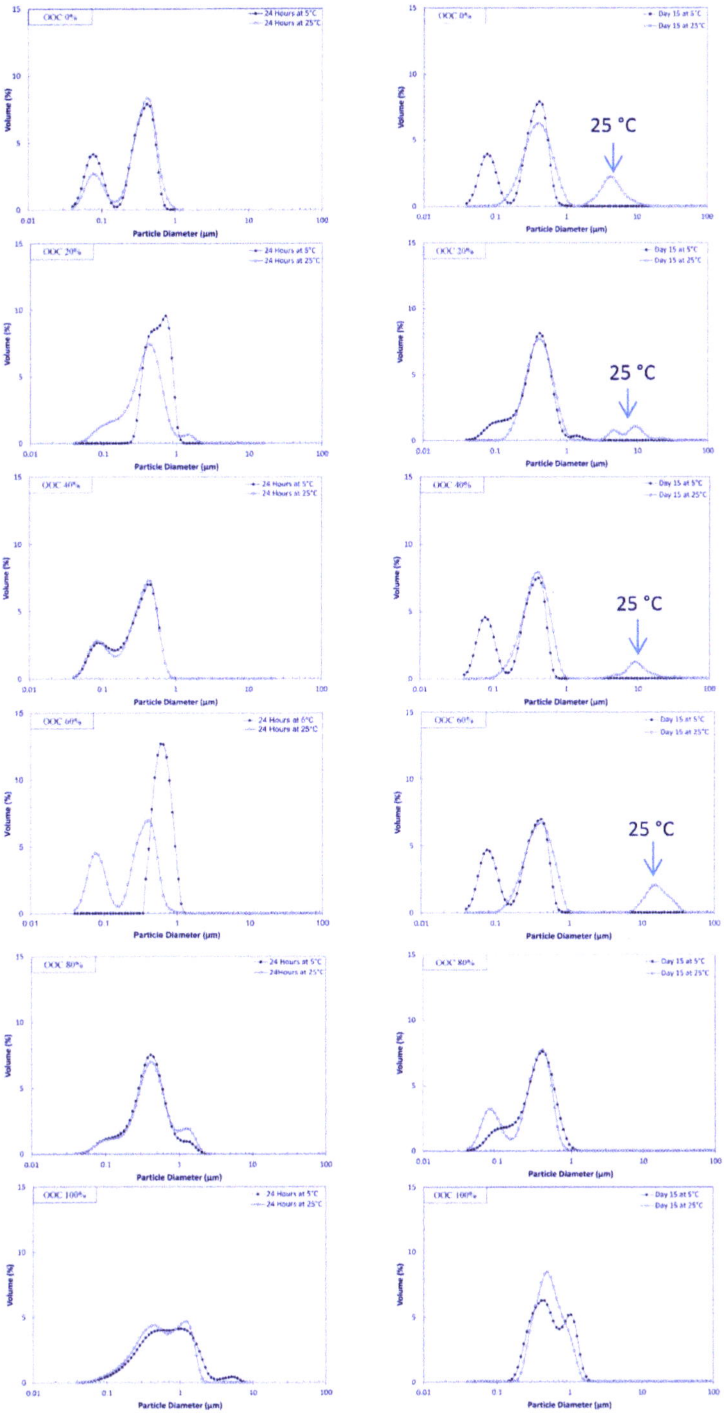

Figure 4. Droplet size distributions of emulsions prepared using 1% (*w/w*) OOC extracts and 5% (*w/w*) soybean oil, by high-pressure homogenization (100 MPa, 4 passes).

Previous studies have evaluated the emulsification capacity of extracts obtained from composted olive oil waste materials [16,17]. The studied emulsifiers produced larger droplets with a main peak around 2 to 3 µm and a relatively constant droplet size distribution for 15 days of storage. The emulsifiers studied in this work encourage, therefore, the use of crude extracts, instead of time-consuming composting materials to produce stable O/W emulsions with smaller droplet sizes and a similar period of storage.

3.2.2. Droplet Surface Charge

The surface charge of emulsion droplets plays an important role in the formation and stabilization of emulsions. In general, a higher ζ-potential means that the emulsifier layer yields more repulsive forces between emulsion droplets, thus preventing their coalescence. A neutral ζ-potential, on the other hand, means that the emulsifier covered interfaces are more likely to destabilize, resulting in the formation of larger particles [33]. As shown in Figure 3a, all emulsions prepared using OOC extracts displayed a strong negative ζ-potential, independently of their stability characteristics. The 80% (*w/w*) aqueous ethanol extract, for example, provided a ζ-potential of −45.9 mV, in agreement with its good emulsion-stabilizing properties. Absolute ethanol extract, on the other hand, presented a ζ-potential of −37.5 mV, despite its limited emulsifying ability.

Carboxylic acid groups with typical pKa values of ~3.5 may be responsible for the negative ζ-potentials of emulsions stabilized by OOC extracts [3]. This group would be fully charged (COO^-) at pH 7, hence the strong negative charge of emulsion droplets. Proteins can be also responsible for the negative surface charge observed in our conditions as they usually show negative ζ-potentials of approximately −40 mV at pH 7 [1,34].

3.3. Effect of OOC 80% Extract Concentration and Oil Mass Fraction on Emulsion Formation

The $d_{4,3}$ of emulsions gradually decreased upon increasing 80% (*w/w*) aqueous ethanolic extract concentration from 0.1 to 1% (*w/w*). Further increase in extract concentration appreciably increased the droplet size of emulsions and resulted in visible creaming at 5% (*w/w*) (Figure 5a,b). The ability of an emulsifier to produce small droplet size emulsions depends on (i) its ability to reduce the interfacial tension, (ii) the speed at which it adsorbs on the droplets' interface and (iii) its effectiveness to produce repulsive forces during homogenization [3,35]. Lower concentrations of OOC extract were insufficient, therefore, to reduce interfacial tension and/or stabilize the newly generated interfaces, while higher concentrations increased the contents of emulsion destabilizing agents (e.g., minerals) in the emulsions.

Increasing oil mass fraction from 2.5 to 10% (*w/w*) also increased the droplet size of OOC-stabilized emulsions (Figure 5c). This suggests that relatively high emulsifier-to-oil ratios are needed to produce small droplet size emulsions using the present extract. Moreover, further increase in oil concentration increases the viscosity of emulsions, which affects their disruption efficiency during homogenization [35]. Overall, small droplet size O/W emulsions can be prepared using 1% (*w/w*) OOC extract and 5% (*w/w*) soybean oil by high-pressure homogenization (100 MPa, four passes). We proceeded, therefore, to evaluate the stability characteristics of these emulsions at different stress conditions to determine their potential application.

3.4. Effect of Stress Conditions on OOC 80% Extract Stabilized Emulsions

Emulsions prepared using OOC 80% (*w/w*) aqueous ethanol extract were stable from pH 5 to 10 with average $d_{4,3}$ of 0.3 µm. Further decrease in pH gradually increased the droplet size of emulsions or resulted in excessive creaming at pH 3 and 2 (Figure 6). Increasing NaCl concentration also reduced the stability of emulsions as indicated by droplet size measurements and visual appearance. The $d_{4,3}$ increased to 6.2 µm at 100 mM and visible creaming was observed from 25 mM (Figure 7). These results agreed with the ζ-potential measurements of emulsions, which gradually decreased upon reducing the pH or increasing NaCl concentration (Figures 6 and 7).

Saponins show a pKa of approximately 3.25 and, therefore, reducing the pH below this value leads to screening the surface charge of emulsions droplets [2]. Moreover, electrostatic screening of the negative charge of saponins can affect their intermolecular interaction with other components, including proteins, affecting the interfacial layer composition of emulsions droplets. The stabilization mechanism of OOC 80% extract-stabilized emulsions depends, therefore, importantly on electrostatic repulsion due to the adsorption of ionizable surface active components with ionizable groups, such as saponins. The contribution of other components (e.g., proteins) to emulsions stability is also suggested, explaining their resistance to complete phase separation (oiling-off) at extreme environmental conditions by providing stronger interfacial coverage.

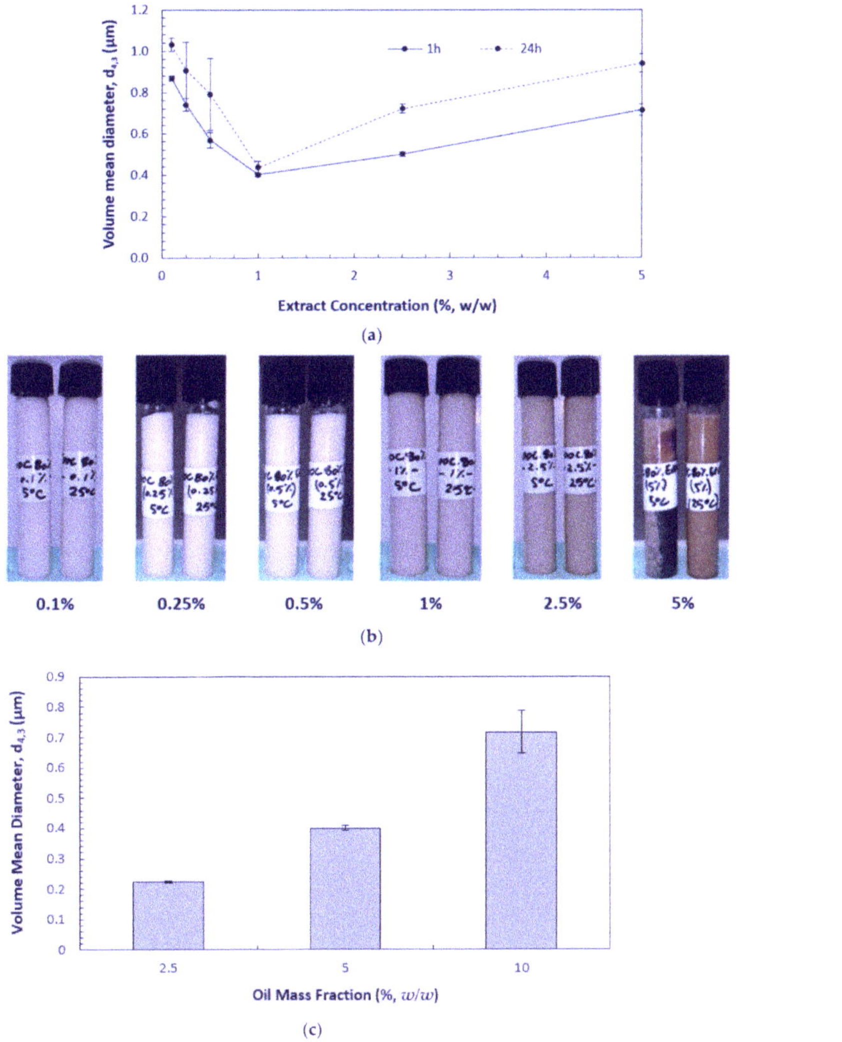

Figure 5. (**a**) Effect of OOC 80% (*w/w*) aqueous ethanolic extract concentration on the volume mean droplet diameter of emulsions after 1 and 24 h of storage at 5 °C; (**b**) effect of OOC 80% extract concentration on the visual appearance of emulsions after 24 h of storage at 5 °C. (**c**) Effect of oil mass fraction on the mean droplet size of emulsions after 24 h of storage at 5 °C.

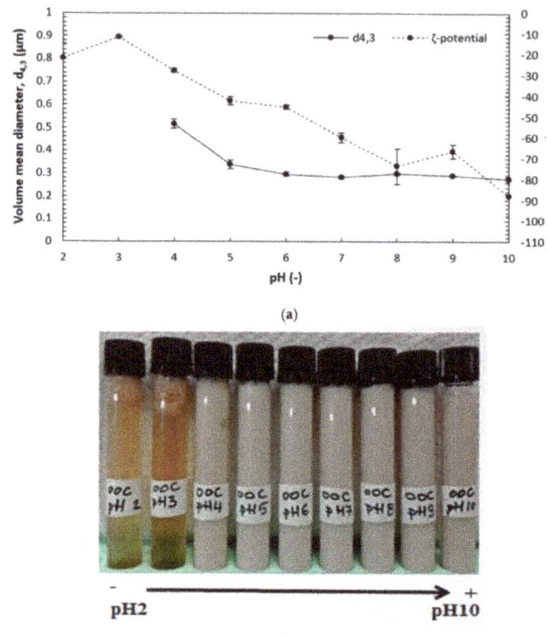

Figure 6. (**a**) Effect of pH on the volume mean droplet diameter ($d_{4,3}$), ζ-potential and (**b**) visual appearance of O/W emulsions prepared using OOC 80% aqueous ethanolic extract.

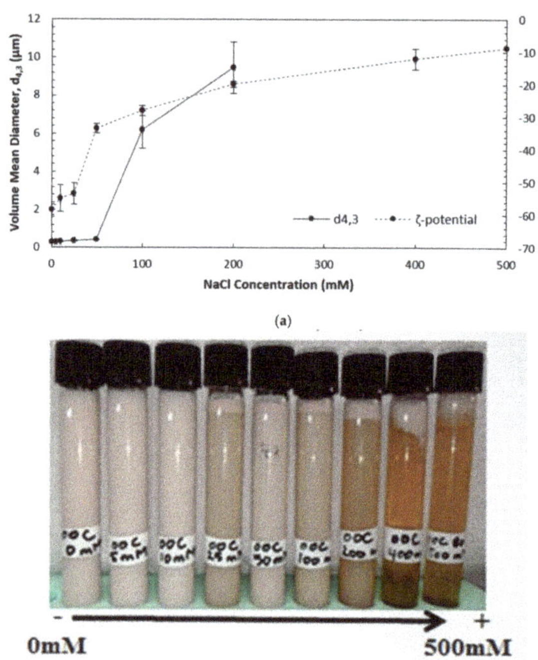

Figure 7. (**a**) Effect of NaCl concentration on the volume mean droplet diameter ($d_{4,3}$), ζ-potential and (**b**) visual appearance of O/W emulsions prepared using OOC 80% aqueous ethanolic extract.

The $d_{4,3}$ of OOC 80% extract-stabilized emulsions did not change after 30 days of storage at different temperatures (Table 2), thus its potential use for the preparation of stable O/W emulsions by high-pressure homogenization.

Table 2. Volume mean droplet diameter (μm) changes of emulsions prepared using 1% (*w/w*) OOC 80% extract and 5% (*w/w*) soybean oil, by high-pressure homogenization.

	5 °C	25 °C	50 °C
Day 1	0.437	0.405	0.401
Day 7	0.416	0.440	0.470
Day 15	0.427	0.405	0.481
Day 30	0.394	0.387	0.426

4. Conclusions

The present work suggests a novel approach for using crude aqueous-ethanolic extracts from untreated olive oil cake to produce stable O/W emulsions with smaller droplet sizes and for an extended period of storage. OOC extracts successfully formed emulsions with average droplet sizes ($d_{4,3}$) of 0.28–0.4 μm. OOC 80% (*w/w*) aqueous ethanolic extract was selected, as it provided the best storage stability for up to 15 days at 25 °C. Our results clearly showed that this extract could produce emulsions with good physical stability at a wide range of pH (5–10) and when stored at 5, 25 and 50 °C for 30 days. However, the emulsions were highly unstable at extreme acidic pH (<4), and at increased ionic strengths ($\geq$25 mM). The main stabilization mechanism is believed to be mainly electrostatic, likely due to the presence of surface-active compounds with ionizable groups such as saponins. In future studies, it would be helpful to extend this work by evaluating other parameters, such as the taste profile, the toxicity, the cost, and the reliability of supply to encourage the application of this extract as a food emulsifier.

Author Contributions: Conceptualization, F.F., A.H. and M.N.; methodology, F.F.; software, F.F.; validation, A.H., M.A.N., and M.N.; formal analysis, F.F.; investigation, F.F. and F.B.; resources, M.N.; data curation, F.F.; writing—original draft preparation, F.F.; writing—review and editing, N.T.; visualization, F.F.; supervision, N.T.; project administration, M.N., A.H. and H.I.; funding acquisition, M.N. All authors have read and agreed to the published version of the manuscript.

Funding: This research was funded by the Japan Science and Technology Agency and the Japan International Cooperation Agency under the framework of SATREPS project [grant number JPMJSA1506].

Institutional Review Board Statement: Not applicable.

Informed Consent Statement: Not applicable.

Data Availability Statement: Data sharing not applicable.

Acknowledgments: This research was financially supported by the Science and Technology Research Partnership for Sustainable Development–SATREPS project, from JICA/JST, Japan.

Conflicts of Interest: The authors declare no conflict of interest.

References

1. McClements, D.J. *Food Emulsions Principles, Practices, and Techniques*, 3rd ed.; CRS Press: Boca Raton, FL, USA, 2015; ISBN 9781498726696.
2. Ozturk, B.; McClements, D.J. Progress in natural emulsifiers for utilization in food emulsions. *Curr. Opin. Food Sci.* **2016**, *7*, 1–6. [CrossRef]
3. Yang, Y.; Leser, M.E.; Sher, A.A.; McClements, D.J. Formation and stability of emulsions using a natural small molecule surfactant: Quillaja saponin (Q-Naturale®). *Food Hydrocoll.* **2013**, *30*, 589–596. [CrossRef]
4. Melanie, H.; Taarji, N.; Zhao, Y.; Khalid, N.; Neves, M.A.; Kobayashi, I.; Tuwo, A.; Nakajima, M. Formulation and characterisation of O/W emulsions stabilised with modified seaweed polysaccharides. *Int. J. Food Sci. Technol.* **2020**, *55*, 211–221. [CrossRef]

5. Chen, Z.; Shu, G.; Taarji, N.; Barrow, C.J.; Nakajima, M.; Khalid, N.; Neves, M.A. Gypenosides as natural emulsifiers for oil-in-water nanoemulsions loaded with astaxanthin: Insights of formulation, stability and release properties. *Food Chem.* **2018**, 261. [CrossRef]
6. Bouhoute, M.; Taarji, N.; de Oliveira Felipe, L.; Habibi, Y.; Kobayashi, I.; Zahar, M.; Isoda, H.; Nakajima, M.; Neves, M.A. Microfibrillated cellulose from Argania spinosa shells as sustainable solid particles for O/W Pickering emulsions. *Carbohydr. Polym.* **2021**, *251*, 116990. [CrossRef]
7. Ralla, T.; Salminen, H.; Edelmann, M.; Dawid, C.; Hofmann, T.; Weiss, J. Sugar Beet Extract (Beta vulgaris L.) as a New Natural Emulsifier: Emulsion Formation. *J. Agric. Food Chem.* **2017**, *65*, 4153–4160. [CrossRef]
8. Ralla, T.; Herz, E.; Salminen, H.; Edelmann, M.; Dawid, C.; Hofmann, T.; Weiss, J. Emulsifying Properties of Natural Extracts from *Panax ginseng* L. *Food Biophys.* **2017**, *12*, 479–490. [CrossRef]
9. Ralla, T.; Salminen, H.; Edelmann, M.; Dawid, C.; Hofmann, T.; Weiss, J. Oat bran extract (*Avena sativa* L.) from food by-product streams as new natural emulsifier. *Food Hydrocoll.* **2018**, *81*, 253–262. [CrossRef]
10. Taarji, N.; Rabelo da Silva, C.A.; Khalid, N.; Gadhi, C.; Hafidi, A.; Kobayashi, I.; Neves, M.A.; Isoda, H.; Nakajima, M. Formulation and stabilization of oil-in-water nanoemulsions using a saponins-rich extract from argan oil press-cake. *Food Chem.* **2018**, *246*, 457–463. [CrossRef]
11. Bouhoute, M.; Taarji, N.; Vodo, S.; Kobayashi, I.; Zahar, M.; Isoda, H.; Nakajima, M.; Neves, M.A. Formation and stability of emulsions using crude extracts as natural emulsifiers from Argan shells. *Colloids Surf. A Physicochem. Eng. Asp.* **2020**, *591*, 124536. [CrossRef]
12. Vodo, S.; Taarji, N.; Bouhoute, M.; de Oliveira Felipe, L.; Neves, M.A.; Kobayashi, I.; Uemura, K.; Nakajima, M. Potential of bagasse obtained using hydrothermal liquefaction pre-treatment as a natural emulsifier. *Int. J. Food Sci. Technol.* **2020**. [CrossRef]
13. Fernández-Bolaños, J.; Rodríguez, G.; Gómez, E.; Guillén, R.; Jiménez, A.; Heredia, A.; Rodríguez, R. Total recovery of the waste of two-phase olive oil processing: Isolation of added-value compounds. *J. Agric. Food Chem.* **2004**, *52*, 5849–5855. [CrossRef]
14. Mirabella, N.; Castellani, V.; Sala, S. Current options for the valorization of food manufacturing waste: A review. *J. Clean. Prod.* **2014**, *65*, 28–41. [CrossRef]
15. Filotheou, A.; Ritzoulis, C.; Avgidou, M.; Kalogianni, E.P.; Pavlou, A.; Panayiotou, C. Novel emulsifiers from olive processing solid waste. *Food Hydrocoll.* **2015**, *48*, 274–281. [CrossRef]
16. Koliastasi, A.; Kompothekra, V.; Giotis, C.; Moustakas, A.K. Emulsifiers from Partially Composted Olive Waste. *Foods* **2019**, *8*, 271. [CrossRef]
17. Koliastasi, A.; Kompothekra, V.; Giotis, C.; Moustakas, A.K.; Skotti, E.P.; Gerakis, A.; Kalogianni, E.P.; Georgiou, D.; Ritzoulis, C. Food Hydrocolloids Novel emulsi fi ers from olive mill compost. *Food Hydrocoll.* **2020**, *99*, 105373. [CrossRef]
18. Zhaobao, X.; Chunhong, T.; Gang, C.; Zhisong, S. Studied on corlorimetric determination of oleanolic acid in Chinese quince. *Nat. Prod. Res. Dev.* **2001**, *13*, 23–26.
19. Rabelo, C.A.S.; Taarji, N.; Khalid, N.; Kobayashi, I.; Nakajima, M.; Neves, M.A. Formulation and characterization of water-in-oil nanoemulsions loaded with açaí berry anthocyanins: Insights of degradation kinetics and stability evaluation of anthocyanins and nanoemulsions. *Food Res. Int.* **2018**, 106. [CrossRef]
20. Oleszek, W.; Hamed, A. Saponin-Based Surfactants. In *Surfactants from Renewable Resources*, 1st ed.; Wiley: Hoboken, NJ, USA, 2010; pp. 239–249.
21. Lam, R.S.H.; Nickerson, M.T. Food proteins: A review on their emulsifying properties using a structure-function approach. *Food Chem.* **2013**, *141*, 975–984. [CrossRef]
22. Souilem, S.; Kobayashi, I.; Neves, M.A.; Jlaiel, L.; Isoda, H.; Sayadi, S.; Nakajima, M. Interfacial characteristics and microchannel emulsification of oleuropein-containing triglyceride oil–water systems. *Food Res. Int.* **2014**, *62*, 467–475. [CrossRef]
23. Di Mattia, C.D.; Sacchetti, G.; Mastrocola, D.; Sarker, D.K.; Pittia, P. Surface properties of phenolic compounds and their influence on the dispersion degree and oxidative stability of olive oil O/W emulsions. *Food Hydrocoll.* **2010**, *24*, 652–658. [CrossRef]
24. Piotrowski, M.; Lewandowska, J.; Wojciechowski, K. Biosurfactant–Protein Mixtures: Quillaja Bark Saponin at Water/Air and Water/Oil Interfaces in Presence of β-Lactoglobulin. *J. Phys. Chem. B* **2012**, *116*, 4843–4850. [CrossRef]
25. Wojciechowski, K.; Kezwon, A.; Lewandowska, J.; Marcinkowski, K. Effect of β-casein on surface activity of Quillaja bark saponin at fluid/fluid interfaces. *Food Hydrocoll.* **2014**, *34*, 208–216. [CrossRef]
26. Wojciechowski, K.; Piotrowski, M.; Popielarz, W.; Sosnowski, T.R. Short- and mid-term adsorption behaviour of Quillaja Bark Saponin and its mixtures with lysozyme. *Food Hydrocoll.* **2011**, *25*, 687–693. [CrossRef]
27. Bandyopadhyay, P.; Ghosh, A.K.; Ghosh, C. Recent developments on polyphenol–protein interactions: Effects on tea and coffee taste, antioxidant properties and the digestive system. *Food Funct.* **2012**, *3*, 592–605. [CrossRef]
28. Mitra, S.; Dungan, S.R. Micellar Properties of Quillaja Saponin. 1. Effects of Temperature, Salt, and pH on Solution Properties. *J. Agric. Food Chem.* **1997**, *45*, 1587–1595. [CrossRef]
29. Stamkulov, N.S.; Mussabekov, K.B.; Aidarova, S.B.; Luckham, P.F. Stabilisation of emulsions by using a combination of an oil soluble ionic surfactant and water soluble polyelectrolytes. I: Emulsion stabilisation and Interfacial tension measurements. *Colloids Surf. A Physicochem. Eng. Asp.* **2009**, *335*, 103–106. [CrossRef]
30. Taarji, N.; Bouhoute, M.; Fainassi, F.; Hafidi, A.; Kobayashi, I.; Neves, M.A.; Tominaga, K.; Isoda, H.; Nakajima, M. Interfacial and emulsifying properties of purified glycyrrhizin and non-purified glycyrrhizin-rich extracts from liquorice root (Glycyrrhiza glabra). *Food Chem.* **2021**, *337*, 127949. [CrossRef]

31. McClements, D.J.; Gumus, C.E. Natural emulsifiers—Biosurfactants, phospholipids, biopolymers, and colloidal particles: Molecular and physicochemical basis of functional performance. *Adv. Colloid Interface Sci.* **2016**, *234*, 3–26. [CrossRef]
32. Ogino, K.; Onishi, M. Interfacial action of natural surfactants in oil/water systems. *J. Colloid Interface Sci.* **1981**, *83*, 18–25. [CrossRef]
33. Lu, G.W.; Gao, P. CHAPTER 3-Emulsions and Microemulsions for Topical and Transdermal Drug Delivery. In *Personal Care & Cosmetic Technology*; William Andrew Publishing: Boston, MA, USA, 2010; pp. 59–94. ISBN 978-0-8155-2025-2.
34. Karaca, A.C.; Low, N.; Nickerson, M. Emulsifying properties of chickpea, faba bean, lentil and pea proteins produced by isoelectric precipitation and salt extraction. *Food Res. Int.* **2011**, *44*, 2742–2750. [CrossRef]
35. Baret, J.-C.; Kleinschmidt, F.; El Harrak, A.; Griffiths, A.D. Kinetic Aspects of Emulsion Stabilization by Surfactants: A Microfluidic Analysis. *Langmuir* **2009**, *25*, 6088–6093. [CrossRef] [PubMed]

processes

Article

Microbial Population Dynamics during Unstable Operation of a Semicontinuous Anaerobic Digester Fed with a Mild-Treated Olive Mill Solid Waste

Juan Cubero-Cardoso, África Fernández-Prior, Javier Ramiro-Garcia, Angeles Trujillo-Reyes, Belén Caballero-Guerrero, Guillermo Rodríguez-Gutiérrez and Fernando G. Fermoso *

Instituto de la Grasa (IG), Spanish Scientific Research Council (CSIC), Carretera de Utrera, km 1, 41013 Seville, Spain; j.cubero@dqcm.uhu.es (J.C.-C.); mfernandez26@us.es (Á.F.-P.); jramiro@ig.csic.es (J.R.-G.); angeles.trujillo@ugr.es (A.T.-R.); cabaguer@ig.csic.es (B.C.-G.); guirogu@ig.csic.es (G.R.-G.)
* Correspondence: fgfermoso@ig.csic.es; Tel.: +34-954611550

Abstract: This research evaluates process instability together with microbial population dynamics of the startup of an anaerobic digestion of a mild pretreated solid olive oil waste. The pretreatment consisted of a mild thermal treatment called thermo-malaxation and a subsequent dephenolized process of the olive mill solid waste. The anaerobic digestion process of the mild pretreated and partially dephenolized biomass was studied for three Hydraulic Retention Times (HRTs), with 21 days each HRT, with an organic load rate of 1 g VS/L d, carried out at mesophilic temperature (35 ± 1 °C). The average value of methane yield decreased from 204 ± 9 mL CH_4/g VS d on day 21, the last day of the first HRT, to 87 ± 24 mL CH_4/g VS d on day 60, the last day of the third HRT. The alkalinity decreased drastically, indicating instability of the anaerobic digestion process. Although phenolic compounds were partially extracted in the pretreatment, the observed increase in phenolic compounds during reactor operation might be contributed to the methane production decay. Interestingly, volatile fatty acids decreased with time, indicating that not only the methanogenic stage but also the hydrolysis stage was affected. Indeed, the microbial analysis showed that the abundance of hydrolytic bacteria decreased over time. It is also worth noticing that hydrogenotrophic methanogens, while present during the first two HRTs, were not observed at the end of the last HRT. This observation, together with the increase in the relative abundance of acetoclastic methanogens, showed a shift in the methane production pathway from hydrogenotrophic methanogenesis to acetotrophic methanogenesis.

Keywords: anaerobic digestion; solid waste; thermal treatment; phenol; valorization

Citation: Cubero-Cardoso, J.; Fernández-Prior, Á.; Ramiro-Garcia, J.; Trujillo-Reyes, A.; Caballero-Guerrero, B.; Rodríguez-Gutiérrez, G.; Fermoso, F.G. Microbial Population Dynamics during Unstable Operation of a Semicontinuous Anaerobic Digester Fed with a Mild-Treated Olive Mill Solid Waste. *Processes* **2023**, *11*, 2724. https://doi.org/10.3390/pr11092724

Academic Editors: Sebastián Sánchez Villasclaras and Juan Francisco García Martín

Received: 28 July 2023
Revised: 28 August 2023
Accepted: 5 September 2023
Published: 12 September 2023

1. Introduction

The importance of the olive oil sector in the main producing countries is well known, as is the case of Spain, where 47% of the worldwide consumed olive oil is produced. It should also be borne in mind that the type of cropping system is gradually changing from traditional to intensive or even super-intensive in order to increase production [1]. All this leads to the production of between 5 and 7 million tons of by-products per year in Spain while simultaneously exhausting agronomic resources more rapidly, forcing the use of a larger quantity of fertilizers. This increase makes managing this massive number of by-products even more difficult.

In Spain, the most common system used to extract olive oil is the two-phase extraction system, which produces a semi-solid waste with high humidity (60–70%) [2]. The advantage of this country is the centralization of the industry dedicated to the use of olive mill solid waste, called pomace extractors. The main use is the extraction of pomace oil. Most of it is obtained by extraction with an organic solvent after a drying process that consumes a large amount of energy resources [3]. And once the pomace oil has been extracted, the final solid is mainly used as biomass for electricity generation. The increase in production, the high drying costs, and the growing importance of the bioactive components present in this type of by-product are proliferating the appearance of new alternatives to improve its management [4]. It should be noted that more and more pomace extractors are using a three-phase extraction system. This process allows oil to be recovered during centrifugation, a liquid fraction rich in bioactive components, and a solid with a lower degree of humidity (50–55%). Therefore, this oil extraction facilitates and reduces the costs of subsequent drying to extract the pomace oil remaining in the solid with solvents. This trend makes it possible to study new ways of management and use [5].

The application of bioprocesses is always an alternative to be considered for the valorisation of this type of agri-food by-product. The limitation in this type of treatment is the presence of inhibitory substances, which for decades have limited the use of by-products from the olive oil industry [6]. The application of the three-phase extraction system makes it possible to obtain a detoxified liquid phase after extracting the phenolic components, as they are precisely the main components responsible for this toxicity. But, it also makes it possible to obtain a solid with a lower phytotoxic content since most of the phenolic compounds have been solubilised in the liquid phase [3]. If the growing appearance of these two phases with a less toxic character is combined with the increasing energy demand, the need would arise to look for bioprocesses that allow us to manage the whole of the olive mill solid waste at the same time as obtaining energy and other products of great interest.

Among all types of bioprocesses, anaerobic digestion can be highlighted, which allows taking advantage of this source of organic matter, generating methane and a stabilised digestate for agricultural use, i.e., a source of energy and biomass for use in agriculture [6,7]. The extraction of phenolic components during the application of the three-phase extraction system would be favourable for the subsequent anaerobic digestion process due to the toxic effect of these compounds on anaerobic microorganisms, particularly the methanogens [8]. In this sense, the necessity arises to study the use of this technology from the two fractions, liquid and solid, which allows a better use while allowing us to manage the growing volumes of this by-product, generating energy and improving the conditions of exhaustion in the olive plantations themselves.

The objective of this study is to broaden the knowledge behind the processes of anaerobic degradation of agri-food residues such as the olive oil solid waste under mesophilic conditions operating in a semi-continuous regime. More specifically, the research focuses on evaluating the instability of the degradation process of a complex biomass as the olive oil solid waste, with the objective of determining the relationship between the decreased methane production and the present microbial population. The implications of understanding this link could be of great importance for designing biological systems capable of maximizing methane production.

2. Materials and Methods

2.1. Materials and Reagents

The reagents that have been used to carry out all the experimentation are the following: $NaHCO_3$ (Panreac, Castellar del Vallès, Barcelona, Spain), Na_2CO_3 (Panreac); Folin–Ciocalteu reagent (Panreac); Methanol 99.9% (Panreac); Gallic Acid (Sigma Aldrich, Steinheim, Germany); $H_2SO_4/AgSO_4$ 10g/L (Panreac); Potassium hydrogen phthalate (Sigma Aldrich); $K_2Cr_2O_7$ (Panreac); $(NH_4)_2Fe(SO_4)$ (Panreac); commercial $K_2Cr_2O_7$ 1N (Panreac); Sulfuric Acid 96% (Carlo Erba Reagents, Carrer dels Filadors, Sabadell, Spain); Sodium hydroxide 99% (Labkem, Barcelona, Spain); and pH buffer solution (Hach, Dusseldorf, Germany).

All samples were characterized by the following determination: Total, Mineral and Volatile Solids (TS, MS, and VS, respectively), Soluble Chemical Oxygen Demand (SCOD), Total Chemical Oxygen Demand (TCOD), pH, and alkalinity. These procedures were conducted in accordance with the guidelines provided by the Standard Methods of APHA [9].

The determination of total phenolic compounds was carried out using a colorimetric technique referred to as the Folin–Ciocalteau method [10]. The individual phenolic compounds were identified and quantified utilizing a Hewlett-Packard 1100 series high-performance liquid chromatography (HPLC) system from Agilent, located in Barcelona, Spain, following method described in Fernández-Prior et al. [11]. Volatile fatty acids (VFA) in the range of C2–C5 were assessed using a Shimadzu GC-2010 gas chromatograph, equipped with a 0.25 mm × 25 m column comprised of 100% ethylene glycol composition, along with a flame ionization detector (FID). The temperature of the oven was gradually raised from 100 to 170 °C at a rate of 5 °C·min^{-1}. A carrier gas mixture of nitrogen (30 mL·min^{-1}), hydrogen (40 mL·min^{-1}), and air (399.8 mL·min^{-1}) was employed at a flow rate of 40.1 mL·min^{-1}, maintained at 456 kPa.

2.2. Thermo-Malaxation Pre-Treatment

The raw material for this study was obtained from semi-solid olive mill waste sourced at the Instituto de la Grasa (CSIC, Seville, Spain) premises. The semi-solid olive mill waste was processed using Pieralisis equipment (Pieralisis, Jesi, Italy) with a milling capacity of 1000 kg/h. The resulting olive mill solid waste was acquired through a two-stage extraction process designed to obtain olive oil. The application of subsequent heat treatment or thermo-malaxation consisted of beating the olive pomace at 60 °C for 90 min. It was then centrifuged in a decanter giving rise to three phases: a first solid phase (SP), a liquid phase (LP), and pomace oil (POO). A fourth phase of suspended solids (SS) was obtained from the storage and sedimentation of the SP. Thanks to the sedimentation time (two months), a new liquid phase free of suspended solids (LFP) was obtained with an increase in HT due to the hydrolysis of the HT precursors, as in a previous study [8]. LPF served as an abundant reservoir of phenolic compounds, subjected to a solid–liquid extraction system to capture these compounds. Subsequently, the phenolic compounds were extracted from the column using a solution composed of ethanol and water in comparable volumes. Phenolic compounds were then concentrated by completely eliminating the ethanol. Finally, a dephenolized liquid fraction (DLP) was obtained. The whole set of samples, SP and DLP, were stored at −20 °C until their use in the different assays. Table 1 provides a summary of the analytical characteristics of the distinct phases utilized within the anaerobic digestion process. These phases are a solid phase (SP), dephenolized liquid phase (DLP), and inoculum, as will be explained in the next section.

Table 1. Physicochemical characterization of the different samples used in the assay.

		DLP	SP	Inoculum
pH		4.9 ± 0.1	4.6 ± 0.1	7.4 ± 0.1
TS	mg/kg	43,135 ± 188	428,811 ± 6716	57,126 ± 437
MS	mg/kg	9403 ± 285	16,827 ± 605	19,988 ± 252
VS	mg/kg	33,732 ± 461	411,984 ± 7023	37,138 ± 388
VS/TS		0.78 ± 0.01	0.96 ± 0.02	0.65 ± 0.01
tCOD	mg O_2/kg	98,148 ± 565	576,024 ± 40,220	77,792 ± 1738
sCOD	mg O_2/L	92,650 ± 1322	60,076 ± 4755	3115 ± 212
Total phenolics	mg gallic acid eq./kg	2396 ± 84	3935 ± 155	106 ± 4
C2	mg O_2/L	-	-	1316 ± 9
C3	mg O_2/L	-	-	1113 ± 8
i-C4	mg O_2/L	-	-	41 ± 1
n-C4	mg O_2/L	-	-	487 ± 4
i-C5	mg O_2/L	-	-	142 ± 1
n-C5	mg O_2/L	-	-	476 ± 4
Total VFA	mg O_2/L	-	-	3576 ± 10
Alkalinity	mg $CaCO_3$/L	-	-	7044 ± 175

Acetic acid (C2), Propionic acid (C3), Isobutyric acid (i-C4), Butyric acid (n-C4), Isovaleric acid (i-C5), and Valeric acid (n-C5).

2.3. Semi-Continous Anaerobic Process Procedure

Microorganisms used in the anaerobic assay were a mixed culture from the anaerobic digester of the "Copero" wastewater treatment plant in Sevilla, Spain. Physicochemical characterization from inoculum is shown in Table 1.

Two glass reactors with a capacity of two litres have been used. 10 g VS/L of inoculum was introduced in the reactors together with water until 1.7 L of working volume at the start of the assay. The reactors were fed every day with 81 mL of SP + DLP + Water, were continuously stirred with a cylindrical magnetic bar, and were thermostated at 35 °C. Biogas was transported by rubber tube until a gas bubbler, with a NaOH solution (3 N), was used for CO_2 removal. Methane production displaced water from a closed tank and was quantified with a graduated cylinder under standard temperature and pressure conditions (25 °C and 1 atm). In the preceding 7 days leading up to this experiment, both reactors were prepared and acclimated using mixtures of a synthetic solution (SS) containing glucose (50 g/L) and sodium acetate (25.2 g/L), with an organic load rate (OLR) set at 1 g VS/L d. Following the acclimation phase, the reactors were operated for 63 days. They were supplied with a combination of solid phase (SP) and dephenolized liquid phase (DLP) after undergoing pre-treatment. During assay, an OLR of 1 g VS/L d was applied over three stages, corresponding with the three hydraulic retention times (HRTs), each spanning 21 days.

2.4. DNA Extraction and Library Preparation

DNA amplification targeted the V4 region of the 16S rRNA using barcoded primers 515f (5′-GTGCCAGCMGCCGCGGTAA) and 806r (5′-GGACTACHVGGGTWTCTAAT). The amplicons were produced via one-step PCR employing the barcodes specified in Ramiro-Garcia et al. [12]. A DNA template (2 μg) was used in the 40 μL PCR reaction, including 10 μL of HF buffer (Thermo Fisher Scientific, Madrid, Spain), 1 μL of dNTP Mix (10 mM; Bioline, London, UK), 1 U of ADN polymerase Phusion™ Plus (Thermo Fisher Scientific), and 10 nM of each barcoded primer. PCR cycles were conducted using an Alpha cycler 1 (PCRmax, Biorad, Watford, UK) following the conditions outlined in Lara M. Paulo et al., [13]. Purification of PCR products was performed using HighPrep™ (Magbio Genomics, Gaithersburg, MD, USA), with elution using 20 μL of Nuclease Free Water (Bioline). Quantification was achieved using a nanodrop (Thermo Fisher Scientific). Subsequently, the purified products were equimolarly combined to form the library pool, which was then subjected to sequencing on the Illumina NovaSeq 6000 platform (Eurofins Genomics, Ebersberg, Germany). Sequence data have been deposited in the European Nucleotide Archive under accession number [PRJEB59213].

2.5. Bioinformatics Analysis

Data from all samples (with a total of 2,984,943 pair-end reads) were analysed using the Silva 138 database [14] to assign taxonomy and NG-Tax [12], a validated pipeline for 16S rRNA analysis, under default parameters. Alpha diversity and beta diversity were calculated using the R packages phyloseq 1.32.0 [15] and picante 1.8.2 [16]. Their plots and the heatmap plot were generated using ggplot 2 23.3.2 [17].

3. Results and Discussion

3.1. Methane Production along Operation Time

The methane production yield was evaluated daily after feeding a mixture obtained after the treatment of olive mill solid waste (SP + DLP) (Figure 1). The anaerobic digestion process exhibits a variable methane production pattern across three HRTs. During the initial HRT, methane production reached 204 ± 9 mL CH_4/(g VS d). Notably, this methane production yield under an OLR of 1 g VS/(L d) closely resembles that achieved for the hydrothermal treatment of olive mill solid waste at 170 °C, accompanied by phenolic compound extraction, which yielded 172 ± 60 mL CH_4/(g VS d) [6]. Between the first and the second HRT, 34% of methane production decreased in the degradation of SP + DLP (Table 2). During the second HRT, the anaerobic digestion process seems to have been more stable. However, the production decreased by 35% from 135 ± 18 to 87 ± 24 mL CH_4/(g VS d) in the third HRT. The stability of olive mill solid waste in the anaerobic digestion was maintained at an OLR of approximately 1 g VS/(L d) according to Serrano et.al. [18]. Biodegradability was also reduced in the anaerobic digestion process, reaching only 27%. This effect is also shown in solid concentrations in Table 2. During the third HRT, the TS decreased, but the VS increased by 5%. In similar studies with a strawberry extrudate treated with more severe thermal treatments and using solid and dephenolized liquid phase, the same tendency of decrease in methane production is observed with 1 g VS/(L d) [19].

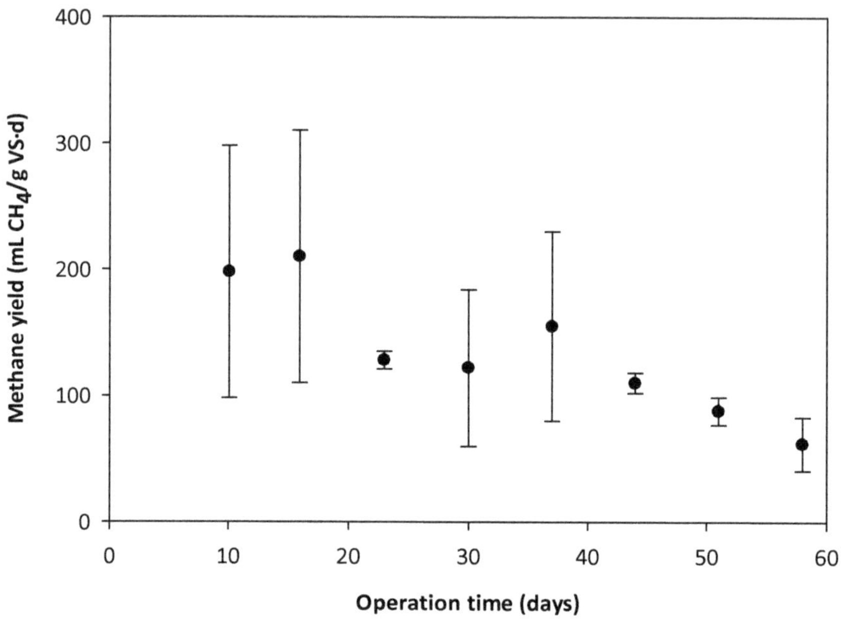

Figure 1. Variation of the methane production yield with their corresponding standard deviations along operation time.

Table 2. pH, alkalinity, TS, VS, sCOD, total VFA, total phenolic compound, methane production yield and biodegradability during the different hydraulic retention times (HRTs).

	1 HRT	2 HRT	3 HRT
OLR (g VS/L d)	1	1	1
Days	0–21	22–42	43–60
pH	7.5 ± 0.1	7.2 ± 0.1	6.8 ± 0.1
Alkalinity (mg $CaCO_3$/L)	7967 ± 2000	4065 ± 747	2270± 709
TS (mg/L)	18,940 ± 2316	14,223 ± 887	12,908 ± 891
VS (mg/L)	9832 ± 628	9823 ± 398	10,389 ± 575
sCOD (mg O_2/L)	3695 ± 1125	1912 ± 540	1968 ± 250
Total VFA (mg O_2/L)	3280 ± 2083	751 ± 500	210± 182
Total phenols (mg gallic acid eq./L)	149 ± 30	243 ± 17	244 ± 8
Methane production yield (mL CH_4/g VS d)	204 ± 9	135 ± 18	87 ± 24
Biodegradability CH_4 (%)	74 ± 40	48 ± 33	27 ± 15

3.2. Control Parameters along Operation Time

pH, alkalinity, sCOD, and VFAs were measured to monitor the anaerobic reactor performance (Figure 2). The high sCOD concentration came mainly from the inoculum and increased in the first days. But during the first HTR, sCOD concentration decreased drastically. Similar to sCOD, VFAs decreased during the first HRT, giving a relationship of about 90% VFAs/sCOD. The sCOD presented an average value of 1960 ± 250 mg O_2/(L d) along the second and third HRTs. However, the ratio VFA/sCOD decreased from 90 to 10% along the second and third HRTs, when the VFA concentration reached 210 ± 182 mg O_2/(L d) at the end of the third HRT. The pH did not significantly fluctuate along the operation time, i.e., approximately 6.8–7.5, the value recommended for the anaerobic digestion process [20]. But, alkalinity decreased during the operation time, triggering the instability of the process. The alkalinity decreased from 7967 ± 2000 to 2270 ± 709 mg $CaCO_3$/L. Alkalinity around 2000 mg $CaCO_3$/L was similar at 1 g VS/(L d) in similar studies with olive mill solid waste treatment [6,21].

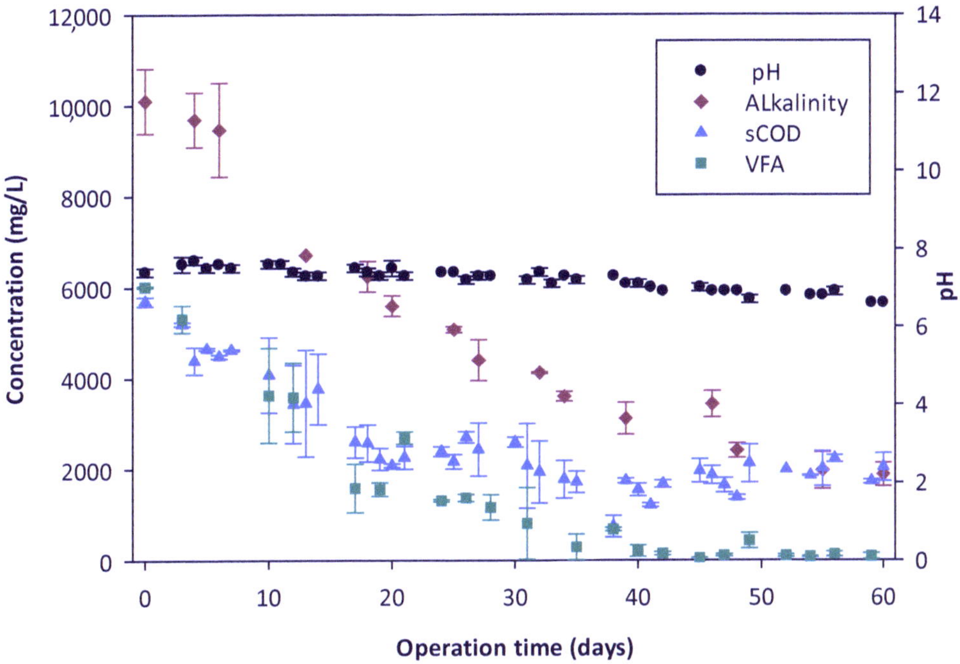

Figure 2. Variation of organic matter measured with sCOD (▲) and VFA (■), and variation of stability measured with pH (•) and Alkalinity (♦) values with their standard deviations along operation time.

Individual VFAs, between C2–C5, were measured along operation time and expressed in mg O_2/L (Figure 3). The inoculum that was used for this test was observed to have a large amount of total VFA with more than 1000 mg O_2/L acetic and propionic acids (Table 1). After two days, a large increase in these acids was observed, reaching more than 6000 mg O_2/L of total VFA (Figure 3). Individual VFAs, mainly Acetic and propionic acids, decreased in the first HRT to less than 2000 and 1000 mg O_2/L, respectively. Individual VFA concentrations were decreased around 704 ± 727 mg O_2/L, propionic acid being the most abundant at the end of third HRT. This parameter is important to evaluate since different studies have reported the inhibitory effect of AGV, such as acetic and propionic acid, with highly varied concentrations [22,23]. High acetic and propionic acid concentrations suggest the instability of the reactor due to the acetogenesis and methanogenesis stages [24].

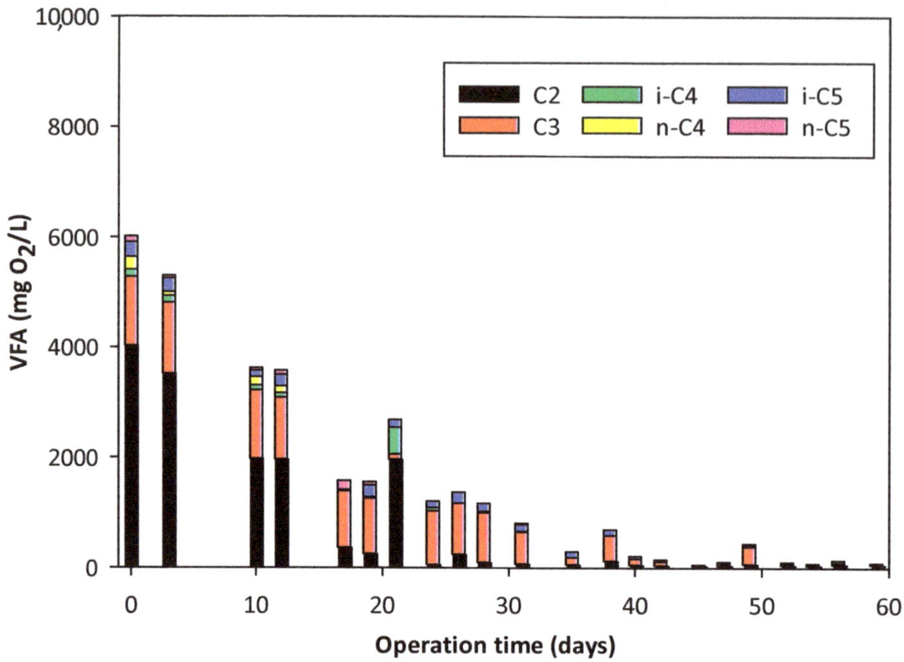

Figure 3. Variation of the individual VFAs values along operation time.

3.3. Variation of the Concentration of Phenolic Compounds along Operation Time

Total and individual phenolic compounds were measured to monitor the anaerobic reactor performance (Figure 4). Total phenolic compound concentration, expressed as mg gallic acid equivalent per litre, increased along the operation time with an average value from 149 ± 30 mg gallic acid/L to 244 ± 8 mg gallic acid/L. According to previous research, phenolic compounds are known inhibitors of the anaerobic digestion process [25]. Microbial growth might be influenced by elevated phenolic compound concentrations [26]. However, the phenolic compound levels observed in these experiments consistently remain below the threshold recognized for inhibitory effects on the anaerobic digestion process, which is typically around 2000 mg/L [27]. Moreover, these phenolic concentrations were even lower than those documented in a study involving olive mill solid waste treated using the identical process as applied in this investigation [21]. Individual phenolic compounds detected with a lower concentration in the reactor were vanillic acid, 4-hydroxybenzoic acid, catechin, and 4-ethylphenol (Table 3). These phenolic compounds were also detected in other studies with the same substrate with other thermal pre-treatment [6,21]. Substrate characterization generally identifies all of these compounds except 4-ethylphenol, an intermediate metabolite indicating incomplete phenolic degradation [6]. Very few simple phenolic compounds remain in the reactors and in very low concentrations. The amount of individual phenolics is far from the amount of total phenolics quantified, which increases with time. This must be due to a polymerization of the phenolics that would remain forming condensed phenols so that although their content is high, their toxicity might be significant. Among simple phenolics, 4-ethylphenol and vanillic acid are the only two that decrease until they disappear, since they are fermentation products. However, catechol and 4-p-hydroxybenzoic acid are intermediate products of the chemical and enzymatic degradation of the main phenolics, and their concentration increases slightly. Since they are compounds with high phytotoxicity, their slight increase could influence the instability of the fermentation [28].

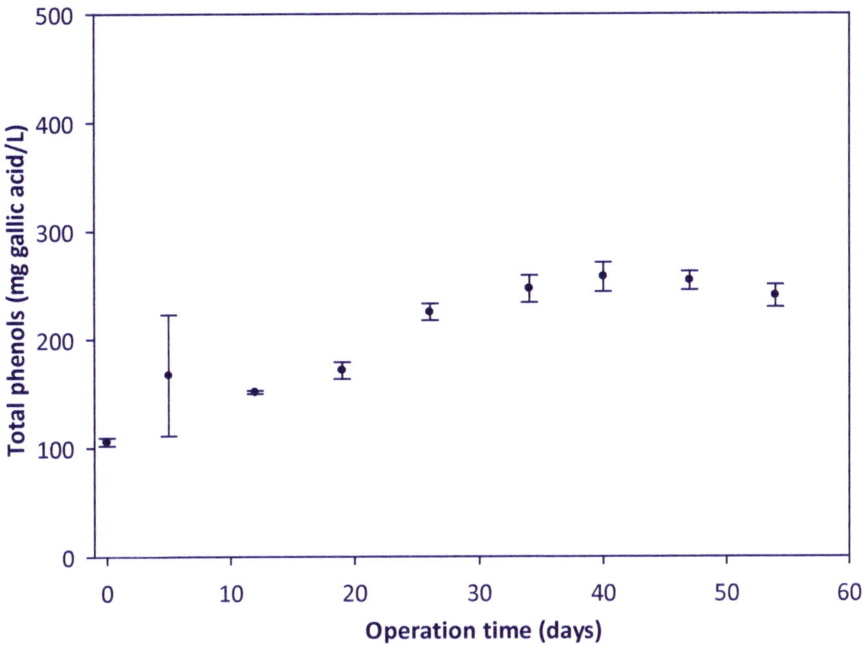

Figure 4. Variation of the total phenolic concentrations with their standard deviation along operation time.

Table 3. Individual phenolic concentration with their standard deviation along operation time.

		Days	12	19	26	34	40	47	54	61
	R.T.	λmax (nm)	Concentration μg/L							
vanillic acid	17.8	254	825 ± 102.5	720 ± 2.5	372.5 ± 12.5	307.5 ± 20	N.D.	N.D.	N.D.	N.D.
4-hydroxybenzoic acid	34.1	254	262.5 ± 10	285 ± 5	260 ± 30	240 ± 15	165 ± 7.5	85 ± 15	395 ± 12.5	285 ± 2.5
catechin	37.6	280	N.D.	2037.5 ± 72.5	2342.5 ±	2832.5 ± 360	2647.5 ± 72.5	4297.5 ± 90	5210 ± 80	4012.5 ± 67.5
4-ethylphenol	49.2	280	traces	traces	traces	traces	traces	traces	traces	traces

R.T., retention time; λmax, maximum wavelength; traces < 0.01 μg/L; N.D., not detected.

3.4. Microbial Population Dynamics along Operation Time

Figure 5 shows the β-diversity and PCoA to determine the general trends of differences and similarities between the biological duplicate of the reactors in inoculum at zero time and the last day of each HRT through cluster analysis. β-Diversity levels include the Weighted Unifrac ASV level, Bray–Curtis ASV level, Weighted Unifrac Genus level, and Bray–Curtis Genus level. Analysing the quantitative data in inoculum at zero time and the last day of each HRT does not show a β-diversity difference between the biological reactor duplicates. During operation time, bacteria analysis showed a difference in β-diversity along each HRT. Also, archaea analysis showed that β-diversity changed slightly in the third HRT (Figure 5).

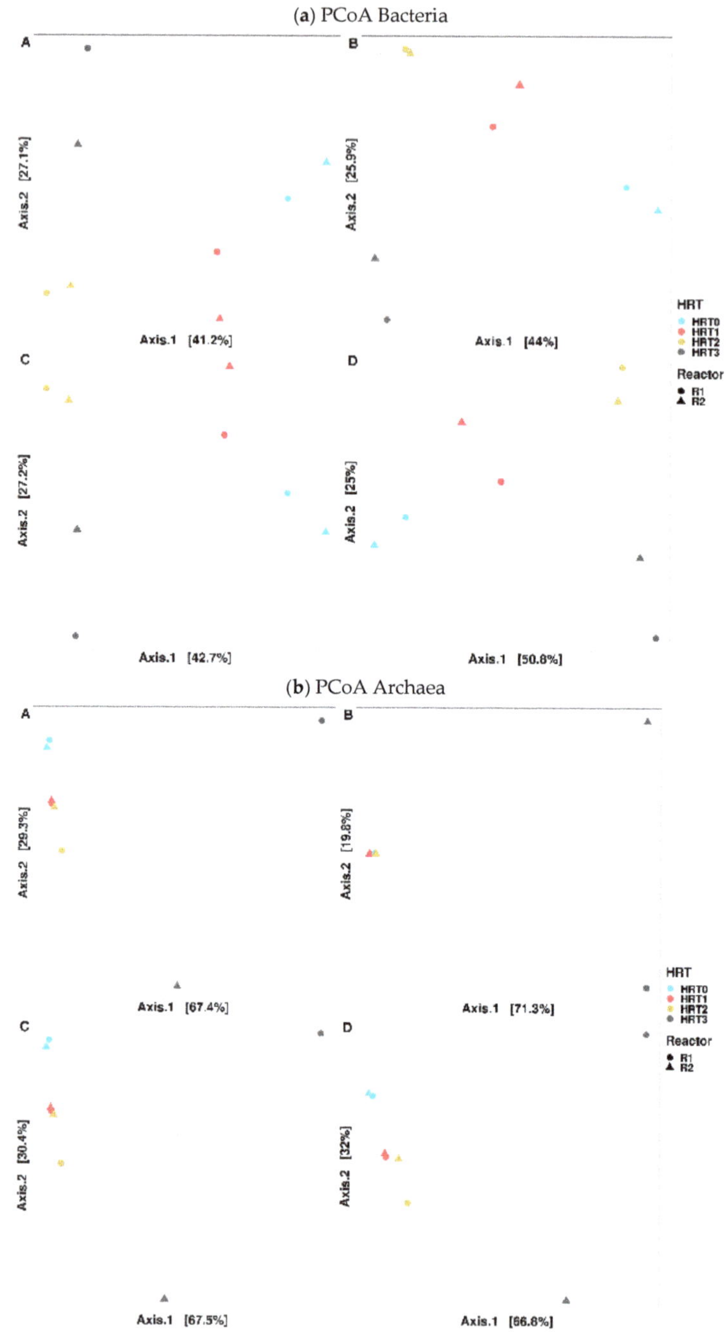

Figure 5. The Beta Diversity PCoA plots for Bacteria (**a**) and Archaea (**b**) are presented, showcasing the inoculum at time zero and the last day of each hydraulic retention time (HRT). These plots are divided into four different levels: A. Weighted Unifrac ASV level. B. Bray–Curtis ASV level. C. Weighted Unifrac Genus level. D. Bray–Curtis Genus level.

The included alpha diversity indices were FaithPD, Shannon, and Simpson, which showed a linear trend (Figure 6). The short measure distance between samples indicates their similarity in the microbial community. Alpha diversity from bacteria has demonstrated a short-range measure, without significant fluctuations, with phylogenetics ranging in FaithPD from 2.0 to 3.6 throughout the digestion progress. Archea fluctuations in FaithPD had slight measure differences, ranging from 0.2 to 0.55. The microorganism of reactors had a similar measure in alpha diversity on the last day of each HRT without having any relevant fluctuations. Regarding the measure ranges in the Shannon and Simpson index, it can be seen that they are similar to other studies on anaerobic digestion [29].

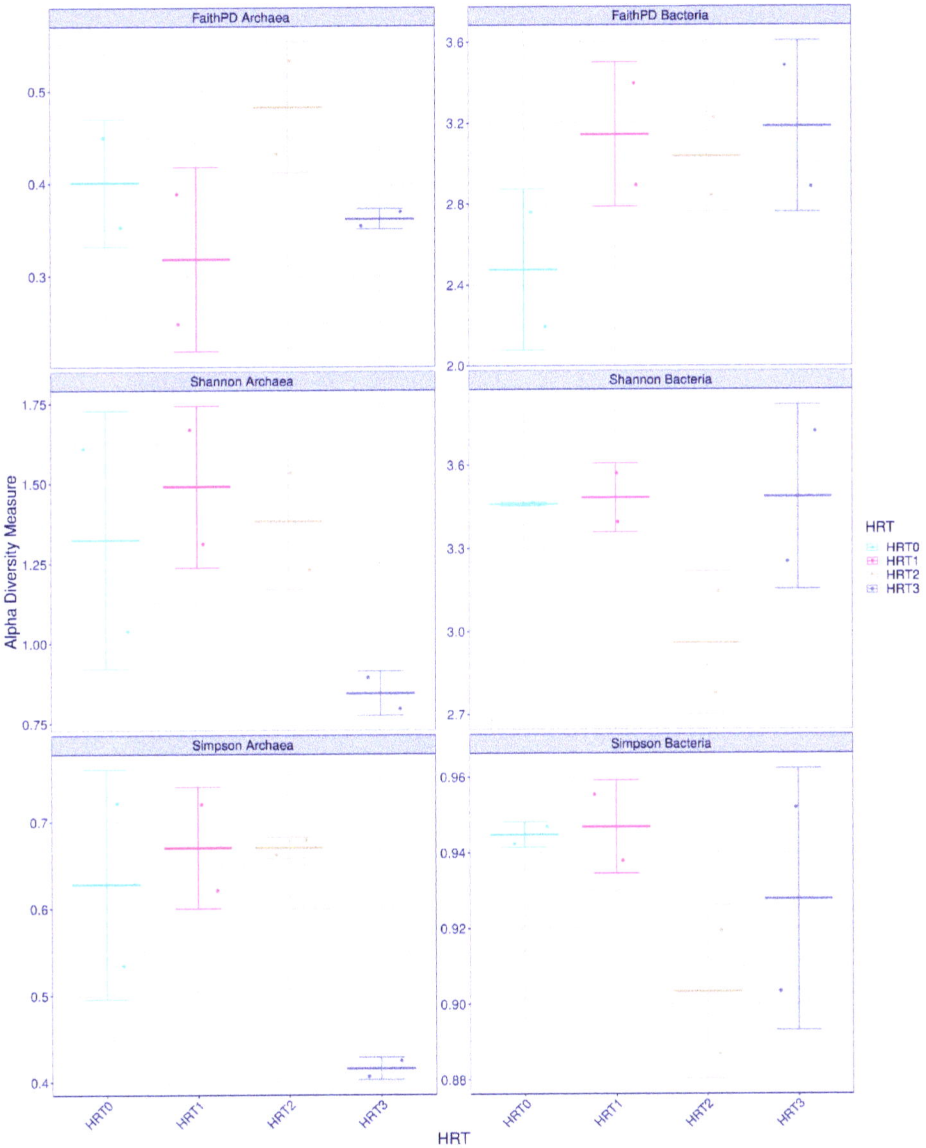

Figure 6. Alpha diversity in inoculum at zero time and the last day of each HRT.

The bacterial microbiome relative composition at the phylum and genus levels is presented in the heatmap abundance (Figure 7). Examination of the bacterial microbiome composition unveiled Bacteroidota as the predominant phylum throughout the experimental period, with various genera undergoing shifts in abundance over time. The *Alistipes* and *dgA-11_gut_group* (family *Rikenellaceae*) genera decreased in relative abundance in favour of genus *Proteiniphilum* (family *Dysgonomonadaceae)* through the experiment, but especially on the last day of the second and third HRT (Figure 7). *Alistipes* has been associated with the production of VFAs and hydrogen through protein and carbohydrate degradation [30]. The decrease in its relative abundance through experimentation would explain imbalances in the AD process, resulting in a reduction of methane production as well as in the accumulation of VFAs [24,31].

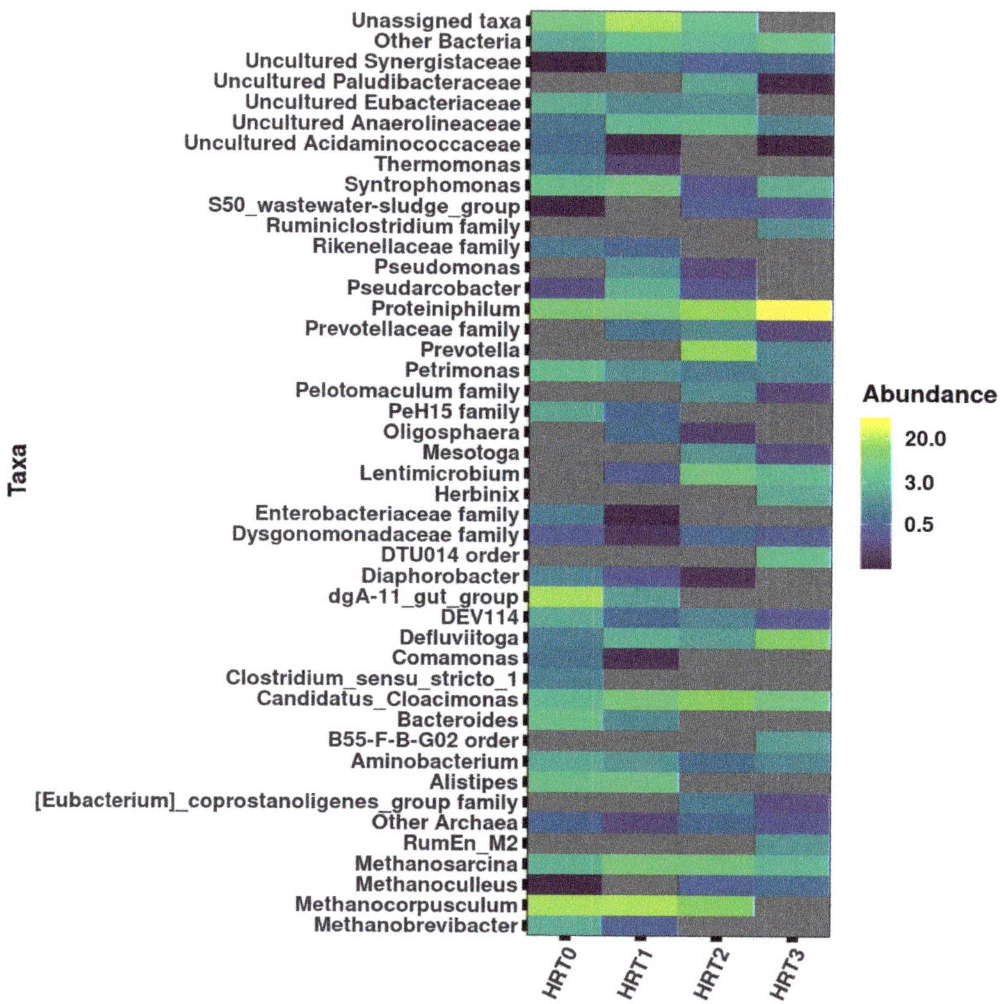

Figure 7. Heatmap abundance samples grouped in inoculum at zero time and the last day of each HRT.

In relation to the archaeal microbial communities, Figure 7 depicts the existence of two primary methanogenic groups crucial for methane production in anaerobic digestion: acetoclastic and hydrogenotrophic methanogens. Acetoclastic methanogens utilize acetate for generating methane and CO_2, whereas hydrogenotrophic methanogens employ H_2 or formate to transform CO_2 into methane [32,33]. The predominant pathway of methane production from in the inoculum at zero time to the last day of the second HRT was the hydrogenotrophic methanogenesis pathway. However, during the experiment, a decline was observed of hydrogenothophic methanogens of the Halobacterota and Euryarchaeota phylums, e.g., the *Methanocorpusculum* and *Methanobrevibacter* genera, until their almost complete disappearance on the last day of the third HRT (Figure 7). The development of an acetoclastic methanogen classified as *Methanosarcina* genus was promoted for all digesters from in the inoculum at zero time to the last day of the second HRT, which would explain the drastic decrease in VFAs at the beginning of the experiment, as well as the stability in methane production (Figures 1–3) [34]. Furthermore, the drastic decrease in methanogens was reflected in a decrease in methane production at the end of the experiment (Figure 7).

4. Conclusions

Olive solid waste thermo-malaxation treated and dephenolizated was evaluated in a semicontinuous anaerobic digestion process. The average value of methane yield decreased from 204 $\pm$ 9 mL CH_4/g VS d on the first HRT, to 87 $\pm$ 24 mL CH_4/g VS d on the third HRT. Different control parameters were measured to observe the stability of the biodegradation of the complex biomass, with a high phenolic compound concentration, in the anaerobic digestion process. The soluble organic matter, measured by sCOD and VFA, declined over time, showing that the hydrolysis stage was affected. The disappearance of hydrolytic bacteria and increased acetoclastic methanogens confirmed the process's instability.

Author Contributions: Conceptualization, G.R.-G. and F.G.F.; methodology, J.C.-C., B.C.-G. and Á.F.-P.; formal analysis, J.C.-C., A.T.-R., J.R.-G. and Á.F.-P.; resources, G.R.-G. and F.G.F.; writing—original draft preparation, J.C.-C., A.T.-R., J.R.-G. and Á.F.-P.; writing—review and editing, G.R.-G. and F.G.F.; supervision, F.G.F.; funding acquisition, G.R.-G. All authors have read and agreed to the published version of the manuscript.

Funding: This research was supported by the incentives for agents of the Andalusian knowledge system, aid under the Andalusian research, development and innovation plan (PAIDI 2020) (project P18-TP-616). This research has received funding from the European Union's Horizon 2020 research and innovation program under grant agreement N° 101033411.

Data Availability Statement: Data will be made available on request.

Conflicts of Interest: The authors declare no conflict of interest.

References

1. Jiménez Marquez, A.; Hermoso Fernández, M.; Uceda Ojeda, M. Extraction of Virgin Olive Oil by Two-Phase Continuous System. Influence of Different Variables of the Process on Certain Parameters Related to Oil Quality. *Grasas y Aceites* **1995**, *46*, 299–303. [CrossRef]
2. Fermoso, F.G.; Serrano, A.; Alonso-Fariñas, B.; Fernández-Bolaños, J.; Borja, R.; Rodríguez-Gutiérrez, G. *Valuable Compound Extraction, Anaerobic Digestion, and Composting: A Leading Biorefinery Approach for Agricultural Wastes*; American Chemical Society: Washington, DC, USA, 2018; Volume 66, pp. 8451–8468.
3. Lama-Muñoz, A.; Rubio-Senent, F.; Bermúdez-Oria, A.; Fernández-Bolaños, J.; Prior, Á.F.; Rodríguez-Gutiérrez, G. The Use of Industrial Thermal Techniques to Improve the Bioactive Compounds Extraction and the Olive Oil Solid Waste Utilization. *Innov. Food Sci. Emerg. Technol.* **2019**, *55*, 11–17. [CrossRef]
4. Berbel, J.; Posadillo, A. Review and Analysis of Alternatives for the Valorisation of Agro-Industrial Olive Oil By-Products. *Sustainability* **2018**, *10*, 237. [CrossRef]
5. Fernández-Prior, Á.; Bermúdez-Oria, A.; Fernández-Bolaños, J.; Espejo-Calvo, J.A.; López-Maestro, F.; Rodríguez-Gutiérrez, G. Evolution of Hydroxytyrosol, Hydroxytyrosol 4-β-D-Glucoside, 3,4-Dihydroxyphenylglycol and Tyrosol in Olive Oil Solid Waste or "Alperujo". *Molecules* **2022**, *27*, 8380. [CrossRef] [PubMed]

6. Serrano, A.; Fermoso, F.G.; Alonso-Fariñas, B.; Rodríguez-Gutiérrez, G.; López, S.; Fernandez-Bolaños, J.; Borja, R. Performance Evaluation of Mesophilic Semi-Continuous Anaerobic Digestion of High-Temperature Thermally Pre-Treated Olive Mill Solid Waste. *Waste Manag.* **2019**, *87*, 250–257. [CrossRef] [PubMed]
7. Girardi, F.; Mascitti, A.; d'Alessandro, N.; Tonucci, L.; Marzo, G.A.; Remetti, R. Effects of Oxidative Treatments on Biomethane Potential of Solid Olive Residues. *Waste Biomass Valorization* **2023**, *14*, 1525–1538. [CrossRef]
8. Fernández-Prior, Á.; Trujillo-Reyes, Á.; Serrano, A.; Rodríguez-Gutiérrez, G.; Reinhard, C.; Fermoso, F.G. Biogas Potential of the Side Streams Obtained in a Novel Phenolic Extraction System from Olive Mill Solid Waste. *Molecules* **2020**, *25*, 5438. [CrossRef]
9. APHA. *Standard Methods for Examination of Water and Wastewater*; American Public Health Association: Washington, DC, USA, 2017; ISBN 9780875532356.
10. Singleton, V.L.; Rossi, J.A. Colorimetry of Total Phenolics with Phosphomoybdic-Phosphotungstic Acid Reagents. *Am. J. Enol. Viticult.* **1965**, *16*, 144–158. [CrossRef]
11. Fernández-Prior, Á.; Cardoso, J.C.; Bermúdez-Oria, A.; Reyes, Á.T.; Fernández-Bolaños, J.; Rodríguez-Gutiérrez, G. Application of a Cold-Pressing Treatment to Improve Virgin Olive Oil Production and the Antioxidant Phenolic Profile of Its by-Products. *Antioxidants* **2023**, *12*, 1162. [CrossRef]
12. Ramiro-Garcia, J.; Hermes, G.D.A.; Giatsis, C.; Sipkema, D.; Zoetendal, E.G.; Schaap, P.J.; Smidt, H.; Schmidt, T.S.B.; Tremblay, J.; Fouhy, F.; et al. NG-Tax, a Highly Accurate and Validated Pipeline for Analysis of 16S RRNA Amplicons from Complex Biomes. *F1000Research* **2018**, *5*, 1791. [CrossRef]
13. Paulo, L.M.; Castilla-Archilla, J.; Ramiro-Garcia, J.; Escamez-Picón, J.A.; Hughes, D.; Mahony, T.; Murray, M.; Wilmes, P.; O'Flaherty, V. Microbial Community Redundancy and Resilience Underpins High-Rate Anaerobic Treatment of Dairy-Processing Wastewater at Ambient Temperatures. *Front. Bioeng. Biotechnol.* **2020**, *8*, 192. [CrossRef] [PubMed]
14. Quast, C.; Pruesse, E.; Yilmaz, P.; Gerken, J.; Schweer, T.; Yarza, P.; Peplies, J.; Glöckner, F.O. The SILVA Ribosomal RNA Gene Database Project: Improved Data Processing and Web-Based Tools. *Nucleic Acids Res.* **2013**, *41*, D590–D596. [CrossRef] [PubMed]
15. McMurdie, P.J.; Holmes, S. Phyloseq: An R Package for Reproducible Interactive Analysis and Graphics of Microbiome Census Data. *PLoS ONE* **2013**, *8*, e61217. [CrossRef] [PubMed]
16. Kembel, S.W.; Cowan, P.D.; Helmus, M.R.; Cornwell, W.K.; Morlon, H.; Ackerly, D.D.; Blomberg, S.P.; Webb, C.O. Picante: R Tools for Integrating Phylogenies and Ecology. *Bioinformatics* **2010**, *26*, 1463–1464. [CrossRef]
17. Wickham, H. *Programming with Ggplot2*; Springer: Berlin/Heidelberg, Germany, 2016; pp. 241–253. [CrossRef]
18. Serrano, A.; Villa-Gomez, D.; Fermoso, F.G.; Alonso-Fariñas, B. Is Anaerobic Digestion a Feasible Alternative to the Combustion of Olive Mill Solid Waste in Terms of Energy Production? A Critical Review. *Biofuels Bioprod. Biorefining* **2021**, *15*, 150–162. [CrossRef]
19. Cubero-Cardoso, J.; Muñoz-Arjona, A.; Trujillo-Reyes, Á.; Serrano, A.; Alonso-Fariñas, B.; Rodríguez-Gutiérrez, G.; Urbano, J.; Borja, R.; Fermoso, F.G. Mesophilic Semi-Continuous Anaerobic Digestion of Strawberry Extrudate Pretreated with Steam Explosion. *Foods* **2020**, *9*, 1887. [CrossRef]
20. Xu, R.Z.; Fang, S.; Zhang, L.; Huang, W.; Shao, Q.; Fang, F.; Feng, Q.; Cao, J.; Luo, J. Distribution Patterns of Functional Microbial Community in Anaerobic Digesters under Different Operational Circumstances: A Review. *Bioresour. Technol.* **2021**, *341*, 125823. [CrossRef]
21. Serrano, A.; Fermoso, F.G.; Alonso-Fariñas, B.; Rodríguez-Gutiérrez, G.; López, S.; Fernandez-Bolaños, J.; Borja, R. Long-Term Evaluation of Mesophilic Semi-Continuous Anaerobic Digestion of Olive Mill Solid Waste Pretreated with Steam-Explosion. *Energies* **2019**, *12*, 2222. [CrossRef]
22. Montañés, R.; Pérez, M.; Solera, R. Mesophilic Anaerobic Co-Digestion of Sewage Sludge and a Lixiviation of Sugar Beet Pulp: Optimisation of the Semi-Continuous Process. *Bioresour. Technol.* **2013**, *142*, 655–662. [CrossRef]
23. Casallas-Ojeda, M.R.; Marmolejo-Rebellón, L.F.; Torres-Lozada, P. Identification of Factors and Variables That Influence the Anaerobic Digestion of Municipal Biowaste and Food Waste. *Waste Biomass Valorization* **2021**, *12*, 2889–2904. [CrossRef]
24. Nagarajan, S.; Jones, R.J.; Oram, L.; Massanet-Nicolau, J.; Guwy, A. Intensification of Acidogenic Fermentation for the Production of Biohydrogen and Volatile Fatty—A Perspective. *Fermentation* **2022**, *8*, 325. [CrossRef]
25. Battista, F.; Fino, D.; Erriquens, F.; Mancini, G.; Ruggeri, B. Scaled-up Experimental Biogas Production from Two Agro-Food Waste Mixtures Having High Inhibitory Compound Concentrations. *Renew. Energy* **2015**, *81*, 71–77. [CrossRef]
26. Caroca, E.; Serrano, A.; Borja, R.; Jiménez, A.; Carvajal, A.; Braga, A.F.M.; Rodriguez-Gutierrez, G.; Fermoso, F.G. Influence of Phenols and Furans Released during Thermal Pretreatment of Olive Mill Solid Waste on Its Anaerobic Digestion. *Waste Manag.* **2021**, *120*, 202–208. [CrossRef] [PubMed]
27. Calabrò, P.S.; Fòlino, A.; Tamburino, V.; Zappia, G.; Zema, D.A. Increasing the Tolerance to Polyphenols of the Anaerobic Digestion of Olive Wastewater through Microbial Adaptation. *Biosyst. Eng.* **2018**, *172*, 19–28. [CrossRef]
28. Palmqvist, E.; Hahn-Hägerdal, B. Fermentation of Lignocellulosic Hydrolysates. II: Inhibitors and Mechanisms of Inhibition. *Bioresour. Technol.* **2000**, *74*, 25–33. [CrossRef]
29. Mao, C.; Wang, Y.; Wang, X.; Ren, G.; Yuan, L.; Feng, Y. Correlations between Microbial Community and C:N:P Stoichiometry during the Anaerobic Digestion Process. *Energy* **2019**, *174*, 687–695. [CrossRef]
30. Liu, Q.; Ren, Z.J.; Huang, C.; Liu, B.; Ren, N.; Xing, D. Multiple Syntrophic Interactions Drive Biohythane Production from Waste Sludge in Microbial Electrolysis Cells. *Biotechnol. Biofuels* **2016**, *9*, 162. [CrossRef]

31. Vázquez-Fernández, A.; Suárez-Ojeda, M.E.; Carrera, J. Review about Bioproduction of Volatile Fatty Acids from Wastes and Wastewaters: Influence of Operating Conditions and Organic Composition of the Substrate. *J. Environ. Chem. Eng.* **2022**, *10*, 107917. [CrossRef]
32. Roy, C.K.; Hoshiko, Y.; Toya, S.; Maeda, T. Effect of Different Concentrations of Sodium Selenite on Anaerobic Digestion of Waste Sewage Sludge. *Environ. Technol. Innov.* **2022**, *27*, 102403. [CrossRef]
33. Zhou, M.; Yang, H.; Zheng, D.; Pu, X.; Liu, Y.; Wang, L.; Zhang, Y.; Deng, L. Methanogenic Activity and Microbial Communities Characteristics in Dry and Wet Anaerobic Digestion Sludges from Swine Manure. *Biochem. Eng. J.* **2019**, *152*, 107390. [CrossRef]
34. Azarmanesh, R.; Zarghami Qaretapeh, M.; Hasani Zonoozi, M.; Ghiasinejad, H.; Zhang, Y. Anaerobic Co-Digestion of Sewage Sludge with Other Organic Wastes: A Comprehensive Review Focusing on Selection Criteria, Operational Conditions, and Microbiology. *Chem. Eng. J. Adv.* **2023**, *14*, 100453. [CrossRef]

processes

MDPI

Article

Anaerobic Digestion of the Residue (Combination of Wastewater and Solid Waste) from a New Olive-Oil Manufacturing Process Based on an Olive Cold-Pressing System: Kinetic Approach and Process Performance

Mª José Fernández-Rodríguez [1,2], Juan Cubero-Cardoso [1], David de la Lama-Calvente [1], África Fernández-Prior [1], Guillermo Rodríguez-Gutiérrez [1] and Rafael Borja [1,*]

[1] Spanish National Research Council (CSIC)—Instituto de la Grasa (IG), Department of Food Biotechnology, Campus Universidad Pablo de Olavide, Edificio 46, Carretera de Utrera, km 1, 41013 Seville, Spain
[2] Department of Vegetal Biology and Ecology, Faculty of Biology, University of Seville, 41080 Seville, Spain
* Correspondence: rborja@ig.csic.es; Tel.: +34-954611550; Fax: +34-954616790

Abstract: This research evaluates the anaerobic digestion (AD) process of the residue generated in a new olive-oil manufacturing process for cold-pressed olive, a residue consisting of a mixture of the wastewater and solid waste obtained from this process. Additionally, in order to assess the possible influence of the level of ripening of the olives on the performance of anaerobic processing, olives of the Picual variety were collected at two stages, i.e., green olives and olives in veraison. The AD processes of the residues obtained from the cold-pressing process and the process without pressure (control) were comparatively assessed by means of biochemical methane potential (BMP) assays conducted at mesophilic temperature (35 ± 1 °C). Maximum values for methane yield (390 ± 1 NL CH_4/kg VS_{added}) and biodegradability (84.5%) were obtained from the cold-pressed green olive residues. For the rest of the wastes studied, biodegradability also reached high values, ranging from 79.1 to 79.6%. The logistic model adequately fit the experimental data and allowed for the assessment of the anaerobic biodegradation of these wastes and for obtaining the kinetic parameters for each case studied. The theoretical values for ultimate methane production predicted from this model showed less than a 1% deviation from the experimental values. A decrease was detected for both types of olives tested in the rate of maximum methane production, R_m, during the cold-pressing process, from 44.3 ± 0.1 to 30.1 ± 1.3 L CH_4/(kg VS·d) (green olives) and from 43.9 ± 1.5 to 38.7 ± 1.6 L CH_4/(kg VS·d) (olives in veraison). Finally, the highest energy output result was detected in the waste from cold-pressed green olives (15.7 kJ/g $VS_{removed}$), which coincided with its high methane yield.

Keywords: anaerobic digestion; mixture waste; wastewater and solid waste; cold-pressing olive-oil manufacturing process; kinetics; process performance

Citation: Fernández-Rodríguez, M.J.; Cubero-Cardoso, J.; de la Lama-Calvente, D.; Fernández-Prior, Á.; Rodríguez-Gutiérrez, G.; Borja, R. Anaerobic Digestion of the Residue (Combination of Wastewater and Solid Waste) from a New Olive-Oil Manufacturing Process Based on an Olive Cold-Pressing System: Kinetic Approach and Process Performance. *Processes* **2022**, *10*, 2552. https://doi.org/10.3390/pr10122552

Academic Editor: Sebastián Sánchez Villasclaras

Received: 28 October 2022
Accepted: 28 November 2022
Published: 1 December 2022

1. Introduction

The olive oil industry is among the most important sectors in the food industry in Spain, where 80% is concentrated in the southern region of the country, called Andalusia. Spain is responsible for about 45% of olive oil production worldwide, which means that an average of 7 million tons per year of waste, or by-product, is generated [1]. In Spain, this by-product is mainly generated through the two-phase, continuous extraction method and is called alperujo. In spite of the difficulty of its treatment, technologies have already been implemented that serve to use all of its components. One such technology is thermal treatment [2]. Thermal treatment improves the management of the by-product but does not have a direct impact on the quality of the oil, which is obtained in previous stages in the mill. One of the main changes being made to improve the quality of the oil is to process decreasingly ripe olives, which promotes the production of oils with high organoleptic and functional quality [3]. Gone are the years when in Spain the priority was to obtain a greater quantity of oil by searching for olives with a high degree of ripeness, which in most cases meant harvesting olives that were damaged, fermented, or susceptible to fermentation. Nowadays, the aim is to harvest green olives to obtain a high-quality oil with the detriment of quantity, since fat recovery is much lower than in the case of riper olives. For this reason, technologies are being sought to increase the percentage of oil recovery from green olives without altering the quality of the main product. In this sense, the application of new systems for beating the olive paste once it has been milled, or the application of vacuum, ultrasound-assisted extraction [4], among others, are being studied. More recently, the use of cold pressure has been employed as a novel system that allows for increasing oil recovery yield while improving the use of its by-products, mainly alperujo (olive pomace). Previous studies have shown that the application of a cold-pressure treatment does not improve the subsequent application of anaerobic digestion (AD) of the olive pomace for the integral utilization of the by-product, although it is also possible [5]. These studies were carried out using the solid fraction obtained after olive-oil extraction with or without cold-press treatment.

As already indicated, the evolution of the sector is focused on the application of thermal treatments such as thermo-malaxation in the pomace extractors for better utilization of the olive pomace. To this end, the use of a beating at temperatures between 50 and 60 °C [2] prior to a three-stage centrifugation has been implemented. An oily phase, a liquid phase, and a solid phase with 55–60% moisture are obtained. The oily phase is the so-called crude pomace oil that must be refined to be marketed as olive pomace oil. The liquid phase is usually concentrated and its final destination is as a source of bioactive components, mainly phenols, or for the formation of fertilizers. The solid fraction can be further extracted from the pomace oil, or used as fuel. This work presents a novel study on how a combination of cold-pressure treatment with thermo-malaxation and three-phase centrifugation influences the subsequent AD of the mixture of the wastewater and solid waste generated. This work studies the application of AD as a final step for the integral utilization of mixture residue through the union of the solid and liquid phases obtained after applying a first step of cold pressure to improve the extractability of the olive oil and a second step of extraction in three phases after thermo-malaxation. The addition of the liquid fraction provides a more accessible source of organic matter in addition to phenols, which are potentially toxic for digestion—factors that depend on previous processing, such as cold pressing.

2. Materials and Methods

2.1. Olive Processing by Cold-Pressing Reactor

The cold pressure treatment was applied in the experimental plants of the Instituto de la Grasa-CSIC. Specifically, a 100 L stainless steel reactor working at a maximum pressure of 1.2 MPa and manual top closure was used. Between 5 and 10 kg of both green and olives in veraison samples were introduced into the reactor. Pressure was applied through an air compressor at up to 7 kg/cm^2 for 10 min. Then, in order to depressurize the reactor, the upper valve was opened and the olives were collected through the lower valve. The treated olives and the untreated control olives were subjected to an oil extraction process using the Abencor equipment [2]. The olives were ground in a hammer mill below 4 mm and the resulting paste beaten at 29 °C for 45 min, adding 100 mL of water and talcum powder halfway through the beating process. In this way, two fractions were obtained, one liquid and one solid. The Abencor system with water addition resembles a three-phase system in the industry. The liquid, which included the oil and water, was separated by decantation. The aqueous and the solid fractions were then stored at −20 °C until further use. Finally, both fractions (liquid and solid) were mixed and subjected to the AD process in the present research.

In order to assess the possible influence of the level of ripening of the olives on the AD performance of the produced wastes, olives of the Picual variety were collected at two stages, i.e., green olives and olives in veraison. The olives were harvested from the local area "Valle de los Pedroches, Pozoblanco" (Cordoba, Spain). The harvest date for both varieties was 5 December 2020.

Therefore, four mixed residues (mixtures of wastewaters and solid waste) were tested: residue from cold-pressed green olives, residue from olives that were not cold-pressed (control), residue from cold-pressed olives in veraison, and their residue without pressing (control).

The main characteristics pertaining to the four residues are presented in Table 1. In addition, Table 2 shows the individual phenolic composition contained in the wastewater fractions and solid wastes, both components (at 50%) of the final residues studied in this research.

Table 1. Principal characteristics of the four residues (mixtures of the wastewaters and solid wastes) subjected to AD experiments. Values represent means ± standard deviations. Different superscripted letters (a, b) mean values are significantly different.

Parameters	Cold-Pressed Green Olives	Green Olives Control	Cold-Pressed Olives in Veraison	Olives in Veraison Control
TS (g/kg)	147 ± 2 ^{a}	186 ± 2 ^{b}	143 ± 2 ^{a}	146 ± 1 ^{a}
VS (g/kg)	171 ± 1 ^{a}	132.2 ± 0.3 ^{b}	130.6 ± 0.5 ^{b}	132 ± 1 ^{b}
VS/TS	0.90	0.92	0.91	0.90
Moisture content (%)	85.6	82.7	85.5	85.3
Total phenols (g gallic acid/L)	10.00	10.97	10.99	11.27
tCOD (g O_2/L)	240 ± 30 ^{a}	200 ± 50 ^{a}	200 ± 80 ^{a}	190 ± 50 ^{a}
sCOD (g O_2/L)	140 ± 50 ^{a}	140 ± 40 ^{a}	160 ± 50 ^{a}	150 ± 30 ^{a}

TS: total solids; VS: volatile solids; tCOD: total chemical oxygen demand; sCOD: soluble chemical oxygen demand.

Table 2. Composition of the main individual phenolics in the liquid fractions (wastewaters) and the solid waste fractions as determined by HLPC. n.d.: not determined.

Phenolics Composition	Wastewaters (mg/L)				Solid Wastes (mg/kg)			
	Green Olives Control	Cold-Pressed Green Olives	Olives in Veraison Control	Cold-Pressed Olives in Veraison	Green Olives Control	Cold-Pressed Green Olives	Olives in Veraison Control	Cold-Pressed Olives in Veraison
3,4-Dihydroxyphenylglycol	107 ± 1	100 ± 3	114 ± 7	172 ± 8	43 ± 2	37 ± 2	69 ± 2	41 ± 3
Hydroxytyrosol glucoside	59 ± 1	5 ± 0	108 ± 6	7 ± 1	10 ± 0	29 ± 1	34 ± 0	41 ± 2
Hydroxytyrosol	440 ± 10	810 ± 10	840 ± 15	510 ± 20	62 ± 2	128 ± 8	209 ± 1	210 ± 10
Tyrosol	127 ± 8	112 ± 5	149 ± 3	154 ± 5	10 ± 1	17 ± 0	24 ± 3	23 ± 1
Syringic acid	19 ± 0	26 ± 0	27 ± 1	29 ± 1	2.6 ± 0.1	31 ± 1	2 ± 0	2 ± 0
Apigenin	n.d.	n.d.	n.d.	n.d.	6.3 ± 0.1	9.2 ± 0.7	7.6 ± 0.8	7.9 ± 0.6
Luteolin	n.d.	n.d.	n.d.	n.d.	25 ± 1	31 ± 1	26 ± 1	30 ± 1
p-coumaric acid	35 ± 2	32 ± 1	31 ± 3	36 ± 2	11 ± 0	15 ± 0	13 ± 1	10 ± 1
Oleuropein	60 ± 1	56 ± 3	53 ± 1	78 ± 1	n.d.	n.d.	n.d.	n.d.
Feluric acid	6.4 ± 0.8	10.3 ± 0.1	9.4 ± 0.7	7.3 ± 0.1	n.d.	n.d.	n.d.	n.d.
Comsegoloside	610 ± 10	610 ± 10	620 ± 10	680 ± 10	244 ± 6	257 ± 9	224 ± 9	261 ± 8

2.2. Biochemical Methane Potential Assays

A mesophilic granular sludge from a full-scale UASB reactor treating brewery wastewaters was used as anaerobic inoculum. The main characteristics of the inoculum were as follows: pH: 7.5 ± 0.2, total solids (TS): 25.0 ± 1.1 g/kg and volatile solids (VS): 19.9 ± 1.2 g/kg. Reactors of 250 mL with a 210 ± 2 mL working volume were used for the AD tests carried out in batch mode. A mesophilic temperature of 35 ± 2 °C was selected for the experiments, which was controlled by placing the reactors in thermostatic baths under constant stirring (400 rpm). The reactors were then filled with an inoculum to a substrate (ISR) ratio of 2 (VS), reaching a final concentration of 24 g VS/L, along with a micronutrient solution [5,6]. At the outset of the experiment, nitrogen gas was flushed through the reactors for two minutes (40 mL headspace volume) with the aim of maintaining anaerobic conditions. The reactors were activated for three days before the addition of each substrate. The reactors were replicated three times for each substrate. In addition, three blanks, with no substrate, were placed in order to obtain the inoculum's endogenous methane production, which will be subtracted from the final yield of each test. The resultant biogas was then passed through a 2N NaOH solution in order to retain the CO_2. The displacement volume was determined as methane, which was expressed according to standard or normalized (N) conditions of pressure and temperature (N: 0 °C, 1 atm).

2.3. Analytical Methods

The different mixtures tested were analyzed by determining the following parameters: total chemical oxygen demand (tCOD), soluble chemical oxygen demand (sCOD), TS, VS, and total phenol concentration. After 26 days of operation, sCOD, TS and VS, pH, total alkalinity, and volatile fatty acids (VFA) were measured in the resultant anaerobic digestates. The analyses were carried out in accordance with the methods previously described elsewhere [6].

2.4. Extraction and Analysis of Individual Phenolics by HPLC-DAD

The preparation of the aqueous samples for the determination of individual phenols was carried out by filtration through a 0.45-micron syringe filter. In the case of the solid phase samples, three sequential extractions were performed using an Ultra Turrax IKA T25 digital blender for 60 s at 1000 rpm, with a methanol:water solution (4:1 (v/v)) at a 1:1 ratio of solid:methanol and water mixture (w/v). The three extractions were combined and brought to dryness under vacuum at 40 °C and dissolved in a ten-times smaller volume of the hydroalcoholic mixture. Identification and quantification of the main individual phenols were made by HPLC with a UV Diode array detector (DAD). The equipment used was a Hewlett-Packard 1100 liquid chromatograph, and the individual phenols were quantified at wavelengths of 254, 280, and 340 nm. The column used was a C-18 Teknokroma Mediterranea Sea 18, 250 mm × 4.6 mm, i.d. 5 µm. The mobile phase was Milli-Q water acidified with trichloroacetic acid (0.01%) (A) and acetonitrile (B). The process was performed in gradient mode described as follows: 95% A, 75% A (30 min), 50% A (45 min), 0% A (47 min), 75% A (50 min), and 95% A (52 min) until completion of the run (55 min). Identification was made according to retention times and absorption spectra for each compound, and quantification was carried out by calibration of a curve with external standards.

2.5. Kinetic Evaluation

The substrate degradation during the AD process was determined by the mathematical modeling of the kinetics derived from experimental methane production. The obtained parameters aid in designing and optimizing full-scale anaerobic plants [7]. Kinetic parameters serve to determine the necessary time for microorganisms to acclimate to their new environment, the length of the digestion period, and the ability of the substrate to biodegrade. Therefore, these kinetic parameters serve as indicators for assessing the performance of the anaerobic reactor.

The logistic function model (LM) was used to estimate performance parameters and kinetic constants in the anaerobic digestion of the four residues tested. This model proved the fit of the experimental data shape of methane production kinetics [8]: an initial exponential increase after a small lag stage with final stabilization at the maximal level of production.

The logistic model is provided by the following equation:

$$B = P/[1 + \exp(4R_m(\lambda - t)/P + 2] \quad (1)$$

where B is the cumulative specific methane production (L CH_4/kg VS_{added}), P is the ultimate methane production (L CH_4/kg VS_{added}), R_m is the maximum methane production rate (L CH_4/(kg VS_{added}·d), t (days) is the digestion time, and λ is the lag time (days).

This model presupposes that the rate of methane production will be influenced by the amount of gas previously produced and that the R_m and maximum capacity for methane production will also affect the process [9]. The Logistic model also estimates the delay in λ and the R_m together with the potential for methane production of the substrates tested. This model has already been used for the anaerobic digestion of different organic substrates, and for estimating methane production in leachate from landfills [9].

2.6. Energy Output

The heat energy output corresponding to the BMP tests was determined by using the experimental data according to Equation (2) [5,10]:

$$E_0 = (P_{CH4} \times \mathcal{E} \times \lambda_m)/VS_{removed} \quad (2)$$

where

E_0 is the energy output in (kJ/g $VS_{removed}$);
P_{CH4} is the cumulative methane production after digestion time (m^3);
$\mathcal{E}$ is the lowest heating value for methane (35,800 kJ/m^3 CH_4);
λ_m is the energy conversion factor of methane (0.9);
$VS_{removed}$ is the grams of VS removed at the end of the BMP test (g/L).

2.7. Statistical Analysis

All analyses and tests were performed in triplicate. The statistical analyses were carried out using the SigmaStat software (Palo Alto, CA, USA). A one-way analysis of variance (ANOVA) test was used to determine levels of confidence among various results. The kinetic mathematical models were adjusted from the experimental data using the Sigma-Plot software (version 11). All the results were expressed as means ± standard deviations.

3. Results and Discussion

3.1. Substrate Characterization

Table 1 presents the main characteristics of the four residues used in this experiment (wastewaters and solid-waste mixture from cold-pressed green olives, residue from olives without cold-pressing (control), residue from cold-pressed olives in veraison, and residue from olives without pressing (control)). The VS contents in the residues from the green olive control, cold-pressed olives in veraison, and olives in veraison control were around 132 ± 1 g/kg, with no significant differences among them (Table 1). In contrast, the residues from cold-pressed green olives presented a VS content of 171 ± 1 g/kg, which is higher and statistically different from the other mixtures studied. The TS/VS ratio was similar for the four residues studied (0.90, 0.92, 0.91, and 0.90 for residues from cold-pressed green olives, residue from green olive control, residue from cold-pressed olives in veraison, and residue from olives in veraison control, respectively). The high values for the VS/TS ratio denote the marked organic character of the residues studied and were considered optimal for the AD process.

One of the main drawbacks of the olive-oil processing residues regarding anaerobic digestion performance is their phenols content, which can be decisive and even inhibit methane production [11]. Table 2 shows the phenolic compounds found in each residue used in this experiment.

Alperujo is a wet–solid waste, which is mainly composed of polysaccharides, proteins, fatty acids, pigments, and polyphenols [12]. Literature shows that alperujo is rich in polyphenols such as hydroxytyrosol but also contains important amounts of other compounds with high added value (e.g., vanillic acid, rutin, caffeic acid, oleuropein, tyrosol, p-coumaric acid, elenolic acid, catechol, and verbascoside) [12]. The main polyphenol found in the studied solid fractions was comsegoloside (Table 2) in concentrations between 224 and 261 mg/kg, followed by hydroxytyrosol, with a lower concentration in the green olive residues. The solid fraction from the control green olives presented a hydroxytyrosol concentration of 62 mg/kg, which increased to 128 mg/kg when the samples were treated (Table 2). In contrast, the solid residues of the olives in veraison presented a hydroxytyrosol concentration of between 209 and 210 mg/kg (Table 2).

The process wastewater or liquid residue consists of lipids, polyphenols, pectins, soluble sugars, and polyalcohols, among other minor compounds [13]. As in the solid fractions, the wastewater or liquid fraction main phenolic compound is hydroxytyrosol but it is also rich in others (e.g., gallic acid, tyrosol, vanillic acid, oleuropein, luteolin, verbascoside, and caffeic acid, among others) [13].

In the olive wastewater studied it is worth highlighting the presence of hydroxytyrosol with 440 ± 10–840 ± 15 mg/L, comsegoloside with concentrations between 610 ± 10–680 ± 10 mg/L, and tyrosol with a significantly higher concentration in the liquid residue from olives in veraison (112 ± 5–154 ± 5 mg/L) (Table 2).

Similar results were reported by Nunes et al. and Mallamices et al. [14,15], where hydroxytyrosol and comsegoloside represented around 79% of the total phenolic compounds present in the olive residues.

3.2. Organic Matter Removal and Methane Yield Coefficients

Graphs showing the production of biogas versus time for the four mixtures assayed (residue from cold-pressed green olives, residue from the green olives control, residue from cold-pressed olives in veraison, and residue from olives in veraison control) are shown in Figure 1. The methane yield of three of the four samples studied was very similar until day 12, except for the residues from the control of the green olive, which produced up to 10% more methane on day 10 after the start-up of the assays. Finally, from day 12, the methane production stabilized at around 300 NL CH_4/kg VS in the residues from olives in veraison, both control and cold-pressed, and the residues from the green olive control. The residues from cold-pressed green olives continued to produce methane until day 23, obtaining the highest methane yield (390 NL CH_4/kg VS). The overall methane production of the other three residues was between 18 and 23% lower (residue from the green olive control, residue from cold-pressed olives in veraison, and residue from olives in veraison control at 320, 302, and 320 NL CH_4/kg VS, respectively). The methane production from the residue of the cold-pressed green olives was significantly higher than those obtained for the other three residues tested. These results are in accordance with the higher biodegradability value found for the residue from cold-pressed green olives, i.e., 84.5%, compared with the others.

Tsigkou et al. [16] reported higher methane yield for raw three-phase olive mill wastewater (OMW) (472 mL CH_4/g VS_{added}). On the other hand, they obtained similar values to those obtained in the present work for the centrifuged OMW (391 NL CH_4/kg VS). The biodegradability reported by Tsigkou et al., 2019, was greater than 90% in all cases. The presence of alperujo in the mixtures studied in the present work could provide more difficult-to-degrade compounds, which was confirmed by the obtained biodegradability (82–86%) as well as by the methane yield (302–390 NL CH_4/kg VS). In another study carried out by Donoso-Bravo et al. [17], lower values for final methane yield were reported (274 NL CH_4/kg VS) after subjecting the olive pomace (OP) to an enzymatic maceration pre-treatment under mesophilic conditions with an ISR of 2.

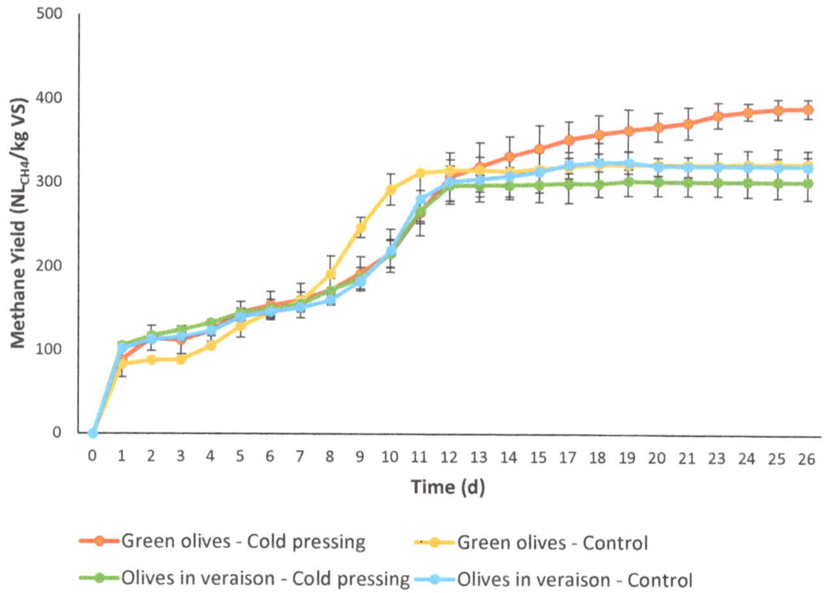

Figure 1. Curve of methane production against time for each test.

3.3. *Characterization of the Anaerobic Digestates*

The final pH of the different tests carried out was within the established optimal values for methane production by the methanogenic Archea [18] (7.91 ± 0.04, 8.07 ± 0.01, 7.91 ± 0.03 and 7.85 ± 0.03 for residue from cold-pressed green olives, residue from the green olive control, residue from cold-pressed olives in veraison, and residue from olives in veraison control, respectively; Table 3). Another very important parameter for the development of methanogenic microorganisms is the buffer capacity system. Total alkalinity values between 2500 and 5000 mg $CaCO_3$/L are considered optimal values for AD. Total alkalinity values ranged from 5900 ± 400 mg $CaCO_3$/L to 7490 ± 20 mg $CaCO_3$/L at the end of the experiment (residue from olives in veraison control and residue from cold-pressed green olives, respectively). Volatile fatty acids (VFAs) are good indicators of process stability in the AD system [19]. The main chemical equilibrium that controls alkalinity is carbonic acid-bicarbonate when pH values are between 6 and 8. The VFAs were measured in each reactor at the end of the experiment in order to avoid acidification inside the reactor. In all the residues studied, the presence of acetic acid was detected with values between 82 ± 2 and 150 ± 10 mg/L. The propionic acid concentration found was <15 mg/L and of isobutyric acid < 25 mg/L. Butyric and isovaleric acid were only found in residues from cold-pressed green olives and the control, at 74.2 ± 0.5 and 27.2 ± 0.1 mg/L, respectively (Table 3). No volatile fatty acids with longer chain accumulation were found. These results indicated that acidification processes did not occur during the anaerobic processes, which is indicative of the high stability of the four systems investigated in this work.

3.4. *Kinetic Modeling*

Table 4 presents a summary of the different parameters determined by the application of LM to the methane production experimental data against time, as can be seen in Figure 1. Errors are defined as the difference between predicted and measured methane yield, and they were <1% for all tests. The high determination coefficients (R^2 > 0.99 across the board) and the low standard errors of estimates show the remarkable fit of the model to the experimental results. As an example, Figure 2 shows the superposition of the experimental points of methane production against time with the theoretical curves of the model applied to the residues from green olives with cold-pressing and the controls.

Table 3. Principal characteristics of the four digestates or anaerobic effluents resulting from the BMP experiments of the mixture residues tested. Values represent means ± standard deviations. Different superscripted letters (a, b) mean values are significantly different. n.d.: not determined.

Parameters	Cold-Pressed Green Olives	Green Olives Control	Cold-Pressed Olives in Veraison	Olives in Veraison Control
pH	7.91 ± 0.04^{a}	8.07 ± 0.01^{b}	7.91 ± 0.03^{a}	7.85 ± 0.03^{a}
TA (g $CaCO_3$/kg)	7490 ± 20^{a}	6900 ± 700^{a}	6500 ± 300^{b}	6000 ± 400^{b}
Acetic acid (mg/L)	150 ± 10^{a}	86 ± 1^{b}	90 ± 10^{b}	82 ± 2^{b}
Propionic acid (mg/L)	n.d.	14.9 ± 0.1^{a}	n.d.	14.1 ± 0.1^{a}
Isobutyric acid (mg/L)	n.d.	25.3 ± 0.1^{a}	n.d.	23 ± 1^{b}
Butyric acid (mg/L)	74.2 ± 0.5^{a}	n.d.	n.d.	n.d.
Isovaleric acid (mg/L)	27.2 ± 0.1^{a}	n.d.	n.d.	n.d.
SCOD (mg O_2/L)	2800 ± 400^{a}	1700 ± 100^{b}	2400 ± 200^{a}	2500 ± 700^{a}
TS (g/kg)	35.3 ± 0.4^{a}	35.9 ± 0.4^{a}	36.4 ± 0.7^{a}	36 ± 1^{a}
VS (g/kg)	26.4 ± 0.4^{a}	27 ± 1^{a}	27.2 ± 0.6^{a}	27.7 ± 0.9^{a}

Table 4. Values corresponding to the parameters obtained from the logistic model (Sigmoidal 4 parameters) for the four mixture residues studied. Values represent means ± standard deviations. Different superscripted letters (a–c) mean values are significantly different.

Residue	P (NL CH_4/kg VS)	R_m (LCH_4/(kg VS·d))	λ (d)	R^2	S.E.E.	Error (%)
Residue from cold-pressed green olives	387 ± 9^{a}	30 ± 1^{a}	10.3 ± 0.2	0.9958	10.32	0.8
Residue from green olives–control	323 ± 5^{b}	44.3 ± 0.1^{b}	7.8 ± 0.1	0.9965	8.53	0.3
Residue from cold-pressed veraison olives	303 ± 5^{c}	39 ± 2^{c}	9.5 ± 0.1	0.9954	7.65	0.3
Residue from veraison olives–control	320 ± 4^{b}	44 ± 2^{b}	9.6 ± 0.1	0.9961	8.12	0.2

P: ultimate or maximum methane production; R_m: maximum methane production rate; λ: lag time. S.E.E.: standard error of estimate; R^2: determination coefficient; Error (%): difference (in percentage) between the experimental and calculated ultimate methane production.

Regarding the ripening stage of the olives, the values of R_m were significantly higher for the residues from green olives than for olives in veraison. Parallelly, the pressing process caused a decrease in the R_m parameters of the green olives from 44.3 ± 0.1 to 30 ± 1 L CH_4/(kg VS·d); similarly, although less severe, the R_m from the residues from olives in veraison also decreased (from 44 ± 2 to 39 ± 2 L CH_4/(kg VS·d)). This fact may be attributed to the lower phenolic compound content observed in the residues from cold-pressed olives compared with those obtained from the controls, especially in the case of green olives, for which a decrease of 8.8% in the total phenolic compound concentration was observed.

A similar trend in the R_m values was observed when only the solid waste derived from a similar cold-pressing was subjected to batch anaerobic digestion [5]; although, in this case, higher R_m values were found compared with those obtained in the present research. This fact is attributed to the higher phenolic compound concentration found in the mixture residue as a consequence of the addition of wastewater to the solid waste. The previously mentioned research in which single olive pomace was digested also revealed a reduction in the R_m value from 87 ± 7 to 73 ± 6 L CH_4/(kg VS·d) when the green olives were cold-pressed compared with the control (without cold-pressing).

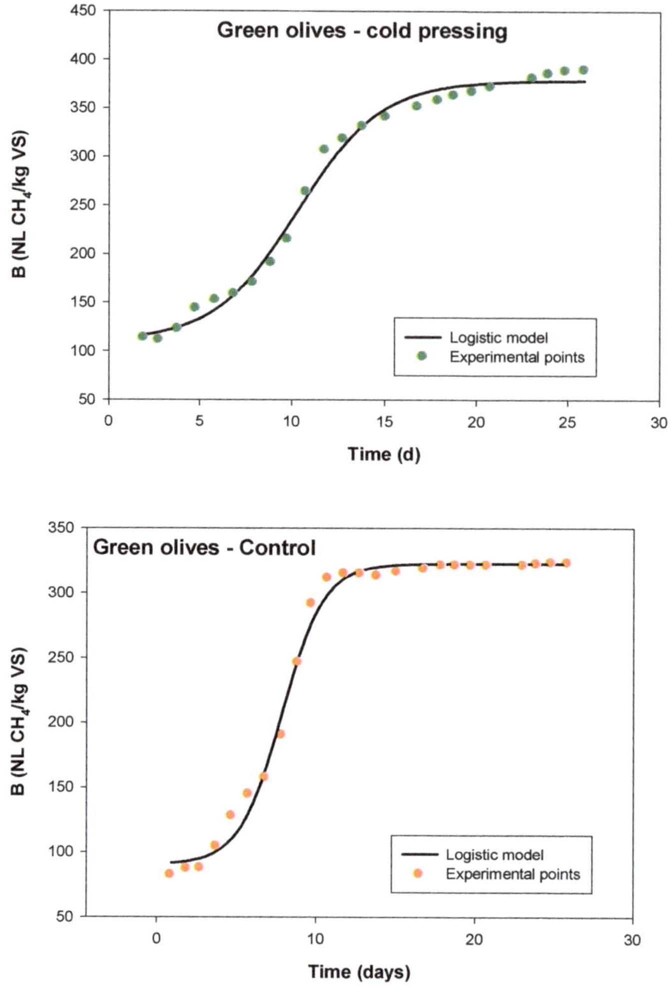

Figure 2. Variation in the experimental values for methane production and the theoretical values obtained from the logistic model (solid lines) for the residues derived from the cold-pressed green olives and the control.

A decrease in the R_m was also observed for two-phase olive mill solid waste (TPOMSW) subjected to a BMP test after pre-treatment by steam explosion at temperatures ranging from 138 to 171 °C and times varying between 5 and 30 min [17]. In this case, the maximum R_m (24.2 ± 0.7 L CH_4/(kg VS·d)), obtained from the pre-treatment conditions of 141 °C and 30 min, is much lower than the results reported in this study, especially in the case of both green and veraison olives subjected to cold-pressing. In addition, Donoso-Bravo et al. [18] also revealed that the direct enzyme addition pre-treatment did not enhance either the rate or the maximum methane production. Similarly, steam explosion showed no increment in the biodegradability of TPOMSW; however, thermal hydrolysis performed at 148 °C for 30 min without rapid depressurization notably enhanced both R_m (50%) and the methane yield (70%) [17].

Considerably lower R_m values than those found in this study were reported when olive agro-food by-products (composed basically of olive-pomace and small proportions of straw) were co-digested with animal manure at ratios of 1:1, 1:2, and 2:1 [20]. In this case, the R_m values ranged from 1.06 to 1.83 L CH_4/(kg VS·d), and these authors determined that biogas production was slower and lower when the agro-food by-product load increased. However, a higher animal manure ratio increased the process kinetics [20].

In contrast, R_m values of 35 ± 3 and 31 ± 2 L CH_4/(kg VS·d) were found in BMP tests of raw olive mill wastewater (OMW) and centrifuged OMW (at 4000 rpm for 15 min), respectively [16], using an anaerobic inoculum derived from an anaerobic digester that treats urban wastewater. This difference may be due to the compositions of VS in the raw and centrifuged OMW, which contribute to a higher R_m. Moreover, the raw OMW presented more highly biodegradable carbohydrates in addition to oils and grease, which resulted in a higher R_m value.

Furthermore, the maximum P value was observed for the residue derived from cold-pressed green olives (387 ± 9 NL CH_4/kg VS), in contrast with that observed for R_m. This may be because of the higher organic matter proportion (VS: 171 ± 1 g/L; tCOD: 236 ± 33 g O_2/L) detected in the residue when compared with others (VS ranged from 130.6 to 132.2 g/L and tCOD from 190 to 202 g O_2/L). The highest P value reached for this residue coincided with the maximum biodegradability (VS removed) value obtained (84.5%) compared with the others, whose VS removal values were between 79.0 and 79.6%.

On the other hand, when only the single solid waste derived from cold-pressed olives or not subjected to this process was anaerobically digested, the maximum value for ultimate methane production was found for the solid waste from green olives that were not cold-pressed (319 ± 6 NL CH_4/kg VS). This result may be due to the soluble matter (sCOD: 113 g/L) detected in the single solid waste, which was higher in comparison with the others (sCOD: 105–107 g/L). In this case, the highest maximum methane yield reached for this solid waste coincided with the highest biodegradability obtained (90.8%) in comparison with the other sole solid wastes, with values within 74.5 and 86.4% [5].

It is also worth pointing out that the P achieved for the mixture residue from the cold-pressed green olives (387 ± 9 NL CH_4/kg VS) was 31.6% higher than that obtained for an olive pomace previously subjected to steam explosion treatment (200 °C for 5 min with rapid decompression) (294 NL CH_4/kg VS) [11]. This behavior may be explained by the fact that the steam explosion process generated undesirable compounds such as furan and complex phenols, which could inhibit the AD process [17,21].

Maximum methane yield values of 320 and 325 L CH_4/kg VS were reported for the anaerobic co-digestion processes of agro-food by-products, composed basically of olive pomace with pre-treated waste sludge and animal manure, respectively, [20]. As seen in Table 4, these values were very similar to the values obtained in the present work when the olives (both green and in veraison) were not subjected to cold-pressing.

The maximum methane yields obtained in batch anaerobic digestion experiments of raw and centrifuged OMW using different types of anaerobic inoculum were higher in every case for the untreated OMW (325–472 L CH_4/kg VS) in comparison with the same OMW after a subsequent centrifugation step (219–391 L CH_4/kg VS [16]. This result may be due to the higher biodegradable total carbohydrates, oils, fats, and soluble COD contents in the raw OMW compared with the centrifuged wastewater [16].

The lag periods found for the four mixture residues used in this work were much higher (between 7.8 and 10.3 days) than the lag periods found in the anaerobic digestion of single solid residues, whose values varied between 0.20 and 0.23 days [5]. This difference may be attributed to the higher phenolic compound content present in the mixture residues compared with those contained in the single solid waste. The shape of the curves of methane production–time observed in the present research clearly indicates a fast increase in methane generation after an initial lag period of 6–7 days, during which the hydrolysis of the slowly digestible substrate components took place for the four mixture residues. Higher lag period values (11.9 days) than those obtained in the present work were reported for the anaerobic co-digestion processes of the mixtures of agro-food by-products, composed mainly of olive pomace, with animal manure at a ratio of 3:1 [20]. These authors demonstrated that a higher proportion of olive residue in the co-digested mixture led to higher lag periods. On the contrary, similar lag period values (7.5–9.0 days) to those obtained in the present work were revealed in the BMP tests of two-phase olive mill solid wastes previously subjected to steam-explosion processes at temperatures ranging from 140 to 170 °C for 5–30 min [17].

3.5. Energy Assessment

Despite being a resource-efficient process, anaerobic digestion reduces the organic matter and considerably decreases the contamination power of wastes. The high amount of methane produced along with the energy output generated during the process are remarkable. Equation (2) was applied to determine [5,10] the energy output from the BMP data from the experiments. The energy yield from the process, or viability, is a key factor in the scaling up of any AD process run at the lab scale. Additionally, it should be considered that the inoculum was not acclimatized to the new residues before the anaerobic experiments; hence, the generated methane had been, presumably, underestimated. The energy output values for mixture residues from cold-pressed green olives were 15.7 kJ/g VS, and 12.3 kJ/g $VS_{removed}$ for the control; meanwhile, for the mixture residues from cold-pressed olives in veraison the values were 12.2 kJ/g VS and 13.0 kJ/g $VS_{removed}$ for the control.

The highest energy output found for the cold-pressed green olives mixture was 27.6% higher than that obtained for the residue from control green olives. By contrast, the values observed for the wastes from olives in veraison were very similar. The highest energy output value found for the residue from cold-pressed green olives is in accordance with the highest methane production and biodegradability values determined in this residue compared with the others.

A previous study [22] revealed that the energy output of a mixture containing 16% olive pomace with corn silage (17%), citrus pulp (25%), whey (18%), cattle manure (4%), and poultry litter (8%) was 1.2 kWh-e/kg dry feedstock mixture. This fact allows for valorizing the most important agricultural wastes and by-products from southern Italy (Sicily), including olive pomace.

Pasalari et al. [10] reported a range of energy output values from 9.4 to 25.5 kJ/g $VS_{removed}$ for the AD process applied to pretreated landfill leachate by an electrochemical oxidation process. The values obtained in this study were of the same order of magnitude as the values reported in the present research. High energy output results (76.25 kJ/g fed VS) were also observed for the co-digestion of sewage sludge and food waste after a microwave pre-treatment [23]. The study concluded that the use of microwave as a pre-treatment enhanced the solubilization of organic compounds and the hydrolysis of protein to NH_4^+-N, and also increased the methane yield as well as the R_m during the co-digestion process [23].

Author Contributions: M.J.F.-R.: conceptualization, investigation, validation, writing—review and editing; J.C.-C.: investigation conceptualization, validation; D.d.l.L.-C.: investigation, conceptualization, validation; Á.F.-P.: investigation, conceptualization, validation; G.R.-G.: supervision, resources, funding acquisition, writing—original draft; R.B.: supervision, writing—original draft, writing—review and editing, resources, funding acquisition. All authors have read and agreed to the published version of the manuscript.

Funding: This research was funded by the regional government of Andalucía, Junta de Andalucía, Project 18-616, "Consejería de Economía, Conocimiento, Empresas y Universidad", Andalucía, Spain, and Project PID2020-114975RB-100/AEI/10.13039/501100011033 was financed by the Spanish Ministry of Science and Innovation. The project FEDER UPO-380782 was financed by the regional government of Andalucía, Junta de Andalucía, "Consejería de Trasformación Económica e Industria", providing financial support to Dr. Fernández-Rodríguez.

Institutional Review Board Statement: Not applicable.

Informed Consent Statement: Not applicable.

Data Availability Statement: Data will be made available on request.

Conflicts of Interest: The authors declare no conflict of interest.

References

1. Fernández-Lobato, L.; López-Sánchez, Y.; Blejman, G.; Jurado, F.; Moyano-Fuentes, J.; Vera, D. Life cycle assessment of the Spanish virgin olive oil production: A case study for Andalusian region. *J. Clean. Prod.* **2021**, *29025*, 125677. [CrossRef]
2. Lama-Muñoz, A.; Rubio-Senent, F.; Bermúdez-Oria, A.; Fernández-Bolaños, J.; Fernández Prior, A.; Rodríguez-Gutiérrez, G. The use of industrial thermal techniques to improve the bioactive compounds extraction and the olive oil solid waste utilization. *Innov. Food Sci. Emerg. Technol.* **2019**, *55*, 11–17. [CrossRef]
3. Beltrán, G.; Del Río, C.; Sánchez, S.; Martínez, L. Seasonal changes in olive fruit characteristics and oil accumulation during ripening process. *J. Sci. Food Agric.* **2004**, *84*, 1783–1790. [CrossRef]
4. Clodoveo, M.L.; Durante, V.; La Notte, D.; Punzi, R.; Gambacorta, G. Ultrasound-assisted extraction of virgin olive oil to improve the process efficiency. *Eur. J. Lipid Sci. Technol.* **2013**, *115*, 1062–1069. [CrossRef]
5. Fernández-Rodríguez, M.J.; Cubero-Cardoso, J.; de la Lama-Calvente, D.; Fernández-Prior, A.; Rodríguez-Gutiérrez, G.; Borja, R. Performance and kinetic evaluation of the anaerobic digestion of olive pomace derived from a novel manufacturing process based on an olive cold-pressing system: Influence of the fruit ripening level. *Biomass Convers. Biorefin.* 2022, *in press*. [CrossRef]
6. Fernández-Rodríguez, M.J.; De la Lama-Calvente, D.; Jiménez-Rodriguez, A.; Borja, R.; Rincón, B. Anaerobic co-digestion of olive mil solid Waste and microalga *Scenedesmus quadricauda*: Effect of different carbon to nitrogen ratios on process performance and kinetics. *J. Appl. Phycol.* **2019**, *31*, 3583–3591. [CrossRef]
7. Rose Benish, P.M.; Mozhiarasi, V.; Nagabalaji, V.; Weichgrebe, D.; Srinivasan, S.V. Optimization of process parameters for enhanced methane production from banana peduncle by thermal pretreatment. *Biomass Convers. Biorefin.* 2022, *in press*.
8. Donoso-Bravo, A.; Perez-Elvira, S.I.; Fernández-Polanco, F. Application of simplified models for anaerobic biodegradability tests. Evaluation of pretreatment processes. *Chem. Eng. J.* **2010**, *160*, 607–614. [CrossRef]
9. Pommier, S.; Chenu, D.; Quintard, M.; Lefebvre, X. A logistic model for the prediction of the influence of water on the solid waste methanization in landfills. *Biotechnol. Bioeng.* **2006**, *97*, 473–482. [CrossRef]
10. Pasalari, H.; Esrafili, A.; Rezaee, A.; Gholami, M.; Farzadkia, M. Electrochemical oxidation pretreatment for enhanced methane potential from landfill leachate in anaerobic co-digestion process: Performance, Gompertz model, and energy assessment. *Chem. Eng. J.* **2021**, *422*, 130046. [CrossRef]
11. Serrano, A.; Fermoso, F.G.; Alonso-Fariñas, B.; Rodríguez-Gutiérrez, G.; Fernandez-Bolaños, J.; Borja, R. Phenols recovery after steam explosion of olive mill solid waste and its influence on a subsequent biomethanization process. *Bioresour. Technol.* **2017**, *243*, 169–178. [CrossRef]
12. Nunes, M.A.; Pimentel, F.B.; Costa, A.S.G.; Alves, R.C.; Oliveira, M.B.P.P. Olive by-products for functional and food applications: Challenging opportunities to face environmental constraints. *Innov. Food Sci. Emerg. Technol.* **2016**, *35*, 139–148. [CrossRef]
13. Araújo, M.; Pimentel, F.B.; Alves, R.C.; Oliveira, M.B.P.P. Phenolic compounds from olive mill wastes: Health effects, analytical approach and application as food antioxidants. *Trends Food Sci. Technol.* **2015**, *45*, 200–211. [CrossRef]
14. Nunes, M.A.; Costa, A.S.G.; Bessada, S.J.S.; Puga, H.; Alves, R.C.; Freitas, V.M.; Oliveira, B.P.P. Olive pomace as a valuable source of bioactive compounds: A study regarding its lipid- and water-soluble components. *Sci. Total Environ.* **2018**, *644*, 229–236. [CrossRef]
15. Mallamaci, R.; Budriesi, R.; Clodoveo, M.L.; Biotti, G.; Micucci, M.; Ragusa, A.; Curci, F.; Muraglia, M.; Corbo, F.; Franchini, C. Olive Tree in Circular Economy as a Source of Secondary Metabolites Active for Human and Animal Health Beyond Oxidative Stress and Inflammation. *Molecules* **2021**, *26*, 1072. [CrossRef] [PubMed]
16. Tsigkou, K.; Sakarika, M.; Kornaros, M. Inoculum origin and waste solid content influence the biochemical methane potential of olive mill wastewater under mesophilic and thermophilic conditions. *Biochem. Eng. J.* **2019**, *151*, 107301. [CrossRef]
17. Donoso-Bravo, A.; Ortega-Martínez, E.; Ruiz-Filippi, G. Impact of milling, enzyme addition, and steam explosion on the solid waste biomethanation of an olive oil production plant. *Bioprocess Biosyst. Eng.* **2016**, *39*, 331–340. [CrossRef] [PubMed]
18. Arif, S.; Liaquat, R.; Adil, M. Applications of materials as additives in anaerobic digestion technology. *Renew. Sustain. Energy Rev.* **2018**, *97*, 354–366. [CrossRef]
19. Siegert, I.; Banks, C. The effect of volatile fatty acid additions on the anaerobic digestion of cellulose and glucose in batch reactors. *Process Biochem.* **2005**, *40*, 3412–3418. [CrossRef]
20. Parralejo, A.I.; Royano, L.; González, J.; González, J.F. Small scale biogas production with animal excrement and agricultural residues. *Ind. Crops Prod.* **2019**, *131*, 307–314. [CrossRef]
21. Rincón, B.; Rodríguez-Gutiérrez, G.; Bujalance, L.; Fernández-Bolaños, J.; Borja, R. Influence of a steam-explosion pre-treatment on the methane yield and kinetics of anaerobic digestion of two-phase olive mill solid waste or alperujo. *Process Saf. Environ. Prot.* **2016**, *102*, 361–369. [CrossRef]
22. Valenti, F.; Zhong, Y.; Sun, M.; Porto, S.M.C.; Toscano, A.; Dale, B.E.; Sibilla, F.; Liao, W. Anaerobic co-digestion of multiple agricultural residues to enhance biogas production in southern Italy. *Waste Manag.* **2018**, *78*, 151–157. [CrossRef] [PubMed]
23. Liu, J.; Zhao, M.; Lu, C.; Yue, P. The effect of microwave pretreatment on anaerobic co-digestion of sludge and food waste: Performance, kinetics and energy recovery. *Environ. Res.* **2020**, *189*, 109856. [CrossRef] [PubMed]

processes

Article

Profile Phenolic Compounds in Spanish-Style and Traditional Brine Black Olives ('Gemlik' Cv.) Provided from Different Regions of Türkiye

Cansu Demir [1,2,*], Elif Yildiz [3] and Ozan Gurbuz [4]

1 Graduate School of Natural and Applied Sciences, Bursa Uludag University, Bursa 16059, Türkiye
2 Olive Research Institute, Izmir 35100, Türkiye
3 Department of Food Processing, Keles Vocation School, Bursa Uludag University, Bursa 16740, Türkiye; elifyildiz@uludag.edu.tr
4 Department of Food Engineering, Faculty of Agriculture, Bursa Uludag University, Bursa 16059, Türkiye; ozang@uludag.edu.tr
* Correspondence: cansu.demir@tarimorman.gov.tr

Abstract: The aim of this study was to evaluate the effect of growing regions and processing methods on the composition and the quantity of phenolic compounds in 'Gemlik' variety table olives. Two different processing methods, Spanish-style and traditional brine (naturally processed) olives, were used in the processing of 'Gemlik' table olives. According to the data obtained in this study, the highest concentrations of phenolic compounds were observed for 3-hydroxytyrosol (4.58–168.21 mg/kg), followed by 4-hydroxyphenyl (0.76–97.58 mg/kg), luteolin 7-glucoside (0.32–58.64 mg/kg), tyrosol (1.57–47.24 mg/kg), and luteolin (0.17–53.56 mg/kg) in overall samples. The highest quantity of phenolic compounds was determined in raw olives, and the lowest phenolic compound content was determined in Spanish-style processed olives. Table olives which are produced by the natural process were observed to contain higher concentrations of phenolic compounds compared with the olives, which are produced in the Spanish style. In this sense, statistical results showed that region and processing methods have significant impacts on the phenolic compounds of table olives.

Keywords: Cv. 'Gemlik'; table olives; phenolic compounds; Spanish style; traditional processing

Citation: Demir, C.; Yildiz, E.; Gurbuz, O. Profile Phenolic Compounds in Spanish-Style and Traditional Brine Black Olives ('Gemlik' Cv.) Provided from Different Regions of Türkiye. *Processes* **2023**, *11*, 2412. https://doi.org/10.3390/pr11082412

Academic Editors: Sebastián Sánchez Villasclaras and Juan Francisco García Martín

Received: 21 June 2023
Revised: 1 August 2023
Accepted: 5 August 2023
Published: 10 August 2023

1. Introduction

Olea europaea L. (olive tree) is one of the oldest plants cultivated in the Mediterranean Basin [1]. The olive is known as the most important fruit among other fruits in Mediterranean countries such as Italy, Spain, and Greece (6). Table olive, a traditional Mediterranean food, is of great importance both economically and socially in Türkiye [2]. According to the International Olive Council (IOC, 2017), worldwide production of table olives is around 3020.500 tons [3]. Regarding the averages for the 2012/13–2015/16 seasons, Türkiye was considered the third largest producer of table olives, with more than 15.6% of the world's production (IOC, 2017) [2].

Olive fruit is known for being a rich source of phenolic compounds [4]. The phenolic compounds of table olives are of great importance in terms of nutritional, color, and flavor properties [5] and have strong antioxidant effects and improve the nutritional and organoleptic qualities of olives [6].

Phenolic compounds, which are bioactive components, have gained importance in recent years mainly due to their antioxidant, anti-inflammatory, and antitumor properties [6–9]. They also help preventing certain diseases, such as Alzheimer's and cancer [10]. The phenolic compounds are formed from the metabolism of primary products, including carbohydrates, fats, and amino acids [9]. Factors such as the degree of ripening, growing conditions, fruit size, and processing method affect the quantity of phenolic compounds in olives [11]. Oleuropein, 3-hydroxytyrosol, and rutin are the most abundant phenolic compounds in olive fruit [7]. Oleuropein has a strong antioxidant activity, it is the main compound that gives characteristic bitterness to olives, and its content in fruit flesh is 2–4% [9]. While the oleuropein content of olives is quite high in the early fruiting period, the amount in some cultivars may decrease during the black ripening period [12]. Additionally, Salis et al. [9] detected 3-hydroxytyrosol, tyrosol, verbascoside, rutin, oleuropein, and luteolin in Spanish-style and Greek-style processed olive fruits of cv. 'Kalamata'. Irmak and Irmak [13] determined hydroxytyrosol, tyrosol, apigenin, and luteolin in the raw and processed olives of varieties 'Ayvalık' and 'Domat'. Hydroxytyrosol, rutin, oleuropein, tyrosol, luteolin-7-glucoside, apigenin, luteolin, verbascoside, and apigenin are determined phenolic compounds in olive fruit [9,13,14]. Also, Uylaser [15] found the highest contents of 3-hydroxytyrosol and vanillic acid in 'Gemlik' variety raw olives.

It has been reported that olive processing methods affect the taste of olives and can significantly change the health properties of olive fruit [9]. The processing method of the final products also has a significant effect on the phenolic compound content. It has been reported that olive processing methods affect the taste of olives and can significantly change the health properties of olive fruit [9]. Laruen et al. [14] determined that oleuropein, 3-hydroxytyrosol, verbascoside, tyrosol, luteolin 7-glucoside, and rutin were higher in processed olives.

Since olives contain a high level of bitterness caused by oleuropein after harvest, they must be processed. Three different commercial olive processing methods have international importance: (1) The Spanish style for green olives, (2) The Greek style for natural black olives, and (3) The Californian style for black olives [6,11,16]. Phenolic compounds are distinct for *Olea europaea* and have antioxidative, antimicrobial, antiviral, anti-inflammatory, and anticarcinogenic effects. Table olive processing methods cause severe losses in phenolic compounds, and as a result, the positive effects of table olives on health may change [16]. In addition, the most widely used table olive production processes in Türkiye are those that aim at the production of natural black olives (unprocessed, in brine, and dry-salted), Californian style for black olives, and Spanish style for green olives [12].

Ninety-three olive varieties are known to be present in Türkiye. The percentage of the distributions of olive varieties in all olive-producing regions of Türkiye in 2014 and 2015 was calculated, and it was determined that 48.71% of the total olive tree consists of cv. 'Gemlik' [17]. The 'Gemlik' variety is thin-skinned and adhered to the flesh and has a high flesh/stone ratio, small seeds, and aromatic features. In addition, higher quantities of aromatic compounds result in higher-quality table olive products [18,19].

'Gemlik' olive is the dominant variety of the Marmara region. This variety shows remarkable agronomic characteristics (such as not showing severe alternation, high adaptability, early yield, being a tree that is partially resistant to cold and diseases, and easy reproduction from steel) and dual-purpose technological advantages (such as black table and olive oil varieties). It is a variety that has spread rapidly in all olive-growing regions of Türkiye (including the Aegean region, East-West Mediterranean region, and even the Southeastern Anatolia region) except the Marmara (Bursa) region, which has been the place of origin for the last 25–30 years.

Due to its superior properties, the 'Gemlik' olive was protected by the Turkish Patent Institute on 23 September 2003 with the 'Geographical Indication Registration Certificate' and 'Name of Origin'. The geographical boundaries for 'Gemlik' olives are stated in the registration document as Gemlik, Iznik, and Orhangazi districts of Bursa, which shows that this variety is a local product [20].

'Gemlik' olives are obtained from Gemlik (Gem), Iznik (Izn), Mudanya (Mud), Orhangazi (Orh), and Erdek (Erd) regions in Bursa, where this variety is the most widely grown, and from the Akhisar (Akh) region in Manisa, where cultivation has grown rapidly in recent years. According to the Turkish Food Codex Table Olive Communiqué, processing methods are divided into two as natural and chemical methods. In the table olive sector, some producers deceive consumers by selling chemically processed olives as naturally processed. In this study, in addition to regional differences, differences between processing methods were also determined. According to the phenolic compounds of the processed olives, it will be possible to identify which process was used. As a result, the deception of the consumer can be prevented.

2. Materials and Methods

2.1. Procedure and Collection Date of the Samples

'Gemlik'-type raw olive fruits were obtained from Marmarabirlik (S.S. Marmara Olive Agricultural Sales Cooperatives Union, Bursa, Türkiye) grown in 6 different regions in Türkiye. Gemlik (Gem), Iznik (Izn), Mudanya (Mud), Orhangazi (Orh), Erdek (Erd) and Akhisar (Akh), in the harvest period of 2019–2020 in November for this study. A total of 220 number/kg olives were hand-harvested at 5 maturity index from these regions, and 3 samples were taken from each region. The maturity index (MI) was determined according to the color changes in peel and pulp with the procedures of Vinha et al. (2005) [21] and varied between 0 and 7 in eight categories. Sampling codes are given in Table 1.

Table 1. The regions and olive processing methods of the samples.

Regions	Codes of Regions	Olive Processing Methods Codes		
		Raw	Natural Processing (NP)	Spanish Style (SS)
Gemlik	Gem	GemRaw	GemNP	GemSS
Iznik	Izn	IznRaw	IznNP	IznSS
Mudanya	Mud	MudRaw	MudNP	MudSS
Orhangazi	Orh	OrhRaw	OrhNP	OrhSS
Erdek	Erd	ErdRaw	ErdNP	ErdSS
Akhisar	Akh	AkhRaw	AkhNP	AkhSS

2.2. Chemicals

HPLC grade water, methanol, and orthophosphoric acid solutions, phenolic compound standards 3-hydroxytyrosol (3Hyt, CAS No: 10597-60-1), protocatechuic acid (Prt, CAS No: 99-50-3), keracyanin (Ker, CAS No: 18719-76-1), coumarin (Kum, CAS No: 41044-12-6), catechin (Kat, CAS No: 154-23-4), tyrosol (Tyr, CAS No: 501-94-0), 4-hydroxyphenyl (4Hdf, CAS No: 67914-60-7), epicatechin (Ep, CAS No: 490-46-0), syringic acid (Syr, CAS No: 530-57-4), oleuropein (Ole, CAS No: 32619-42-4), trans-cinnamic acid (Trs, CAS No: 140-10-3), luteolin 7-glucoside (L7g, CAS No: 5373-11-5), luteolin 4-glucoside (L4g, CAS No: 6920-38-3), luteolin (Lt, CAS No: 491-70-3), kaempherol (Kam, CAS No: 520-18-3), and apigenin (Apg, CAS No: 520-36-5) analytical standards were obtained from Merck (Darmstadt, Germany).

2.3. Table Olive Processing

Two types of olive processing are preferred in black table olive production. The first one is the natural fermentation known as traditional brine olive production, also traditionally called Gemlik style in Türkiye. For production, black table olives (17 kg) were washed with tap water to remove dust, placed in 40 L plastic vessels, and processed by natural processing (NP). Then, olive samples were brined (17% NaCl) and fermented for two months at room temperature (18–25 °C) in the dark. The second one is Spanish-style (SS) table olive production. Firstly, black table olives (17 kg) were washed with tap water to remove dust, placed in 40 L plastic vessels, and NaOH solution (1.5%) was added. The olives were kept in NaOH solution overnight, and the NaOH was allowed to penetrate 2/3 of the olive flesh. After the NaOH, olives were kept in water for 18 h, washed, brined (17% NaCl and 1% lactic acid solution), and fermented for two months at room temperature (18–25 °C) in the dark. After the fermentation, the olive samples were analyzed.

2.4. Phenolic Compound Characterization

Extraction

Phenolic compound analysis was performed using the HPLC-DAD technique with the modified method of Ramirez et al. [22]. For the extraction, firstly, olive pits were removed and homogenized with a blender. A total of 10 g of olive pulp was weighed into 250 mL flasks. Next, 50 mL of petroleum ether was added and shaken at 180 rpm for 2 h. The olive pulp and petroleum ether mixture were filtered through filter paper (pore size 22 µm). Afterward, the olive pulps remaining from the filtrate were taken into 50 mL falcon tubes. Petroleum ether was kept for a while with volatile nitrogen and evaporated. Then, 40 mL of methanol/water mixture prepared at 80:20 (v/v) ratio acidified with 0.001% orthophosphoric acid and kept in the refrigerator overnight was added to the olive pulp. After that, the mixture was kept in an ultrasonic bath for 30 min; then, the upper phase was separated by centrifugation at 3500 rpm (15 min). A total of 20 mL of the same mixture was added to the remaining olive pulp, kept in an ultrasonic bath for 30 min, and centrifuged at 3500 rpm (15 min). After the second phase was separated, 2 mL of each of the 1st and 2nd phases was taken and mixed.

The final mixture was filtered into Eppendorf tubes by 0.22 µm membrane filters, transferred to HPLC vials (Agilent, screw tap V9201911A), and injected into HPLC. The analysis of olive samples was carried out using HPLC equipped with ChemStation software, version A.02.14 (2016 Agilent Technologies, California, USA), a 250 × 4.6 mm C18 column, and a column oven (G1316A) using an Agilent 1260 Infinity II HPLC device and DAD. HPLC conditions were established by modifying the method stated by Ramirez et al. [22]. Solutions of methanol (C) and water (pH adjusted to 2.70 with orthophosphoric acid solution) (B) were used as mobile phase. The flow rate was 1 mL/min, the injection volume was 20 µL, and detection wavelengths were 280 nm and 330 nm. While defining the peaks of phenolic compounds, the wavelength at which the phenolic compounds give the maximum absorbance value is taken as a basis. The calibration curves of the standards were obtained by the intermediate stock solutions at the mg/kg levels. The evaluated phenolic compound codes and branch lengths were as follows: 3-hydroxytyrosol (3Hyt), protocatechuic acid (Prt), keracyanin (Ker), coumarin (Kum), catechin (Kat), tyrosol (Tyr), 4-hydroxyphenyl (4Hdf), epicatechin (Ep), syringic acid (Syr), oleuropein (Ole), and trans-cinnamic acid (Trs) were determined at 280 nm; luteolin 7-glucoside (L7g), luteolin 4-glucoside (L4g), luteolin (Lt), kaempherol (Kam), and apigenin (Apg) determined at 330 nm.

The gradient elution program was planned as follows: to 90% (B) and 10% (C) for 0 min, 95% (B) and 5% (C) in 10 min, 70% (B) and 30% (C) in 20 min, 60% (B) and 40% (C) in 40 min, 60% (B) and 40% (C) in 45 min, 50% (B) and 50% (C) in 45.10 min, 40% (B) and 60% (C) in 50 min, 30% (A) and 70% (B) in 55 min, 0% (B) and 100% (C) in 60 min, 0% (B) and 100% (C) in 68 min, 90% (B) and 10% (C) in 73 min. The identification of phenolic compounds was enabled by the analysis of retention times of standards. The linear calibration curves were obtained from the standards (R^2 = 0.98).

3. Statistical Analysis

Analysis of variance was carried out using JMP 7.0 to determine the significant differences at a level of confidence of $p < 0.05$. The degree of differences in the means was compared using the Student's t-test. Among the chemometrics methods in the classification of olives, principal component analysis (PCA) and hierarchical cluster analysis (HCA) were utilized, and the obtained data were evaluated with Minitab (Minitab 16 Statistical Software, Minitab, Inc., New York, NY, USA).

4. Results and Discussion

HPLC Identification and Quantification of Phenolic Compounds

Various factors are known to affect the presence and content of the phenolic compounds of table olives. These are variety, agronomic process, ripening level, and processing steps of the table olive processing methods. Also, there are some significant differences in the phenolic composition of processed table olives, identified as chemical and enzymatic changes in certain phenolics compounds during several processing methods [12].

In this study, phenolic compounds were evaluated in raw, naturally processed (traditional brine olives), and Spanish-style olives in terms of 3Hyt, Ker, Kum, Kat, Tyr, 4Hdf, Trs, L7g, L4g, Lt, Kam, and Apg.

A chromatogram of fourteen phenolic compounds of the standard mixture solution as well as the internal standard is shown in Figure 1. As presented in the chromatogram, all the studied phenolic compounds had responses at 280 nm and 330 nm, where they were efficaciously separated. The peak chromatograms of raw, traditional brine, and Spanish-style olive extracts are given in Figure 2a–f, respectively. The comparison of the retention times of those compounds with the retention times of authentic standards led to their identification.

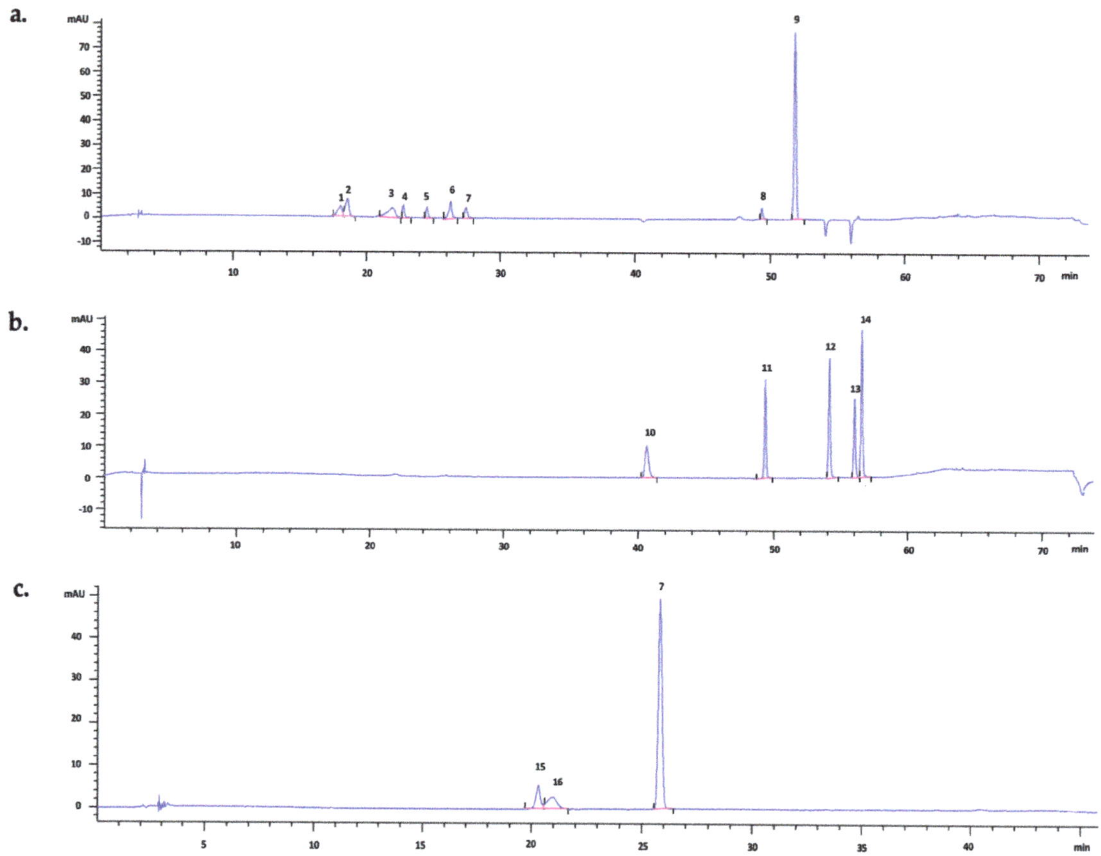

Figure 1. HPLC-UV chromatograms of the standard mixture solutions: (**a**,**b**) 1: 3-hydroxytyrosol; 2: protocatechuic acid; 3: coumarin; 4: tyrosol; 5: 4-hydroxyphenyl; 6: epicatechin; 7: syringic acid; 8: oleuropein; 9: trans-cinnamic acid; 10: luteolin 7-glucoside; 11: luteolin 4-glucoside; 12: luteolin; 13: kaempherol; 14: apigenin; (**c**) 15: catechin; 16: keracyanin.

The high and low concentration calibration curves determined for each phenolic compounds showed a linear response with correlation coefficients of 0.989–0.999. The recovery values of the phenolic compounds ranged from 82.16% to 93.53% with RSD varying from 3.44% to 8.02%.

The quantity of phenolic compounds of raw and processed olives (traditional brine-natural processed olive, NP; Spanish-style processed, SS) obtained from six different regions (Gem, Izn, Mud, Orh, Erd, and Akh) is given in Table 2.

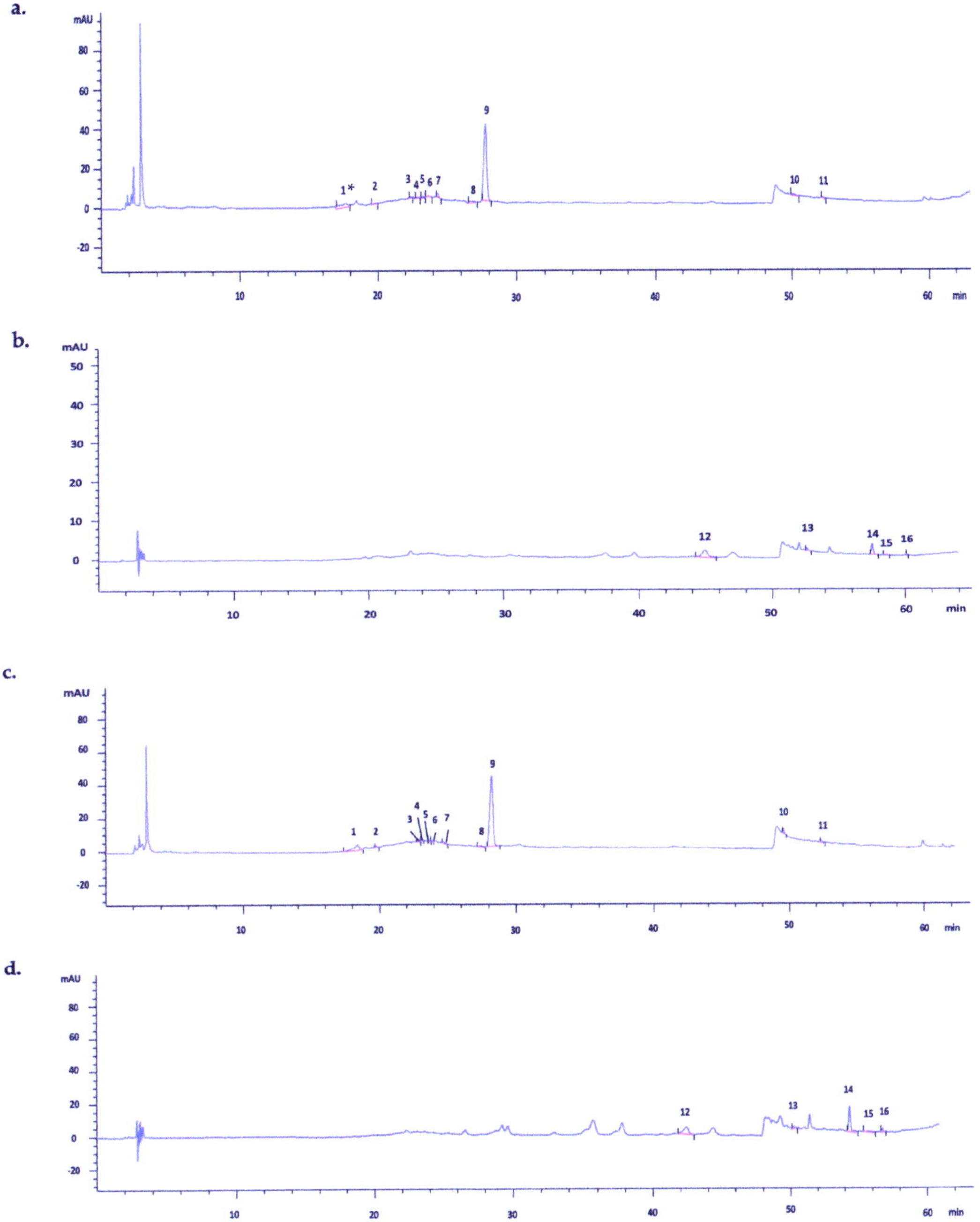

Figure 2. *Cont.*

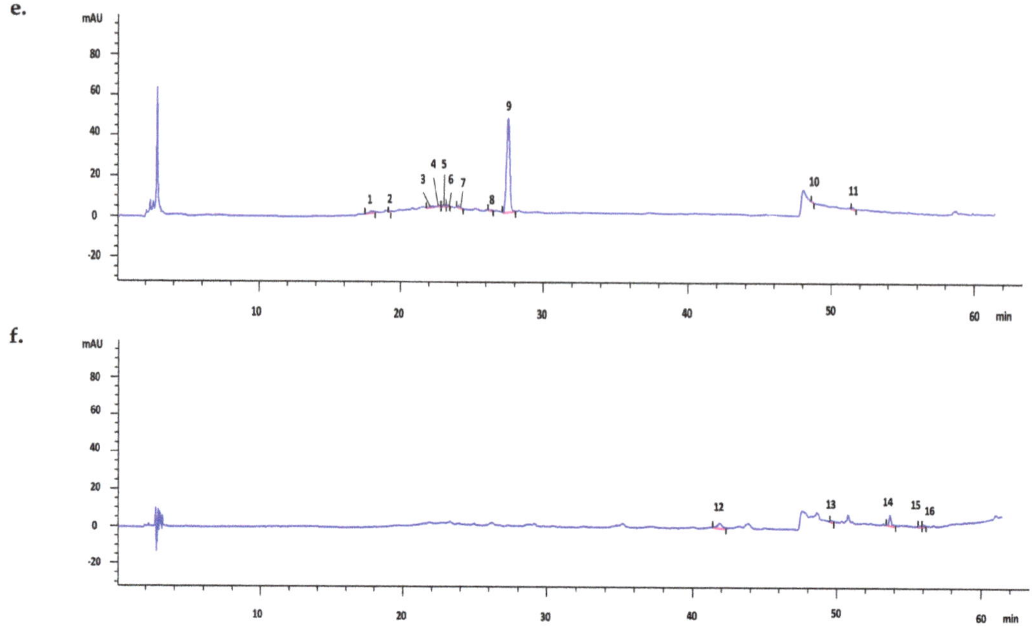

Figure 2. HPLC chromatograms of samples belongs to Gemlik region: (**a**,**b**) Typical HPLC chromatograms of raw olives from Gemlik region; (**c**,**d**) Typical HPLC chromatograms of traditional brine olives from Gemlik region; (**e**,**f**) Typical HPLC chromatograms of Spanish-style olives from Gemlik region. * 1: 3-hydroxytyrosol; 2: protocatechuic acid; 3: keracyanin; 4: coumarin; 5: catechin; 6: tyrosol; 7: 4-hydroxyphenyl; 8: epicatechin; 9: syringic acid; 10: oleuropein; 11: trans-cinnamic acid; 12: luteolin 7-glucoside; 13: luteolin 4-glucoside; 14: luteolin; 15: kaempherol; 16: apigenin.

3Hyt, Tyr, 4Hdf, Lt, and Apg contents were affected by the processing methods, while there was no statistically significant change for L7g and Kam by region ($p < 0.05$, Table 2). The results obtained from previous data confirmed that Hyt and Tyr were identified as the main phenolic compounds of table olives [4,19,23–25].

The main phenolic compound was determined as 3Hyt in raw olives; the content ranged from 104.11 mg/kg (Gem) to 168.21 mg/kg (Erd). These values are much lower than those reported by Uylaser [15] for Hyt ranging from 1.20 mg/kg to 891.80 mg/kg.

In terms of processing, the highest 3Hyt content was observed in SS (49.71 mg/kg); additionally, the content was lower than the results of Sahan et al. (264.56 mg/kg) [12]. The processing methods significantly affected the presence and the content of 3Hyt, which was decreased by the processing. The determined decrease in the SS olives was found to be higher than the NP olives. Contrary to our findings, Salis et al. [9] reported higher amounts of Hyt in processed olives (367.83 μg/g) and determined an important increase in the Hyt content by NP and SS olive processing.

Tyr is a phenolic alcohol, usually present in olives, in lower amounts than Hyt [12]. Try content ranged from 3.66 mg/kg to 8.12 mg/kg in raw olives and from 1.64 mg/kg to 47.24 mg/kg in NP olives. Ozkan et al. [20] reported that Tyr was determined between 33.31 mg/kg and 85.69 mg/kg in raw olive fruits. Accordingly, the amount of Tyr was lower than the findings of Ozkan et al. [20]. Also, higher amounts of Tyr were determined in previous studies [12,26]. These significant differences could be explained by the variety and the applied processes on the olives, especially the use of brine and lye. According to the results of our study, Tyr showed a significant decrease in the SS processing method and might be due to the oxidation of *o*-diphenol during NaOH debittering. Similarly, Salis et al. [9] determined a statistically significant decrease in Tyr.

Table 2. Composition of the phenolic compounds of 'Gemlik' table olives.

	Gemlik Region			Mudanya Region			Iznik Region		
Phenolic Compound (mg/kg)	Raw	Natural Processing	Spanish-Style	Raw	Natural Processing	Spanish-Style	Raw	Natural Processing	Spanish-Style
3Hyt *	104.11 ± 2.66 ^{bcd}**	64.54 ± 1.73 ^{def}	40.86 ± 1.61 ^{efg}	131.69 ± 1.72 ^{abc}	80.32 ± 1.42 ^{de}	42.91 ± 1.36 ^{efg}	145.58 ^{a} ± 2.75	49.71 ± 3.95 ^{ef}	49.71 ± 1.59 ^{ef}
Ker	0.48 ± 0.01	0.29 ± 0.05	0.29 ± 0.05	1.46 ± 0.09	nd ***	1.28 ± 0.07	nd	nd	nd
Kum	1.44 ± 0.04	1.92 ± 0.13	2.13 ± 0.10	5.36 ± 0.55	1.60 ± 0.12	3.36 ± 0.12	2.63 ± 0.12	2.54 ± 0.35	0.23 ± 0.05
Kat	1.44 ± 0.04	1.87 ± 0.12	1.87 ± 0.11	5.36 ± 0.55	5.53 ± 0.35	3.66 ± 0.11	2.63 ± 0.12	4.51 ± 0.66	1.63 ± 0.12
Tyr	8.09 ± 1.28 ^{bc}	4.43 ± 0.60 ^{c}	4.68 ± 0.50 ^{c}	7.26 ± 1.25 ^{c}	3.14 ± 0.28 ^{c}	3.36 ± 0.10 ^{c}	6.87 ± 1.89 ^{c}	16.02 ± 1.21 ^{b}	15.89 ± 1.22 ^{b}
4Hdf	38.89 ± 2.02 ^{bc}	42.77 ± 1.78 ^{bc}	42.77 ± 1.28 ^{bc}	48.12 ± 1.37 ^{bc}	50.95 ± 1.25 ^{bc}	34.63 ± 1.16 ^{bc}	26.86 ± 2.12 ^{cd}	34.03 ± 1.35 ^{bc}	34.04 ± 1.41 ^{bc}
L7g	39.25 ± 2.75	21.46 ± 0.82	0.32 ± 0.05	34.82 ± 1.22	26.30 ± 1.13	14.18 ± 1.55	8.59 ± 1.54	6.80 ± 0.36	8.02 ± 0.93
L4g	3.56 ± 0.02	0.72 ± 0.02	0.09 ± 0.01	0.52 ± 0.01	0.17 ± 0.01	nd	nd	0.96 ± 0.08	0.96 ± 0.09
Trs	0.59 ± 0.03	0.10 ± 0.01	0.32 ± 0.04	3.73 ± 0.20	0.54 ± 0.03	1.05 ± 0.06	0.08 ± 0.01	0.06 ± 0.01	0.06 ± 0.01
Lt	1.60 ± 0.12 ^{lj}	11.28 ± 0.59 ^{efgh}	4.50 ± 0.52 ^{hlj}	13.09 ± 1.14 ^{efg}	0.17 ± 0.01 ^{j}	5.63 ± 0.12 ^{ghlj}	21.60 ± 1.53 ^{cd}	0.96 ± 0.05 ^{lj}	5.62 ± 0.70 ^{ghlj}
Kam	1.60 ± 0.12	0.63 ± 0.02	0.15 ± 0.01	0.79 ± 0.02	0.48 ± 0.01	0.08 ± 0.01	0.32 ± 0.01	0.23 ± 0.08	0.23 ± 0.01
Apg	0.23 ± 0.05 ^{b}	nd	nd	0.52 ± 0.08 ^{b}	0.57 ± 0.01 ^{b}	nd	0.88 ± 0.03 ^{b}	nd	nd
	Orhangazi Region			**Erdek Region**			**Akhisar Region**		
Phenolic Compound (mg/kg)	**Raw**	**Natural Processing**	**Spanish-Style**	**Raw**	**Natural Processing**	**Spanish-Style**	**Raw**	**Natural Processing**	**Spanish-Style**
3Hyt *	155.04 ± 3.42 ^{a}***	32.76 ± 1.43 ^{fg}	4.58 ± 0.12 ^{g}	168.21 ± 2.55 ^{a}	131.01 ± 2.73 ^{abc}	44.97 ± 2.08 ^{ef}	141.38 ± 2.13 ^{ab}	94.32 ± 3.03 ^{cd}	42.52 ± 2.05 ^{efg}
Ker	nd ***	0.20 ± 0.01	nd	nd	nd	nd	nd	0.15 ± 0.01	0.15 ± 0.03
Kum	0.89 ± 0.03	4.17 ± 0.08	1.90 ± 0.08	7.01 ± 0.70	5.90 ± 1.87	4.71 ± 0.17	10.64 ± 1.46	2.30 ± 0.44	1.55 ± 0.44
Kat	7.32 ± 1.21	1.86 ± 0.12	0.09 ± 0.01	5.51 ± 0.55	nd	nd	5.83 ± 0.85	nd	2.85 ± 0.39
Tyr	4.19 ± 0.65 ^{c}	47.24 ± 2.45 ^{a}	1.64 ± 0.04 ^{c}	3.66 ± 0.65 ^{c}	1.95 ± 3.05 ^{c}	1.57 ± 0.44 ^{c}	8.12 ± 1.29 ^{bc}	2.21 ± 0.85 ^{c}	4.50 ± 0.64 ^{c}
4Hdf	30.17 ± 1.84 ^{bc}	27.92 ± 1.84 ^{cd}	4.04 ± 0.07 ^{de}	40.40 ± 1.57 ^{bc}	37.17 ± 11.94 ^{bc}	0.76 ± 0.05 ^{e}	49.41 ± 2.78 ^{bc}	97.58 ± 2.36 ^{a}	54.29 ± 1.80 ^{b}
L7g	15.48 ± 0.57	4.17 ± 0.07	8.02 ± 0.17	19.65 ± 0.93	48.99 ± 3.42	11.26 ± 0.51	40.24 ± 1.11	58.64 ± 1.70	26.21 ± 1.63
L4g	2.39 ± 0.02	4.28 ± 0.09	nd	nd	1.36 ± 0.04	0.39 ± 0.03	nd	1.63 ± 0.29	0.18 ± 0.08
Trs	0.29 ± 0.01	7.37 ± 0.42	0.01 ± 0.00	0.83 ± 0.03	1.36 ± 0.05	3.70 ± 0.64	0.47 ± 0.09	0.66 ± 0.72	nd
Lt	18.92 ± 1.08 ^{cde}	9.09 ± 0.21 ^{fghl}	1.60 ± 0.11 ^{lj}	30.33 ± 1.05 ^{b}	0.60 ± 0.03 ^{j}	14.76 ± 1.74 ^{def}	53.56 ± 2.31 ^{a}	27.18 ± 1.12 ^{bc}	7.38 ± 0.25 ^{fghlj}
Kam	0.39 ± 0.01	nd	0.10 ± 0.01	0.56 ± 0.04	0.66 ± 0.05	0.10 ± 0.01	0.24 ± 0.01	0.73 ± 0.05	0.13 ± 0.01
Apg	0.55 ± 0.02 ^{b}	nd	nd	0.56 ± 0.01 ^{b}	0.42 ± 0.01 ^{b}	0.15 ± 0.03 ^{b}	5.12 ± 0.25 ^{a}	0.46 ± 0.0 ^{b}	nd

* 3Hyt: 3-hydroxytyrosol; Ker: keracyanin; Kum: coumarin; Kat: catechin; Tyr: tyrosol; 4Hdf: 4-hydroxyphenyl; Ep: epicatechin; L7g: luteolin 7-glucoside; L4g: luteolin 4-glucoside; Trs: trans-cinnamic acid; Lt: luteolin; Kam: kaempherol; Apg: apigenin; ** Different letters in each column indicate differences between regions and processing methods ($p < 0.05$). The data are reported as the average of triplicate measurements (n = 3); *** nd: not detected.

4Hdf was the second most abundant phenolic compound in raw olives; contents ranged from 26.86 to 49.41 mg/kg for Izn and Aks, respectively. By the natural processing, the amount of 4Hdf was generally increased, and the highest content was detected in AkhNP (97.58 mg/kg). But SS processing decreased the content of 4Hdf, and the lowest content was detected in ErdSS (0.76 mg/kg).

Apg was detected in NP and raw olives. The highest Apg content was determined in AkhRaw olives (5.12 mg/kg). Content of Apg was decreased dramatically after processing and found under quantification limits in SS olives. Ghorbal et al. [27] reported the A7g content as 1.25–7.79 mg/kg in Gemlik-style processed olives. Our findings are in good agreement with these results.

Additionally, the Lt content varied between 0.17 and 53.56 mg/kg, and after processing, the content was generally decreased by processing. The Lt content in raw and NP olives were found to be higher than in SS-processed ones. Salis et al. [9] reported content of Lt between 92.40 µg/g ft and 118 µg/g in their study, which was higher than our results.

As shown in Table 2, L7g amounts varied between 0.32 and 58.64 mg/kg in olive samples. The lowest L7g content was determined in SS processed olives, while the highest content was obtained in raw and NP olive samples. After processing, the L7g amount was decreased significantly in NP and SS processed olives. In general, L7g amounts were found to be higher than the Lt content (except Orh and Erd). This determined difference may arise from the glucosidase activity of L7g during the processing. Also, our findings are in good agreement with previous studies [28].

Kam was detected in all olive samples except from the OrhNP. The Kam content in raw and NP olives was found to be higher than SS processed ones. Additionally, Kam content (0.08–1.60 mg/kg) was lower than that of Ozcan et al. [29], who reported 1.31 and 4.18 mg/100 g in olive fruit. Ker was only determined in olives obtained from the Gemlik region (both raw and processed). Also, Kum content was determined higher (0.23–7.01 mg/kg) than reported by Ozcan et al. [29] (0.06–0.89 mg/100 g). Protocatechuic acid (Prt) and epicatechin (Ep) were not detected in any of the samples.

For statistical evaluation, the phenolic compounds of raw and processed olives were classified with chemometrics methods (principal component analysis, PCA) according to regions and processing methods. The data matrix of variables (table olive samples and their phenolic compounds) was processed in PCA. Olive samples obtained from six regions were classified according to raw, NP, and SS. The PCA model was constructed with two principal components as 45% of the total variance. PC1 described 28.23% of total variance; PC2 explained 16.21% of total variance, Figures 3–7.

As can be seen from scree plot of eigenvalues, there is a sharp decrease in eigenvalue from PC1 to PC2 and then a relatively flat appearance from PC2 to PC4. This is the possible reason for the low total variance explained with only two PCs (PC1 and PC2). Table 3 shows the eigenvalues along with percent explained variance (% Exp. Var.) and cumulative percent explained variance (Cum. % Exp. Var.).

Table 3. Eigenvalues along with percent explained variance (% Exp. Var.) and cumulative percent explained variance (Cum. % Exp. Var.).

No	Eigenvalues	% Exp. Var.	Cum. % Exp. Var.
1	3.90	28.23	28.23
2	2.24	16.21	44.44
3	2.15	15.59	60.03
4	1.79	12.93	72.95
5	1.16	8.38	81.33
6	1.06	7.67	89.00
7	0.65	4.73	93.73
8	0.47	3.41	97.13
9	0.23	1.67	98.80
10	0.17	1.20	100.00

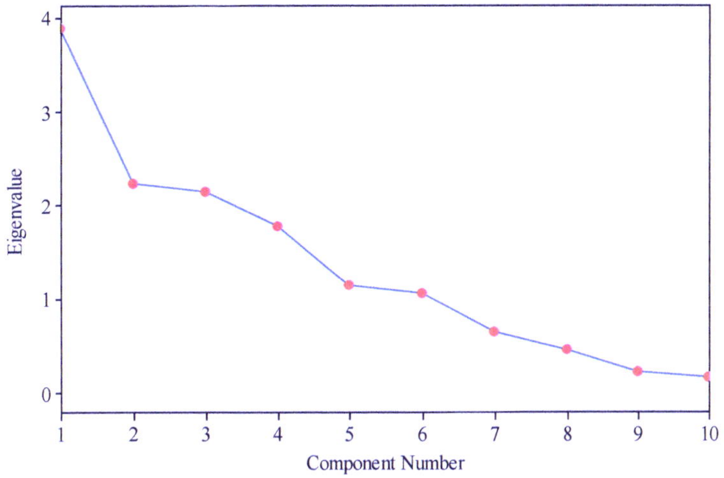

Figure 3. Plot of eigenvalues vs. number of principal components from PCA.

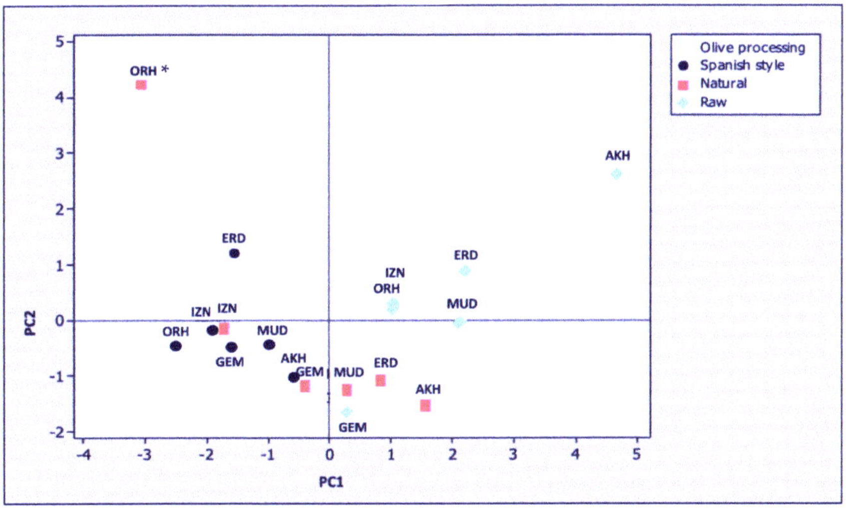

Figure 4. Score plot of PC1 vs. PC2 from principal component analysis. * GEM: Gemlik region; IZN: Iznik region; MUD: Mudanya region; ORH: Orhangazi region; ERD: Erdek region; AKH: Akhisar region.

As seen from Table 3, the sum of the four PC reaches about 73% cumulative explained variance. Figure 4 shows the score plot of PC1 vs. PC2 from principal component analysis. Figure 5 shows the score plot of PC1 vs. PC3 from principal component analysis. Figure 6 shows the three-dimensional score plot of PC1, PC2, and PC3 from principal component analysis.

As can be seen from Figures 4–6 additional third PC is also important to differentiate samples from three processing methods. Especially, the three-dimensional score plot indicates that samples with natural processing are located at the center of the score plot, while raw and Spanish-style samples are on the either side of the natural processed samples.

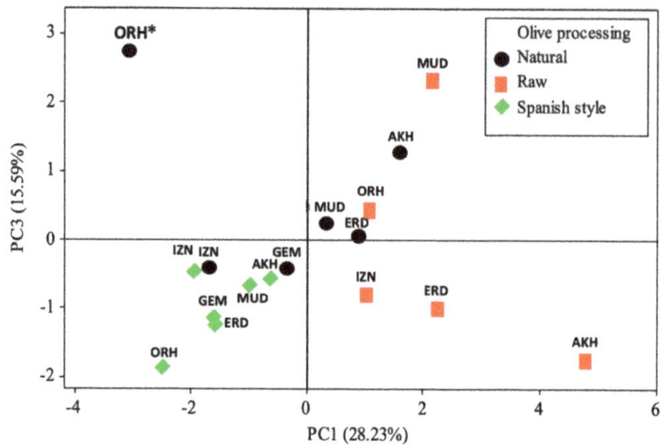

Figure 5. Score plot of PC1 vs. PC3 from principal component C. * GEM: Gemlik region; IZN: Iznik region; MUD: Mudanya region; ORH: Orhangazi region; ERD: Erdek region; AKH: Akhisar region.

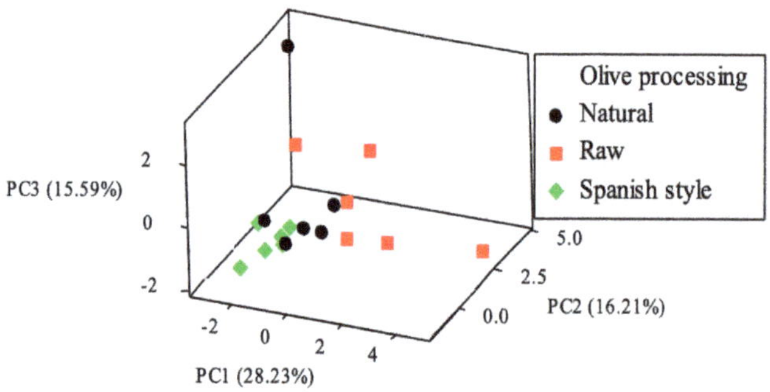

Figure 6. Three-dimensional score plot of PC1, PC2, and PC3 from principal component analysis.

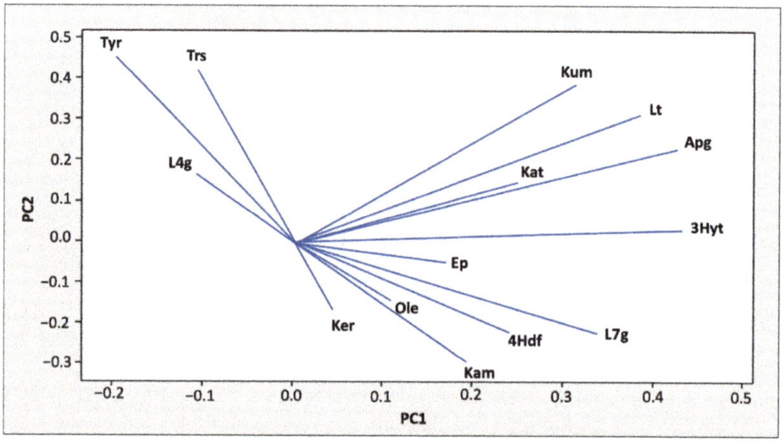

Figure 7. Loading plot of PC1 vs. PC2 from PCA. 3Hyt: 3-hydroxytyrosol; Ker: keracyanin; Kum: coumarin; Kat: catechin; Tyr: tyrosol; 4Hdf: 4-hydroxyphenyl; Ep: epicatechin; L7g: luteolin 7-glucoside; L4g: luteolin 4-glucoside; Trs: trans-cinnamic acid; Lt: luteolin; Kam: kaempherol; Apg: apigenin; Ole: oleuropein.

Olives taken from the Orh are distinguished in the classification of natural processing and are characterized by Trs, Tyr, and L4g phenolic compounds (Figures 4 and 7). Kum and Lt are included in the characterization of raw olives obtained from Orh, Mud, and Gem regions, and 3Hyt, L7g, Kam, Kat, and 4Hyd were characterized in raw olives from Akh, Erd, and Izn (Figures 4 and 7).

According to PCA biplot analysis, the 3Hyt, L7g, Kam, Kat, and 4Hdf were effective in the characterization of 'Gemlik' type black table NP olives obtained from the Akh, Erd, Mud, and Gem regions and processed with natural processing.

On the other hand, raw, NP, and SS processed olives obtained from Erd and Mud were the most similar samples in phenolic compounds. The dendrogram based on the HCA results (Euclidian method) of olive samples could be divided into three main groups of olive processing methods based on their phenolic compounds (Figures 8 and 9).

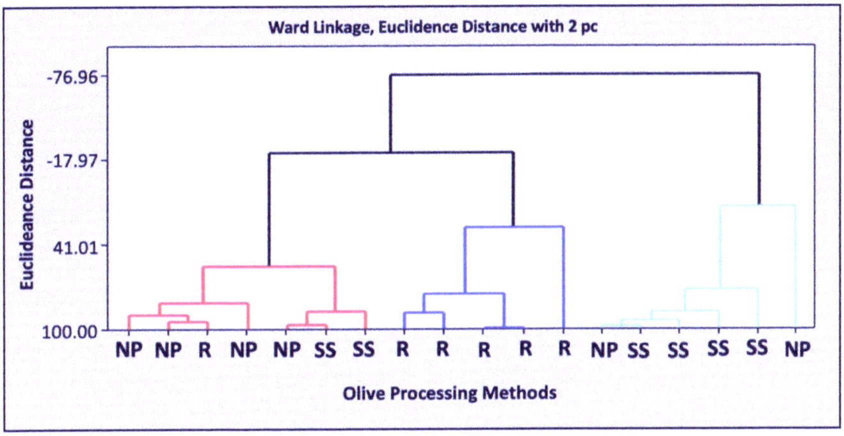

Figure 8. Dendrogram of the samples from hierarchical cluster analysis (HCA) by using two PCs from PCA; NP: natural processing (traditional brine olives); R: raw; SS: Spanish style.

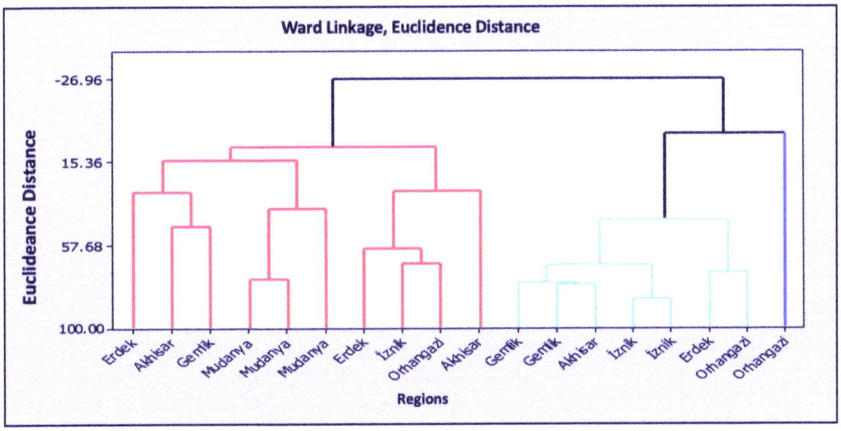

Figure 9. Dendrogram of the samples from hierarchical cluster analysis (HCA) by using standardized data.

A dendrogram was generated by using the Word Linkage method and Euclidean distance measure. As shown in Figure 9, it is made up of three subgroups and includes natural processing, natural processing, raw, and natural processing (subgroup 1); raw, raw, raw, and raw (subgroup 2); and natural processing, Spanish style, Spanish style, Spanish style, and natural processing (subgroup 3) according to the olive processing methods. According to the HCA analysis, given in Figure 9, the study consisted of three subgroups. These are Erd, Mud, Gem, Akh, Gem, Akh, and Mud (subgroup 1) Erd, Mud, Izn, Orh, and Akh (subgroup 2); and Izn, Izn, Gem, Orh, Erd, and Orh (subgroup 3).

There are similar chemometric (including PCA and HCA) investigations based only on fruit profiles (especially total or individual phenolics) data for Türkiye [30] and Algeria olives [31].

The olives obtained from the GemNP and the AkhRaw had the most similar phenolic compound profile to each other. On the other hand, raw olives obtained from Erd and Mud regions and NP and SS processed olives obtained from Izn were the most similar samples. As seen in the dendrogram, Tyr, Trs, and L7g were prominent phenolic compounds in samples such as ErdRaw and OrhNP.

Chemometric analysis of phenolic compound data (PC1, PC2) provided some important indications about the characterization and classification of 'Gemlik' varieties (Figures 4 and 7).

Comparing the changes in the phenolic content of olives after treatment, significant differences were observed between raw and treated olives. After fermentation, changes in the profile and amount of simple phenolic compounds are mainly due to the diffusion of substances from the olives into the brine and vice versa. When alkali is used, sodium hydroxide and components with carboxyl and hydroxyl groups react, and hydrophilic derivatives are washed away [32].

Variations between the amounts of phenolic compounds may vary depending on some factors. Similarly, previous studies have shown that the composition and content of phenolic compounds in olives depend on many factors such as the variety, region, climate, and development conditions of the fruit, degree of maturity, type of harvest, pests, olive processing, and storage method.

It has been observed that traditional brine olive (NP) methods are the most important method for preserving the content of phenolic compounds. It has been determined that olives with the highest content of phenolic compounds were produced by NP olive samples. There are certain disadvantages of natural processes, such as longer processing time, being less applicable in the industry, high sensory bitterness, and therefore less appeal to consumers than SS olives with high salt content [24]. In addition, in NP methods, the fermentation does not take place in a standardized way, so it has adverse effects on food safety, sensory properties, and health effects [4].

5. Conclusions

Among the phenolic compounds identified in 'Gemlik' olives, 3-hydroxytirosol was the most abundant one. The natural (traditional brine) and Spanish-style processing methods differently decreased phenolic content of black table olives. The loss in phenolic compounds was found to be lower in the natural processing method. These results may indicate that natural processing is a more efficient processing method than Spanish-style processing for obtaining phenolic-compound-rich 'Gemlik' table olives and developing healthier fermented foods.

As they grow outside the geographical boundaries, the phenolic profile of the Akhisar-grown 'Gemlik' variety olives was found to be similar to Gemlik, Iznik, and Orhangazi-grown ones.

Additionally, according to chemometric analysis, it was determined that natural and Spanish-style processed 'Gemlik' olives can be identified according to some phenolic compounds (3-hydroxytyrosol, luteolin 7-glicoside, kaempherol, catechin, and 4-hydroxyphenyl). In other words, it will be possible to determine the processing method according to the phenolic compound profile of the 'Gemlik' variety. In this way, deception of the consumer can be prevented by the phenolic compound determination of the olives supplied from markets.

Author Contributions: C.D.: General planning, sample collection; experiments, analysis, tabulation and interpretation of the data, and writing of the manuscript. E.Y.: Writing of the manuscript, review of the manuscript. O.G.: General planning, review of the manuscript. All authors have read and agreed to the published version of the manuscript.

Funding: The authors would like to thank to TAGEM (General Directorate of Agricultural Research and Policies, Ankara, Türkiye) for the financial support (Project No: TAGEM/HSGYAD/18/A3/P1/1158) and Marmarabirlik for technical support of the project.

Data Availability Statement: Data will be available from the corresponding author on reasonable request.

Conflicts of Interest: The authors declare no conflict of interest.

References

1. Iraklia, M.; Chatzopouloua, P.; Ekateriniadou, L. Optimization of ultrasound-assisted extraction of phenolic compounds: Oleuropein, phenolic acids, phenolic alcohols and flavonoids from olive leaves and evaluation of its antioxidant activities. *Ind. Crops Prod.* **2018**, *124*, 382–388. [CrossRef]
2. Ozturk, F.; Ocak, O.; Unal, M.K. Effect of ozone treatment on the physical, microbiological and sensorial properties of Spanish-style table olives. *Grasas y Aceites* **2020**, *71*, e348. [CrossRef]
3. *International Olive Council IOC/T.28/Doc No 1/Rev 1*; Guidelines for Accomplishment of the Requirements of the Norm ISO 17025 by the Laboratories of Sensory Analysis of Virgin Olive Oil. International Olive Oil Council: Madrid, Spain, 2017.
4. Pistarino, E.; Aliakbarian, B.; Casazza, A.A.; Paini, M.; Cosulich, M.E.; Perego, P. Combined effect of starter culture and temperature on phenolic compounds during fermentation of Taggiasca black olives. *Food Chem.* **2013**, *138*, 2043–2049. [CrossRef] [PubMed]
5. Serrano, P.G.; Romero, C.; Brenes, M.; García, P.G. Enrichment in phenolic compounds of black ripe olives through nanofiltration and vacuum evaporation techniques. *Innov. Food Sci. Emerg. Technol.* **2019**, *51*, 73–79. [CrossRef]
6. Rodriguez, P.R.; Barreiro, C.G.; Carballo, M.E.; Perez, C.N.; Otero, R.R.; Gonzales, F.M.; Grande, G.B. Applicability of an in-vitro digestion model to assess the bioaccessibility of phenolic compounds from olive-related products. *Molecules* **2021**, *26*, 6667. [CrossRef] [PubMed]
7. Deng, J.; Xu, Z.; Xiang, C.; Liu, J.; Zhou, L.; Li, T.; Yang, Z.; Ding, C. Comparative evaluation of maceration and ultrasonic-assisted extraction of phenolic compounds from fresh olives. *Ultrason. Sonochem.* **2017**, *37*, 328–334. [CrossRef] [PubMed]
8. Ryan, D.; Robards, K.; Lavee, S. Determination of phenolic compounds in olives by reversed- phase chromatography and mass spectrometry. *J. Chromatogr.* **1999**, *832*, 87–96. [CrossRef]
9. Salis, C.; Papadakis, I.E.; Hagdimitriou, M. Identification and quantification of phenolic compounds in fresh and processed table olives of cv. 'Kalamata'. *Not. Bot. Horti Agrobot. Cluj-Napoca* **2021**, *49*, 12394. [CrossRef]
10. Medinaa, G.; Sanzb, C.; Leónc, L.; Pérezb, A.G.; Rosac, R. Phenolic variability in fruit from the 'Arbequina' olive cultivar under Mediterranean and Subtropical climatic conditions. *Grasas y Aceites* **2021**, *72*, 4–438. [CrossRef]
11. Mettouchia, S.; Sacchib, R.; Moussaa, Z.E.O.; Paduanob, A.; Savaresec, M.; Tamendjaria, A. Effect of Spanish style processing on the phenolic compounds and antioxidant activity of Algerian green table olives. *Grasas y Aceites* **2016**, *67*, 114.
12. Sahan, Y.; Cansev, A.; Gulen, H. Effect of processing techniques on Antioxidative enzyme activities, antioxidant capacity, phenolic compounds, and fatty acids of table olives. *Food Sci. Biotechnol.* **2013**, *22*, 613–620. [CrossRef]
13. Irmak, S.; Irmak, H. The effects of green table olive processing methods on polyphenol content of some Turkish table olive varieties. *J. Food Health Technol.* **2021**, *4*, 296–308.
14. Lauren, M.; Crawforda, D.M.; Holstegeb, S.C.; Wanga, C. High-throughput extraction method for phenolic compounds in olive fruit (*Olea europaea*). *J. Food Compost. Anal.* **2018**, *66*, 136–144.
15. Uylaser, V. Changes in phenolic compounds during ripening in 'Gemlik' variety olive fruits obtained from different locations. *CyTA—J. Food* **2015**, *13*, 167–173. [CrossRef]
16. Johnson, R.L.; Mitchell, A.E. Reducing phenolics related to bitterness in table olives. *J. Food Qual.* **2018**, *12*, 3193185. [CrossRef]
17. Ozaltas, M.; Savran, M.K.; Ulas, M.; Kaptan, S.; Kokturk, H. *Turkiye Olive Growing Sector Report*; Olive Research Institute: Bornova, Izmir, Türkiye, 2016; ISBN 978-605-9175-57-9.
18. Koyuncu, G.; Cabaroglu, T. Determination of quality characteristics and aroma compounds of 'Gemlik' variety olive fruit grown in Adana province. *Food* **2020**, *45*, 1163–1174.
19. Ozkan, G.; San, B.; Akcay, U.Ç.; Dagdelen, A.; Dolgun, O.; Konuskan, D.; Gulsoy, S. *Comparison of Fruit, Oil and Molecular Characteristics of 'Gemlik' Olive Varieties Grown in Marmara, Aegean, Western and Eastern Mediterranean Regions*; TUBITAK Project No: 109O099; The Scientific and Technological Research Council of Türkiye (TUBITAK): Ankara, Türkiye, 2011.
20. Anonymous. Institute Geographical Indication Registration Certificate. Turkish Patent 76, 23 September 2003.
21. Vinha, A.F.; Ferreres, F.; Silva, B.M.; Valentao, G.A.; Pereira, J.A.; Olivera, M.B.; Seabra, R.M.; Andrade, P.B. Phenolic profiles of Portuguese olive fruits (*Olea europaea* L.): Influences of cultivar and geographical origin. *Food Chem.* **2005**, *89*, 561–568. [CrossRef]
22. Ramírez, E.; Medina, E.; Brenes, M.; Romero, C. Endogenous enzymes involved in the transformation of oleuropein in Spanish table olive varieties. *J. Agric. Food Chem.* **2014**, *62*, 9569–9575. [CrossRef]

23. Malheiro, R.; Sousa, A.; Casal, S.; Bento, A.; Pereira, J.A. Cultivar effect on the phenolic composition and antioxidant potential of stoned table olives. *Food Chem. Toxicol.* **2011**, *49*, 450–457. [CrossRef]
24. Lodolinia, E.M.; Bañegilb, M.C.; Fernándezb, A.; Adámezb, J.D.; Ramírezb, R.; Vertedor, D.M. Monitoring of acrylamide and phenolic compounds in table olive after high hydrostatic pressure and cooking treatments. *Food Chem.* **2019**, *286*, 250–259. [CrossRef]
25. Othman, B.N.; Roblain, P.; Thonart and Hamdi, M. Tunisian table olive phenolic compounds and their antioxidant capacity. *J. Food Sci.* **2008**, *73*, C235–C240. [CrossRef] [PubMed]
26. Arslan, D.; Ozcan, M.M. Phenolic profile and antioxidant activity of olive fruits of the Turkish variety "Sarıulak" from different locations. *Grasas y Aceites* **2011**, *62*, 53–461. [CrossRef]
27. Ghorbala, B.A.; Leventdurura, S.; Agirmana, B.; Gunduz, B.P.; Kelebek, H.; Çarşanba, E.; Darıcı, M.; Erten, H. Influence of geographic origin on agronomic traits and phenolic content of cv. 'Gemlik' olive fruits. *J. Food Compost. Anal.* **2018**, *74*, 1–9. [CrossRef]
28. Tsantili, E. Original article quality attributes and their relations in fresh black ripe 'Kalamon' olives (*Olea europaea* L.) for table use-phenolic compounds and total antioxidant capacity. *Int. J. Food Sci. Technol.* **2014**, *49*, 657–665. [CrossRef]
29. Ozcan, M.M.; Fındık, S.; Aljuhaimi, F.; Ghafoor, K.; Babiker, E.E.; Adiamo, O. The effect of harvest time and varieties on total phenolics, antioxidant activity and phenolic compounds of olive fruit and leaves. *J. Food Sci. Technol.* **2019**, *56*, 2373–2385. [CrossRef] [PubMed]
30. Aktas, A.B.; Ozen, B.; Tokatli, F.; Sen, I. Phenolics profile of a naturally debittering olive in comparison to regular olive varieties. *J. Sci. Food. Agric.* **2014**, *94*, 691–698. [CrossRef] [PubMed]
31. Medjkouh, L.; Tamendjari, A.; Alves, C.R.; Laribi, R.; Oliveira, M.B.P.P. Phenolic profiles of eight olive cultivars from Algeria: Effect of *Bactrocera oleae* attack. *Food Funct.* **2018**, *9*, 890–897. [CrossRef] [PubMed]
32. Irmak, S. The effects of processing methods on polyphenol content of some Turkish black table olives. *Riv. Ital. Delle Sostanze Grasse* **2021**, *98*, 205–215.

MDPI AG
Grosspeteranlage 5
4052 Basel
Switzerland
Tel.: +41 61 683 77 34

Processes Editorial Office
E-mail: processes@mdpi.com
www.mdpi.com/journal/processes

www.ingramcontent.com/pod-product-compliance
Lightning Source LLC
Chambersburg PA
CBHW061548210926
51850CB00009B/2038
9783725819720